THE HUNSLET ENGINE WORKS

Over a century and a half
of locomotive building

*Dedicated to my wife Felicity
and to our two sons,
Peter and Christopher,
for their encouragement.*

THE HUNSLET ENGINE WORKS

Over a century and a half
of locomotive building

Don Townsley

PLATEWAY PRESS
ISBN 1 871980 38 0

THE HUNSLET ENGINE WORKS

*Over a century and a half
of locomotive building*

© Don Townsley & Plateway Press 1998

ISBN 1 871980 38 0

Printed by:
Postprint, Taverner House, Harling Road, East Harling
Norwich NR16 2QR

Jacket Design:
Roy C. Link, 1 Station Cottages, Harling Road, East Harling
Norwich NR16 2QP

Published by:
Plateway Press, Taverner House, Harling Road, East Harling
Norwich NR16 2QR

Closed but retained for posterity. The Imposing doorway of the 1882 Grade II Listed office block with its backdated inscription photographed in 1997 after the demolition of the Hunslet Engine Works.

CHRIS NICHOLSON

A classic posed shot of a contractors locomotive with an interesting history. Hunslet no. 202 was built to run on 5'–3" track gauge and supplied on 20 March 1878 to C. Holland at Kilrea, Ireland, (now Northern Ireland), and named 'Maghera'. It presumably worked on the construction of the branch from Macfin to Cookstown. By 1886 it was back at Hunslet where it was rebuilt to standard gauge. It is seen here named 'Cumbria' under the ownership of W. Scott & Co. for whom it worked on contracts including the Great Central Railway's London extension through Buckinghamshire at the turn of the century. A small 10 x 15 in. locomotive it was unusual amongst the 0-6-0 contractors locomotivies in having outside cylinders and a particularly long wheelbase. In this picture it retains the wooden brake blocks to intermediate and trailing wheels only and whilst the original sandboxes and sandgear are still in place there are no sandpipes. A later view of the same locomotive shows replacement cast iron brake blocks and the front sandbox moved back under the nameplate to feed sand between leading and intermediate wheels.

Andrew Neale Collection

Acknowledgements

P. J. O. Alcock

C. P. Atkins

Allan C. Baker

Colin Boocock

A. J. Booth

D. N. Bowden

J. I. C. Boyd

Bradford Metropolitan District
  Council

V. J. Bradley

R. M. Burnley

Sheila Bye

Keith Chester

John Corrie

Paul Cotterell

Neale Coupar

Bob Darvill

G. S. Derrett-Smith

A. R. Etherington

Roger Ford

David Gillan

Oscar Glenn

Chris Hawkins

Philip J. Hindley

G. Horsman

A. O. Hunt

Hunslet-Barclay Ltd

Imperial War Museum

Industrial Railway Society

Institution of Mechanical Engineers

Peter Kewney

R. Kitching

Leeds City Council

M. J. Lyons

Merlin Gerin (Groupe Schneider)

Brian Morrison

National Railway Museum

Andrew Neale

J. A. Peden

D. A. Rayner

R. N. Redman

R. C. (Dick) Riley

Andrew Ross

Rabbi Walter Rothschild

L. W. Rowe

D. Trevor Rowe

Eric Sawford

Mrs M. J. Shackleton

Cliff Shepherd

Brian Smeeton

Martin Smith

Howard Snowden

P. J. Stamper

J. G. Timcke

John Wallis

Weatherall, Green & Smith

D. W. Winkworth

John Wrighton

And to all colleagues, named or unnamed above or in the text,
who all contributed something by virtue of 'being there'.

Contents

New life for the new Millennium

This aerial view was taken on 19 August 1998 and shows completion of the £11 million investment by Groupe Schneider to provide a state-of-the-art factory for the production of electrical switchgear and allied components, thereby breathing new life into the area. The Hunslet and Manning Wardle offices can be picked out, as can the Shepherd and Todd archway and the 1858 gateposts. Trees cover the old track-bed to Hunslet Lane goods yard, now a retail park, and the incline from the Midland main line to the Hunslet Engine works is similarly overgrown.

BRIAN PETER – GROUPE SCHNEIDER

Foreword

Don Townsley has produced a major and very important chapter in the history of the world's locomotive manufacturers.

This definitive volume chronicles almost 150 years of evolution during which the Leeds-based Hunslet Engine Company contributed significantly through bold strategy, vision and innovation to the shaping of what must be one of the most emotive and exciting business activities.

That this relatively small specialist company of six hundred people should produce such an extensive range of successful main-line and industrial locomotives, and not just locomotives but fork-lift trucks, turbines, motor cars and machine tools is a tribute to its founders, successors and employees.

Its continuing quest for the acquisition of complementary companies both in the UK and abroad, together with ongoing development including its revolutionary flameproof rack-and-pinion mining locomotives, ensured a stable business long after the last days of steam. Indeed as recently as 1982 Hunslet Holdings was the seventeenth most profitable British manufacturer in terms of return on equity ahead of many household names.

I can only give a brief insight to whet the appetite for this engrossing and often amusing book, and mention that it was a great pleasure to write these few words for Don who has become a very good friend over the years, and hope I do him and his book justice.

As you will doubtless realize as you progress, Don has a wealth of knowledge in the railway industry generally, and without his like, and his patience in research, the importance of recording this intriguing history of what in essence made Britain Great would be lost for ever as information is misplaced and memories fade.

Peter Kewney

P. E. KEWNEY
Managing Director, Hunslet Barclay Ltd

Author

Don Townsley is a Chartered Engineer by profession, a Member of the Institution of Mechanical Engineers and a Member of the Institute of Management.

Born and educated in Leeds, he spent 40 years with the Hunslet Engine Company and Hunslet Holdings plc,. from apprentice through draughtsman and locomotive designer to General Sales Manager, the latter period of which brought him into contact with most main line, industrial, underground and narrow gauge railways throughout the world.

From 1989 he operated as a freelance railway consultant with clients including London Underground and Docklands Light Railway. As Principal Rolling Stock Consultant for one of Britain's largest multi-disciplinary consultancy organisations he acted as Engineer for the Royal Mail class 325 electric multiple unit trains and drew up specifications and tender documentation for LTS Rail Ltd's class 357 "Electrostar" trains. He continues to be involved with rolling stock procurement, acceptance and commissioning for a number of clients.

A founder member of the Leeds Model Railway Society in 1947, and now the society's President, he was Exhibition Manager for the annual Leeds model railway exhibitions from 1958 until 1992.

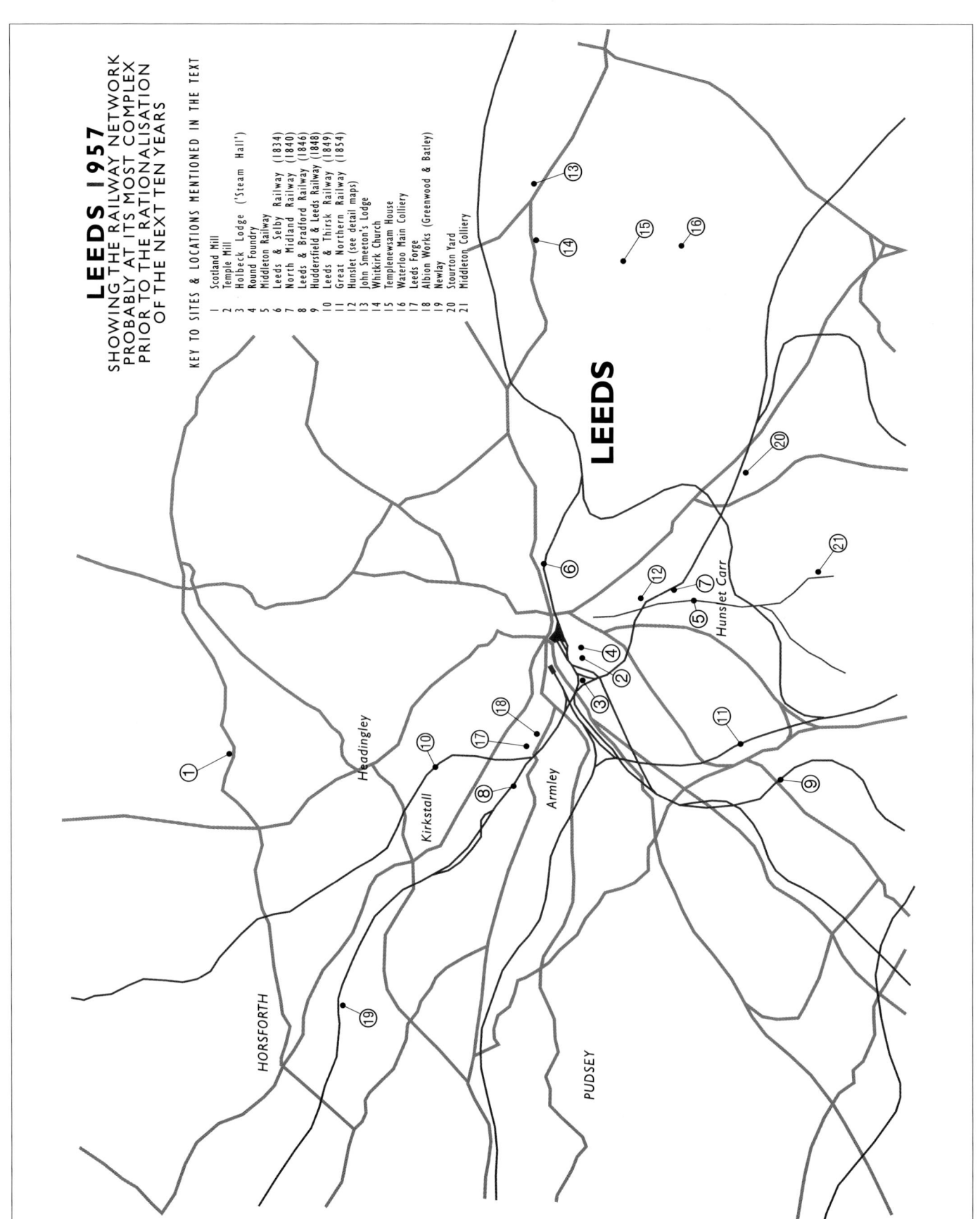

LEEDS 1957

SHOWING THE RAILWAY NETWORK
PROBABLY AT ITS MOST COMPLEX
PRIOR TO THE RATIONALISATION
OF THE NEXT TEN YEARS

KEY TO SITES & LOCATIONS MENTIONED IN THE TEXT

1   Scotland Mill
2   Temple Mill
3   Holbeck Lodge ('Steam Hall')
4   Round Foundry
5   Middleton Railway
6   Leeds & Selby Railway (1834)
7   North Midland Railway (1840)
8   Leeds & Bradford Railway (1846)
9   Huddersfield & Leeds Railway (1848)
10 Leeds & Thirsk Railway (1849)
11 Great Northern Railway (1854)
12 Hunslet (see detail maps)
13 John Smeeton's Lodge
14 Whitkirk Church
15 Templenewsam House
16 Waterloo Main Colliery
17 Leeds Forge
18 Albion Works (Greenwood & Batley)
19 Newlay
20 Stourton Yard
21 Middleton Colliery

LEEDS

HORSFORTH

Headingley

Kirkstall

Armley

PUDSEY

Hunslet Carr

Introduction

To refer to Leeds as a 'railway' town would probably raise incredulous looks from most railway enthusiasts, to whom the term would bring to mind thoughts of Swindon, Derby, Crewe or similar.

The appellation 'railway town' is normally used to describe those centres where, historically, the existence of a railway created the need for a factory to supply its locomotives, carriages and wagons. This factory could well have been away from an existing town on a greenfield site, and thus the railway begat the town, or at least caused it to grow, and became the prime reason for its continued prosperity. Swindon, Crewe, Wolverton and Horwich are four examples, and there are many more.

Leeds does not fit this stereotype, for while, ultimately, it was served by five of the major pre-grouping railways, it did not rely on any one individual railway for its prosperity. Indeed the reverse was true, for the town and its first locomotive builders were established before the railways themselves came, and in consequence local enterprise and labour provided the means by which the railways were able to establish themselves.

It must be acknowledged that other towns that were not tied to a particular railway company – Glasgow, Newcastle and Manchester, for example – also provided a similar service, but none could quite match the unique combination of inventiveness, innovation, variety and longevity over every continent and industry characterized by the manufacturers of Leeds.

The town (city by 1893) fathered the first commercially successful steam locomotive, and, long before the main-line railway companies had even considered the need for their own workshops, produced express passenger and goods locomotives for the Great Western, Great Northern, Midland, London and North Western, North Eastern, Manchester, Sheffield and Lincolnshire, London, Brighton and South Coast, South Eastern and many other railways. Leeds-trained engineers went on to hold high office not only in the railway industry but in other industries also. Men such as David Joy, Charles Parsons, John Chester Craven, Richard Peacock, Luke Longbottom – even

Friedrich Krupp – and, master craftsman and mentor of them all, Matthew Murray collectively shaped the future in a way probably unequalled by any other group of people emanating from such a geographically compact seat of learning.

The Leeds 'railway connection' was not limited to locomotives. Samson Fox set up in business in 1874 on an 18 1/2-acre site at Armley in west Leeds, founding the Leeds Forge Company. By 1896 he had two equally huge plants in America. The names 'Fox' and 'Leeds Forge' became synonymous with pressed-steel railway construction on both sides of the Atlantic, and set designs and design standards for bogies and freight wagons that are very little changed in India and other countries to this day. A story in itself.

Leeds Forge became part of Cammell Laird in 1923 and merged into the Birmingham-based Metropolitan Cammell Carriage and Wagon Company in 1929. The board room of GEC Alsthom Metro-Cammell Ltd at Washwood Heath in Birmingham still boasts leather chairs inscribed 'LEEDS FORGE'.

As an example of nineteenth-century business drive it is difficult to better the 2 ft. 6 in. gauge railway laid for exhibition purposes by Leeds Forge in 1896. Sited at Newlay, five miles north-west of Leeds, alongside the Midland Railway main line to Carlisle, an area of approximately 1000 yards by 100 yards allowed the construction of a substantial railway system, with level track and gradients together with single and reverse curves, points, sidings and loops. Complete goods and passenger trains were provided, the former built throughout of Fox's pressed steel, the latter having bodies provided by the Lancaster Railway Carriage and Wagon Company mounted on pressed-steel underframes and bogies. Kitsons provided an 0–8–4 tank locomotive to E. R. Calthrop's patent, as supplied to the Barsi Light Railway and similar to later 2–6–4 tanks for the Leek and Manifold Valley Railway.

Down the years there were other substantial Leeds contributors to the railway scene world-wide, but perhaps the point has been adequately made.

The United Kingdom was almost alone among the industrial nations of the world in that, once their own workshops had become established, the main-line railway companies were largely self-sufficient in the provision of motive power and rolling stock. The private manufacturers increasingly had to look abroad and to special applications to survive. Some found niche markets with the smaller local railway companies, particularly in South Wales, and there were times when even the big boys such as the Great Central needed a helping hand; but, in the main, the burgeoning Empire was the salvation of many.

Having weaned its child, the British railway system, the proud parent manufacturer found only crumbs at the table of its grown-up offspring. But this is life, and life goes on – compare or contrast as you will this situation with the changes that are taking place in Britain during this last decade of the twentieth century.

This book was originally conceived some fifteen years ago for publication in 1989 as *Hunslet 125*, a sequel to Tom Rolt's *A Hunslet Hundred*, and much material was amassed to that end. Events diverted attention from the task, and its scope has now been widened in an attempt to put on record the contribution made by the City of Leeds in general, and by one Leeds factory in particular, to the growth of railways throughout the world, and to comment on the eventual decline of the fundamental manufacturing processes within the 'golden triangle' of Hunslet. The name Hunslet itself, although widely known abroad, was largely unfamiliar to the British railway enthusiast until the end of main-line steam on British Rail, but it is now posthumously revered by thousands on account of the large numbers of retired locomotives built in Hunslet that are owned and operated by the preservationists and museums at home and abroad.

I have tried to avoid too much repetition of text already published in the four major volumes produced by Kitson-Clark, Rolt, Redman and Lane between 1935 and 1969, but an element of duplication is necessary in order to paint the background and form the framework of our tale. Where any differences of stated fact or interpretation may be evident in comparison with earlier works, this is probably due in the former case to the emergence of additional corroborative evidence, and in the latter to the personal stance and beliefs of this particular writer and to the received wisdom of the intervening sixty years. It is fashionable in the 1990s to rewrite history, and to debunk the views of generations long gone in the pursuit of naïve and self-righteous 'political correctness'. This is not intended to be a terribly fashionable volume, as 1990s fashions go, but rather a tribute to the generations of directors, managers and workers who with pride of purpose believed in what they were doing and were not afraid of calling Britain great. Present-day cynicism might pejoratively consider some of them 'movers and shakers', but what matter, since for the most part they had learned their trade well and knew what they were moving or shaking.

A two-part format has been chosen for the book, to provide initially a historical survey up to 1948 and secondly a personal viewpoint from 1949 to the present day. From the historical point of view, care has been taken when revisiting the previous writings to validate the information wherever possible. In the second part of the book, personal notes predominate, augmented by the recollections of valued friends and colleagues. Press releases and reports have been used where the real story behind them was known, and not by repeating the sales pitch. I was responsible for a very large number of them, latterly, and of course they had to be positive – or, as Alan Clark might have put it, they tended sometimes to be 'economical with the *actualité*'. They were never dishonest.

In this single volume I have not attempted to cover every locomotive, or even every product, of the Hunslet Engine Works. Many locomotives are of course referred to by name or number, either because they typified a strand of the company's activities, or because by design or vicissitude they are noteworthy in their own right. I could say much the same of the many worthy individuals named here who all played their part in the great adventure; but to mention every locomotive, every individual, would be impossible in a work of this size. I should be sorry to think that I have left out anyone's particular favourite, or inadvertently neglected an acquaintance. If the reader can be content with my broad canvas, however, and will indulge me in occasional personally interesting excursions, I believe that, by the end, he will consider that this has been a story worth telling.

I have been asked many times to include a works list of locomotives built; but such were their number and variety that this would be a book in itself. The reader is asked therefore to anticipate, God willing, a second illustrated volume complying with this request.

CHAPTER I

As permanent as the Pyramids themselves

TAKE the siege of Sebastopol, the overthrow of the Khalifa at Omdurman, the Battle of the Somme, the onslaught after the D-Day landings in World War II, the Bridge over the River Kwai and the Falklands dispute, and one can be forgiven for thinking in terms of an epic television serialization of the history of the British and Allied armed forces. Add the foothills of the Himalayas, the highest lake in the world, the Rift Valley, the Atacama and Western Deserts, the North-West Frontier, Snowdonia and more than a Heinz variety of other countries and locations, including the temple of Horus at Edfu, and thoughts move to Civilization in the grand Kenneth Clark manner. Mix in the Channel Tunnel, the Thorp nuclear waste reprocessing plant at Sellafield, and virtually every manufacturing and extractive processing industry known to man and the scope becomes encyclopaedic. Throw in an out-of-work itinerant blacksmith and commuters on a high-tech train stranded on account of 'the wrong kind of ice' on the overhead line and one may be forgiven for suspecting a continuity slip-up – a send-up perhaps, not to be taken too seriously – even high farce.

But no, this is deadly serious, portrayed and enacted in a microcosm within a square mile not fifteen minutes' walk from the heart of England's most central and now its third-largest city.

The itinerant blacksmith is the godfather of it all, one Matthew Murray, who, it is written, in 1789 walked the sixty miles from Stockton on Tees to the leafy, and now affluent residential, suburb of Adel, five miles from the township of Leeds, to offer his services to Thomas Marshall, then pioneering flax-spinning at Scotland Mill – a site now only a decibel away from the flight path to the Leeds and Bradford Airport.

Leeds, or *Leodis,* had been mentioned in the Domesday Book as a parish of 1000 acres tended by less than forty small farmers supported practically by a grinding mill and spiritually by the Church of St Peter. Three hundred years later the settlement had still only grown to fifty families, but a 'cottage industry' of woollen weaving was beginning to emerge.

Much later, in 1626, the Crown bestowed the first Charter of Incorporation making Leeds a Municipal Borough possessed of 'a Parish Church of note'.

In the early years of the eighteenth century Daniel Defoe of Robinson Crusoe fame spoke of £10,000 to £20,000 worth of cloth changing hands in just over one hour on trestle tables out in the open by Leeds Bridge in Lower Briggate. Almost three centuries later the cloth tables have gone, but two miles further along the river bank to the east there is, on Sunday mornings, what was at its inception probably the largest car boot sale in the north of England. Continuity is the theme of this book.

By the end of the eighteenth century, Walpole had accused Leeds of being 'dingy'; others preferred to refer to 'muck and money'. The Industrial Revolution had left its mark, the River Aire was navigable, and by 1816 the Leeds and Liverpool Canal would completely sever the country from west to east in its passage from Liverpool, naturally, through Leeds to Hull via Goole.

In 1790, at the age of 25, Murray had introduced steam-driven colliery pumping engines in addition to patenting a new design of yarn-spinning machine, which Marshall installed in his giant new Mill south of the river in 1792, leaving Adel to the enjoyment of its wealthy residents. This giant mill was in the parish of Holbeck, just off Water Lane. The eponymous Hol Beck (*beck,* Yorkshire vernacular for *stream,* cf. German *bach*) still trickles along not far from the south bank of the river Aire, and the aptly named Water Lane alongside is still a major traffic route, as is Marshall Street at right-angles to it, which right angle still contains some rather dilapidated remains of the original mill.

As Leeds developed as the centre of the world's flax-spinning industry, Marshall's Mills of 1792 had already been extended five times before the vast new addition in the Egyptian style was built in 1838–40. The vast façade was modelled on the Temple of Horus at Edfu, and the chimney was a replica of Cleopatra's needle. The façade, now a listed building, and much of the mill also, stand intact to this day as part of the huge Kays mail order complex.

Murray had moved with Marshall to Holbeck, and by 1795 his new machinery was up and running. He teamed up with David Wood to build a small workshop on nearby Mill Green, and two years later, with extra capital provided by James Fenton and William Lister, the firm of Fenton, Murray & Wood opened extended premises at Camp Fields across Marshall Street and fronting on to Water Lane. In these premises they consolidated the Leeds tradition for mill engines, machinery and plant that John Smeaton had fostered many years previously.

Smeaton was by training a mathematical instrument maker but later turned civil engineer, best remembered for the Eddystone lighthouse. He designed and provided the plant for local foundries which supplied his castings. By the time he retired in 1791 he had reputedly been responsible for fifty mills and thirteen steam pumping engines. Smeaton died in 1792 and is buried in Whitkirk church, five miles east of Leeds, near the lodge in which he worked. There is a commemorative tablet in the shape of a lighthouse on the north wall of the sanctuary of this church, where the author and his family before him have worshipped for the intervening two centuries.

Smeaton had established Leeds's pre-eminence in engineering. Murray consolidated and developed this eminence in a practical way that has not received the recognition it deserved. The Murray family home, Holbeck Lodge, built under his direction half a mile away off Water Lane, and by 1846 marooned in the triangle of lines at the west end of Leeds City station, was reputedly the first dwelling to incorporate underfloor steam central heating. It thereby earned the sobriquet 'Steam Hall', but it was demolished without ceremony in 1959.

In 1810 John Blenkinsop, a colliery agent at Middleton, Leeds, had entrusted Murray with the design and building of a steam locomotive to run on Blenkinsop's patent rack rail system. The track gauge was 4 ft. 1 in. and the locomotive weighed roughly 5 tons. The rack drive was on the left-hand side, and two double-acting cylinders were

The ornate façade of John Marshall's Temple Mills just across the road from Murray's works. A centenary sketch by Mike Peace of Leeds City Council's planning department, reproduced by courtesy of the Council and with acknowledgement to the artist.

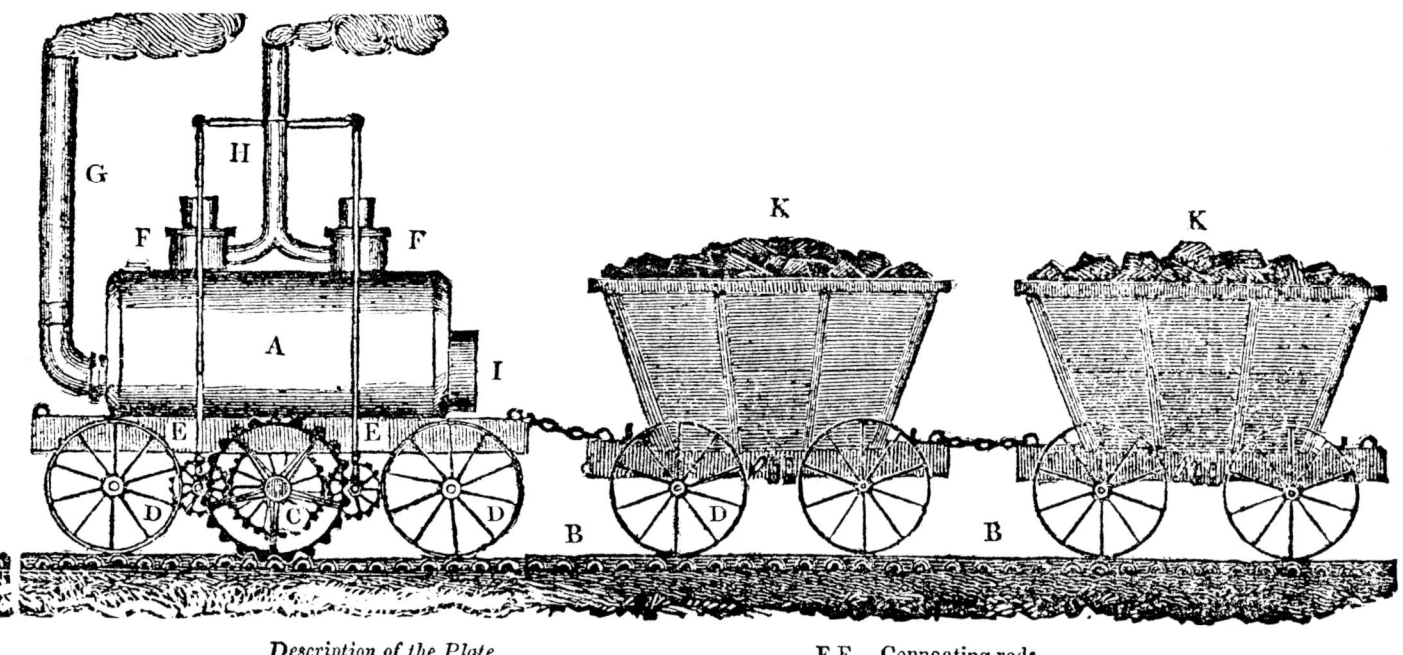

Description of the Plate.

A. Boiler.
B B. Rail road.
C. The propelling wheel, which is put in motion by the agency of steam, or any first mover.
D D. The carriage wheels.

E E. Connecting rods.
F F. Steam cylinders.
G. Smoke chimney.
H. Steam or discharging pipe.
I. Fire place.
K K. Coal waggons, or carriages of any description.

'It only came lumbering on like a cart.' So David Joy described the Middleton Railway rack locomotive built by Murray for John Blenkinsop in 1812. John Blenkinsop's letter of 26 March 1814 and the accompanying engraving appeared in The Monthly Magazine of 1 June 1814. Thanks are due to Mrs Sheila Bye, Editor of the Middleton Railway Trust Ltd's magazine The Old Run, who originally discovered this document.

supplied with steam from a single-flue oval boiler exhausting through a tall chimney of reduced diameter. The boiler pressure of 55 p.s.i. was high for the time and required a royalty payment to the owners of Trevithick's patent for the use of 'strong steam'. The price of the first two locomotives was £350 each.

The *Leeds Mercury* describes the trials of 24 June 1812 in some detail, and comments on the 'thousands' of people who had gathered to witness the successful event. The newspaper recorded the 'iron horse' as fulfilling the duties of 16 horses and drawing 27 wagons, a load of 94 tons, at speeds of up to $3^1/_2$ m.p.h., with 10 m.p.h. being possible with a light load. Named *Prince Regent* and *Salamanca*, the locomotives began regular work on 12 August 1812, becoming the first regularly-employed commercial railway locomotives. Two more were built in 1813, and they were to work until 1835.

The Murray engines attracted the attention of probably the first recorded 'train spotter', David Joy, who wrote years later that *living in Hunslet Lane, on the London Road, the old coal railway from the Middleton Pits into Leeds, ran behind our house a few fields off, and we used to see the steam from the engines rise above the trees. Once I remember going with my nurse, who held my hand (I had to stretch it up to hers, I was so little) while we stood to watch the engine with its train of coal-wagons pass. We were told it would come up like a flash of lightning, but it only came lumbering on like a cart.*

Joy, born in 1825, was the son of Edward Joy of the Leeds Oil Mills situated between the Middleton Railway's town terminus and what was later to be the North Midland Railway passenger station, opened in 1840. The young Joy had no interest in joining the family business, and in 1841 he was apprenticed to Fenton, Murray & Wood. His subsequent diaries, serialized (in places inaccurately) in abridged form in the *Railway Magazine* during 1908, give a racy and often surprisingly humorous account of the fast-changing locomotive building scene of those days, all the more fascinating in that so few other eye-witness accounts were ever committed to paper. Many of the famous names of the early days are there, and reading them more than a century later one is continually astonished by their grasp of theory – necessarily somewhat empirical but none the worse for that – and attention to detail.

Murray did not build any more locomotives, preferring to concentrate on his other activities, and is reputed to have declined an enquiry from the Stockton and Darlington Railway in 1825, on the basis that 'it does not suit with the present arrangements of our business to take orders for high pressure or locomotive engines'. However, in contradiction to this, in the same year Murray recommended to the Stockton and Darlington an articulated locomotive of which a description appeared (post-humously) in *Newton's London Journal* in 1826.

Murray died on 20 February 1826 at the age of 61 and was interred in Holbeck cemetery, where his workmen placed a commemorative cast-iron obelisk. The towns-people of Leeds did not appreciate his significance. Ironically they later erected a statue in what is now City Square to the memory of James Watt, whose only connection with the city had been to buy land adjacent to Camp Fields in order that his son could spy on the activities of Murray and his partners. Murray's endeavours are commemorated publicly only by a neglected plaque on the wall of an insignificant, and much later, building marking the site of his Water Lane foundry.

On Murray's death his share of the company passed to his son-in-law Richard Jackson (hence the firm's change of name to Fenton, Murray & Jackson) under whose superintendence locomotive work recommenced in earnest in 1831 – some three years, be it noted, before the first main-line railway company reached Leeds, this being the Leeds & Selby Railway at its modest terminus at Marsh Lane some distance from the city centre.

The inaugural train on the Leeds & Selby ran on 15 December 1834, when the Fenton, Murray & Jackson locomotive *Nelson* departed at 6.30 a.m. with 150 passengers in three first-class and four open carriages. Slippery rails and much application of ashes allowed only slow progress up the grade to Crossgates, much to the enjoyment of a reported twenty thousand spectators on the outskirts of Leeds, and the train did not arrive at Selby until 9 a.m. The return was much more expeditious, taking 1 hour 15 minutes for the 20 miles, with 30 m.p.h. being sustained over the last seven miles.

By the mid-1830s the Holbeck works was almost certainly the largest locomotive manufactory in the country, if not in the world, supplying not only the emerging British railways but exporting to Belgium, France, Germany and America. It built mainly to subcontract from Robert Stephenson. The records were

destroyed in a fire in 1872 while the works was in the hands of Smith, Beacock & Tannett, machine tool manufacturers, who had renamed it the Victoria Foundry during their occupancy, which lasted until 1894. Samson Fox, born in 1838, was an apprentice in the Smith, Beacock & Tannett days.

It is known, however, that Fenton, Murray & Jackson also built *Vulcan* and *Fury* for the Liverpool & Manchester Railway, a number of 2–2–2s for the North Midland Railway, and what were reputed to be the best twenty of the 62 'Firefly' class 2–2–2 express passenger engines designed by Daniel Gooch for the broad-gauge Great Western Railway. These last were based on Stephenson's *North Star*, and the drawings and templates were prepared by the railway company. The driving wheels were 7 ft. 0 in. in diameter, and the cylinders measured 15 x 18 in. The order for 62 locomotives was split as follows:

20	Fenton, Murray & Jackson, Leeds
16	Nasmyth, Gaskell & Co., Manchester
10	Sharp, Roberts, Manchester
6	Jones, Turner & Evans, Newton-le-Willows
6	R. B. Longridge & Co., Bedlington
2	G. and J. Rennie, London
2	Stothert and Slaughter, Bristol

An interesting pointer to the times is that a small railway engine of this size – the 'Fireflies' weighed 24 tons in working order – could cost in the region of £2000, whereas exactly one hundred years later, a 48 ton steelworks shunter was selling for only £4500. The last of the 'Firefly' class, *Argus*, left the works in August 1842, and David Joy describes the surprisingly advanced testing thus:

After all the work of a day, I well remember staying over hours to see one of these engines (the last) tried in steam. It was placed on special rails with struts in front and behind, and the middle wheels were resting on underground pillars [?rollers], which they drive [sic] round, and to which was attached a counter to show the revolutions per minute or the miles per hour. After that test a dynamometer was yoked on at the trailing buffer, and the hauling power noted.

The most famous of all the Leeds batch of 'Fireflies', *Ixion*, took part in the gauges trial of 1845, when the famous Gauges Commission pronounced on the comparison between the Great Western Railway's broad gauge (7 ft. 0 in.) and the so-called standard gauge (4 ft. 8½ in.).

In the intervening period, Fenton, Murray & Jackson ceased trading; or, as E. L. Ahrons put it, 'in 1843 the oldest locomotive firm in the kingdom, Fenton, Murray &

Great Western Railway broad-gauge 'Firefly' class locomotive Argus, *as originally built. The photograph is believed to have been taken at Chippenham around 1853. The last of the Leeds-built batch, this is the engine witnessed by David Joy when on test at Fenton, Murray & Jackson's works in August 1842.*

FIREFLY SOCIETY

Jackson, dropped out of existence.' Immediately prior to this statement in his book *The British Steam Railway Locomotive from 1825 to 1925*, reference is made to the Liverpool & Manchester Railway's 'Bird' class locomotives built in the railway company's own workshops at Edgehill between 1841 and 1844. This self-build policy was to have a far-reaching effect on the railways of Britain, and indirectly resulted in the formation of the Locomotive Manufacturers' Association in 1875.

The Ordnance Survey sheet of 1846, produced after locomotive building had ceased, shows the former works of Fenton, Murray & Jackson as having three 'fitting-up shops', two foundries, two 'boiler building houses' and sundry other buildings, with the main entrance fronting on to Water Lane. This gloriously random collection of interconnected buildings extends back to fill the right

angle of the Marshall Street/Water Lane junction opposite Marshall Mills culminating at the far end in the famous 'Round Foundry', a three-storey circular structure which, with its rectangular portico, gave the appearance of a large cylinder complete with valve chest.

The 'Round Foundry' itself is long gone, but many of the smaller buildings at the Water Lane end of the complex may still be seen as part of a hotch-potch of old and new engaged in various light industrial activities.

Meanwhile, in 1837, one of Murray's former apprentices, Charles Todd, had formed a partnership with James Kitson to create a locomotive factory in Hunslet with financial backing from David Laird, a wealthy Scottish farmer.

Todd, Kitson & Laird's first order was for six locomotives for the Liverpool & Manchester Railway. The first two, which were 12 x 18 in. 0–4–2 tender 'luggage

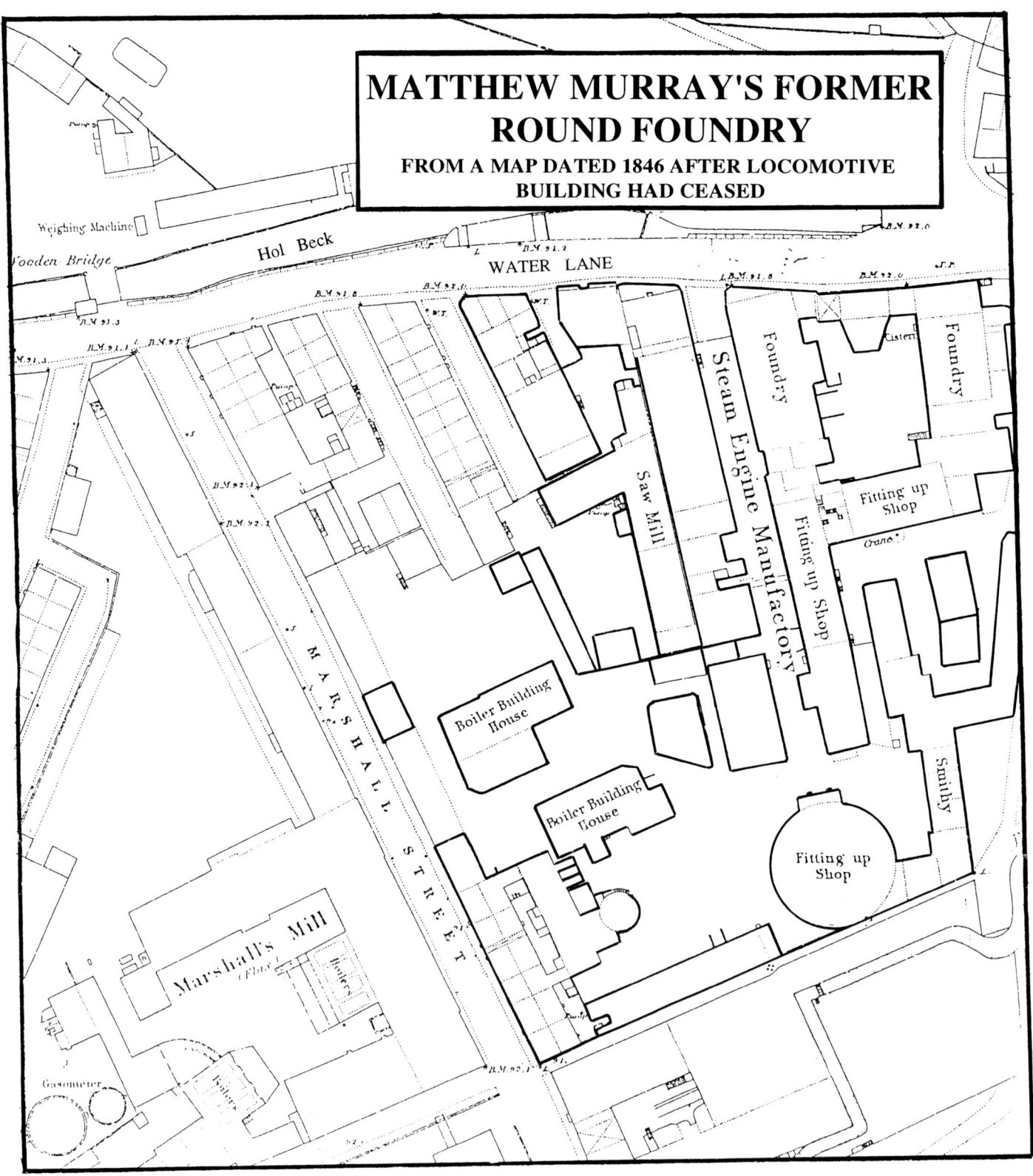

MATTHEW MURRAY'S FORMER ROUND FOUNDRY

FROM A MAP DATED 1846 AFTER LOCOMOTIVE BUILDING HAD CEASED

Two views of buildings on Foundry Street, off Water Lane, Leeds, in November 1995. The brickwork and general construction suggest great antiquity, and the positions match exactly the foundry and 'fitting up' shops of Fenton, Murray & Wood's locomotive manufactory as shown on the 1846 Ordnance Survey sheet. AUTHOR

The Liverpool & Manchester Railway 'luggage engine' Lion pictured after rescue from Princes Dock and restored as a working exhibit at the L&M centenary celebrations at Wavertree in 1930, complete with reproduction original train. Very much pre-Titfield Thunderbolt.   ALLAN C. BAKER COLLECTION

Photographed in October 1995, this arch stands in the centre of the northern boundary of the quadrangle of the original Railway Foundry in Pearson Street. Since the quadrangle was only used for material storage and experimental work after E. B. Wilson's new Railway Foundry was built in 1846, this arch is unlikely to have been built any later than this. It is reasonable to assume from its architectural style that it dates from much earlier, and it may quite conceivably be the Todd, Kitson & Laird gateway out of which the Liverpool and Manchester Railway's Lion first emerged in 1839.   AUTHOR

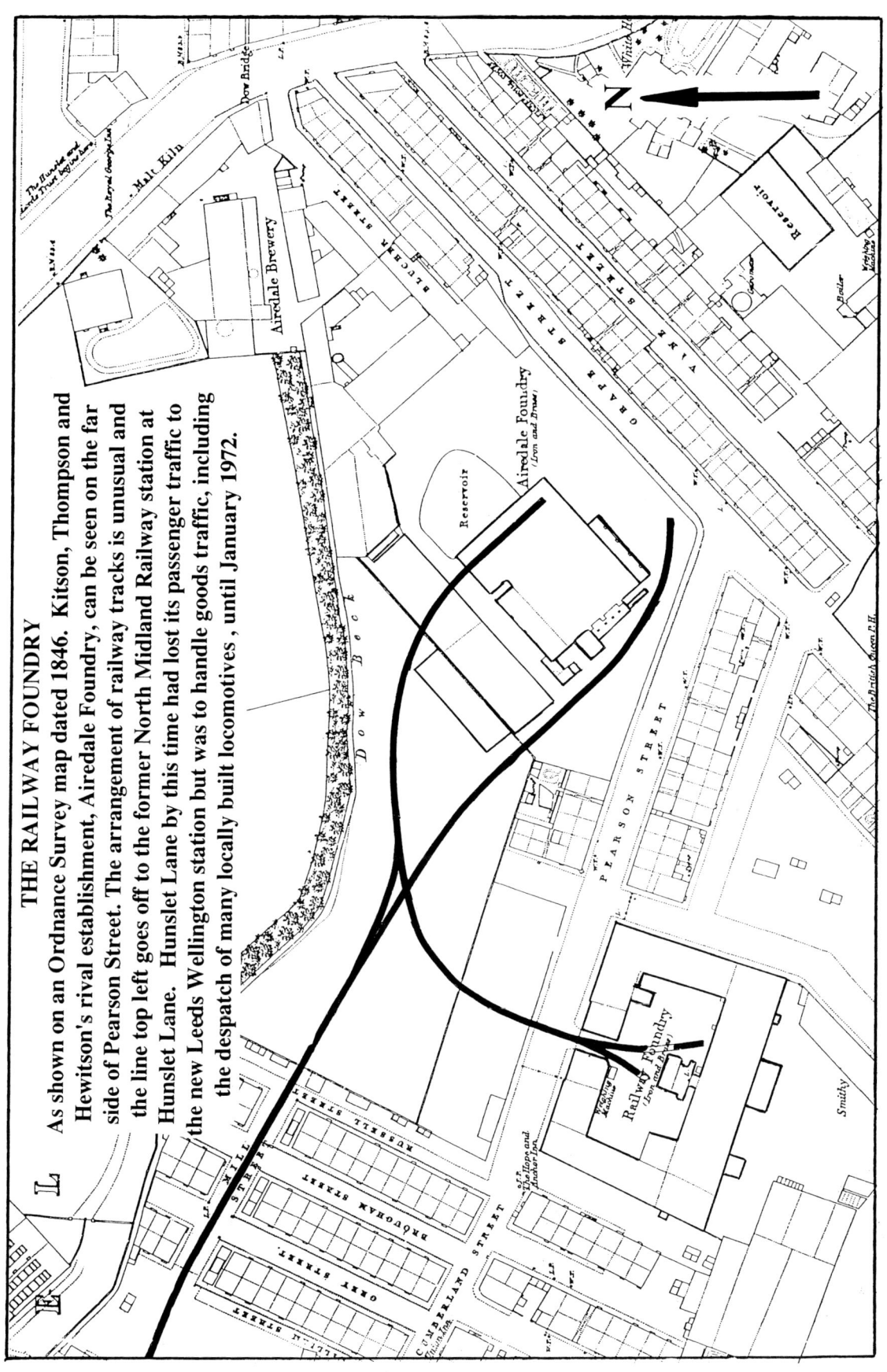

THE RAILWAY FOUNDRY

As shown on an Ordnance Survey map dated 1846. Kitson, Thompson and Hewitson's rival establishment, Airedale Foundry, can be seen on the far side of Pearson Street. The arrangement of railway tracks is unusual and the line top left goes off to the former North Midland Railway station at Hunslet Lane. Hunslet Lane by this time had lost its passenger traffic to the new Leeds Wellington station but was to handle goods traffic, including the despatch of many locally built locomotives, until January 1972.

engines', were named *Lion* and *Tiger*. They were supplied in 1838 at a cost of £1100 each, quite a drop from the price of the Great Western 2–2–2 'Fireflies'.

Lion, the more famous of the two, was taken over by the Grand Junction Railway and later became London & North Western Railway no. 116. In May 1859 it was sold to the Mersey Docks and Harbour Board, where it served, minus its wheels, as a pumping engine at Princes Graving Dock until 1928. The dock Board then presented it to the Liverpool Engineering Society, which had it restored to working order at the Crewe Works of the London, Midland and Scottish Railway. *Lion* was in steam at the 1930 Railway Centenary exhibition and celebration at Wavertree, drawing a representative train, and finally achieved immortality as '*Thunderbolt*' in Ealing Studios' classic comedy *The Titfield Thunderbolt* in 1952.

Kitson and Laird split with Todd in 1839 to form Laird, Kitson & Co., later Kitson, Thompson & Hewitson and finally Kitson & Company. This succession concentrated on main-line locomotives, and roughly 5000 locomotives were built at the Airedale Foundry before Kitsons themselves closed in 1938. But this is outside our particular story and we must concentrate on Todd's succession.

Todd brought yet another partner into the tangled Leeds locomotive web: John Shepherd. Shepherd & Todd were located in a quadrangle of buildings in Pearson Street clearly marked on the 1846 survey as 'Railway Foundry' and with a line of rails crossing Pearson Street to link up with a branch line from the North Midland Railway's Hunslet goods and passenger terminus. This branch also served, indeed more readily served, the embryonic Airedale Foundry, with the Railway Foundry track making a trailing connection.

Railway folklore tells of *Lion* and its fellows being assembled in an old mill and of walls being taken down to release the beasts for despatch. This might just be so, but more likely is the need to remove chimneys to clear a low archway.

Express engine designed by John Gray for the York & North Midland and Leeds & Selby Railways and built by Shepherd & Todd in 1840. Fitted with Gray's patented 'horse leg' gear, these are believed to have been the first locomotives built with expansive valve motion. Antelope and Ariel of the Y&NMR, David Joy's favourites, were of this type.

THE ENGINEER

Both the Railway Foundry and the first part of the Airedale Foundry are described in contemporary accounts as having been converted from old mills. Both enterprises pre-date the North Midland Railway; therefore the more direct access into the Airedale Foundry is of no help in deciding *exactly* where the Liverpool & Manchester Railway engines were built. It would fit nicely into the chronology of our tale to think that they were built in what became Shepherd & Todd's Railway Foundry, but the evidence is circumstantial and no records are to hand either to prove or to gainsay this assumption. If it was the case, the arch that could have been the problem was still standing in 1997, and it does no harm to attempt to conjure up the scene way back in 1838.

For five years Shepherd & Todd produced main-line locomotives, about twenty or so, in the rather cramped premises. Three 0–4–2 tender locomotives appeared in 1839 for the Leeds & Manchester Railway. Bearing a strong resemblance to *Lion*, thereby lending credence to the idea of the Liverpool & Manchester engines being Railway Foundry products, they had larger 14 x 18 in. cylinders and are shown on contemporary sketches as having a more modern boiler and firebox arrangement. The remaining locomotives were mainly 2–2–2 and 0–6–0 tender engines for the Hull & Selby and the York & North Midland Railways, but there were at least two early export examples for the Paris–Orléans Railway. Remember, in passing, that there was no connecting railway at this time, and no road vehicles other than those that could be drawn by horses, and these early locomotives must have been virtually manhandled to the Leeds & Liverpool Canal some ³/₄ mile away at Clarence Dock or some similar point.

Two of the 2–2–2 Y&NMR passenger engines, designed and built by Shepherd & Todd at the Railway Foundry in 1841, were named *Antelope* and *Ariel*. David Joy saw them in 1843 and he described them as handsome and fast. At that time they had John Gray's expansive reversing gear called the 'horse leg' gear, patented in 1838. Joy revisited York engine sheds in 1883 and saw the *Antelope*, which he thought 'a lovely, light, fleet looking engine still' although by then rebuilt and enlarged.

Todd departed in June 1844 to run the nearby Sun Foundry (later Hathorn Davy & Company) in Dewsbury Road, where he built twenty more locomotives before leaving Leeds altogether. Edward Brown Wilson, son of a Hull shipowner, took his place at the Railway Foundry, joining Shepherd in November of the same year.

In August of the following year, six long-boiler four-coupled passenger engines were ordered from the Railway Foundry by Thomas Cabry, Locomotive Superintendent of the Y&NMR. David Joy made the drawings. Of 2–4–0 wheel arrangement, they had 15 x 20 in. cylinders and driving wheels of 6 ft. 0 in. diameter.

We have witnessed the beginnings of an industry that three or four generations later must have seemed as permanent and immemorial as the Pyramids themselves. In setting the scene we have met the young David Joy and introduced Edward Brown Wilson. We shall see in the next chapter how this powerful if quarrelsome double-act did much to ensure that innovation and enterprise turned over the years into a tradition.

CHAPTER 2

E. B. Wilson – Strength through Joy

Wilson saw the inadequacy of the old Railway Foundry only too clearly. He was an ambitious man with plans for expansion which obviously led to disagreements with Shepherd. He appears to have left the company temporarily during 1846, when the trading name became Fenton & Craven for a while. Of these two, James Fenton was the son of Murray's partner of the same name at the Round Foundry; and John Chester Craven, later to become what Tom Rolt has called the 'formidable martinet' in charge at Brighton, was an alumnus of the Round Foundry. Shepherd's name dropped out of the picture completely from this time.

Before his temporary absence Wilson had seen the talent in the 20-year-old David Joy. Like Craven, the young Joy had served some time as a pupil of Fenton, Murray & Jackson, joining the Railway Foundry when the Holbeck empire tumbled in 1843; he was promoted by Wilson to the post of acting chief draughtsman in 1845. Joy was later to design the 'Jenny Lind' class of passenger locomotive. E. B. Wilson returned to effect a complete take-over of the company late in 1846.

The impact that the 'Jenny Linds' had on the subsequent development of motive power should not be underestimated; nor should the contribution that E. B. Wilson, along with Robert Stephenson, Hawthorns, Sharp Brothers and others, made in the provision of main-line locomotives to the various railways of Britain in these early years. That the success of the firm of E. B. Wilson was due to the prodigious efforts and strength of character of two individuals, Wilson and Joy, is beyond doubt. The latter in particular deserves far greater recognition than he has hitherto received.

In writing the history of Hudswell, Clarke & Co. Ltd, the author's contemporary and friend Ronald Nelson Redman expertly covered the story of E. B. Wilson in his book *The Railway Foundry* in 1968. Redman, naturally enough, was looking on it from the Hudswell Clarke viewpoint; but it has to be said that whereas Hudswells occupied a site on the periphery where the old paint shop had stood, the Hunslet Engine Company's works of 1864 was built very much at the business end of the Railway Foundry. Indeed, the Hunslet erecting shop was on the exact spot where the E. B. Wilson locomotives had been erected, and the Hunslet Engine Works eventually covered virtually the whole of the Railway Foundry estate. At the risk of duplication it is therefore necessary to cover the products of E. B. Wilson again in some detail.

Extracts from Joy's diaries are invaluable in piecing together the history of locomotives and understanding the relationships in these pioneering yet stabilizing days, In January 1845, he mentions an order for ten engines for the Manchester, Bury & Rossendale Railway.

Dutton, our manager, knew nothing of locomotives, so I had to take all particulars from the railway's engineer, Mr Cawbey [Cabry?], and I made all the drawings of the engine, going to the station to copy details from Stephenson's engines chiefly.

Mr Cawbey was very particular about his water level and steam space, as the engines had to work on an incline of 1 in 33.

Early in this spring [1845] Stephenson's Great A came out, and was the engine which ran in the battle of the gauges against the Great Western Railway's [Leeds-built] engine Ixion.

Apart from the general interest in locomotive matters excited by the 'Battle of the Gauges', these trials gave Joy his first real ride on a locomotive. This is how he describes this epoch-making circumstance:

We pupils used to frequent the railway station very much, and one afternoon, watching the 4 p.m. York express start, the driver, Sid. Watkins, asked me if I would like a ride. (Will a duck swim? Rather.) No coat, nothing on, I popped on to the engine, and away we went, so jolly. This was my first fair run on an engine with a train; only to Castleford, still it was fine. I had many another like it. And this was Great A.

Together with his fellow-pupils and others, Joy appears to have had many opportunities of riding upon the footplates of locomotives during the autumn of 1845.

At this time we lads were always on the engines of the Y. & N. Midland Railway (as it then was called), taking little trips to Castleford, Milford, and sometimes to York. And just now we got an order from this railway for six engines – Thomas Carby [Cabry?], engineer. I had all these drawings to make.

Joy presented his shaded drawings of this engine to the local Mechanics' Institute. He illustrates the Y&NMR locomotive No. 10, of which he gives the following dimensions:

Wheels four-coupled, 6 ft. diameter; cylinders, 15 in. by 20 in.; boiler, 12 ft. by 3 ft. 4 in.; steam, 90 lbs.

The point of interest concerning this Y&NMR engine is disclosed in the next paragraph.

This is the first I distinctly recall of the link gear, though our Manchester, Bury and Rossendale engine had it on (but on the Manchester, Bury and Rossendale engine it was the box link). The next run I remember was with Joe Elliott, on Zetland, *a little engine, like the Y. & N. Midland Railway engine, but smaller – 5 ft. 6 in. wheels.*

Here is an entry for 28 February 1846.

The first Manchester, Bury and Rossendale Railway engine was finished, and we took her for a run to Normanton. A crowd of us was on the footplate (Fenton was not named in my original tour at all) and all over her. Then we had a big dinner at Normanton, and lots of wine. After having had a bottle, or near it, I passed the bottle. And E. B. Wilson, to whom I sat next, said: 'Joy, you'll never make an engineer if you don't drink your bottle of wine.' I said: 'I will, nevertheless.' We all returned to Leeds on the engine, being hooked on to a Leeds and Manchester train.

There was not a sober man on the footplate but myself. The driver of the Manchester and Leeds train not only made us pull the train, but he put on his brake for mischief. Didn't our engine 'spit fire'!

On March 3rd we read:

My birthday and, free of my apprenticeship, had a spread at home.

Joy lost no time in seeking employment, for the next day he went *to Railway Foundry – to ask for a berth. No, they wanted no paid hands. To Wilsons' at night for a big dinner. Myers, the Russian, there. Twice as many bottles of wine drunk as guests.*

A month later, on April 5th, he went *to Manchester on executors' business, also to seek a draughtsman's berth. Called at Leeds and Manchester Railway offices – on Sharpe [sic], Roberts & Co. – and others. No go. Stayed at Uncle John's. Home on Thursday, 6th, and found a letter asking me to call at Railway Foundry. Went and found a 'break-up'. Wilson out of it, and Fenton and Craven waiting. They offered me a draughtsman's berth at 31s. 6d. (guinea and a half), took it and began, disgusted. One engine to build like Manchester, Bury and Rossendale Railway – but with 5 ft. wheel. A drunken draughtsman (Archer) making drawings exactly the same as*

Manchester, Bury and Rossendale Railway – but only wheel dropped to suit rail-centre; the line of the cylinder the same!!! I took up drawings and put them right. But that man could shade gloriously when he was drunk – so I learnt from him – I was not too proud.

With E. B. Wilson 'out of it' and Alex B. Wilson in charge but knowing nothing about locomotives, Joy obviously had his work cut out – so much so that he went on doctor's orders to the Isle of Wight for two or three weeks in the autumn of 1846, 'to recruit', as he put it.

Returning home 'recruited' on 29 November he found a letter from the Railway Foundry urging him back to work. The following morning, which was a Saturday, Joy went and found E. B. Wilson back in power.

I was to go same night to London and on to Brighton on Sunday to see John Gray about an order for ten engines, of which I was to take all particulars. [Gray was the Locomotive Superintendent of the London, Brighton & South Coast Railway.]

Arrived at Brighton on Sunday night. Spent three weeks taking tracings all day, and receiving instructions from Gray after 7 at night. He gave me an engine pass, so I went all over the line on the various engines – to Chichester on Satellite, *a little engine by Rennies…*

After describing a number of trips he concludes:

Returned to Leeds and started drawings for the new engines, [LBSC nos.] 60–69 inclusive. We had got the new drawing offices – over the entry – and Jimmy Fenton for manager.

How that one nickname makes the whole thing come alive.

Locomotive superintendentships were not positions for life in those days, for Joy says:

I had hardly started the Gray drawings when word came that Gray was 'out', and we were to design a new engine. I, as chief draughtsman, had it to do – so set off scheming by order of Fenton, of course, on the lines of the last engine we had, Leeds and Dewsbury, short boiler (11 ft. 6 in.), outside cylinders, drivers far back, trailing wheels behind the firebox, and as much heating surface as possible in tubes, and 60 sq. ft. in firebox, if possible. Got out 10 or 12 schemes in a week, and threw all aside – after dissension. Then – 12 noon, Saturday – Fenton came to me and said: 'Try another, and give inside cylinders 15 in. by 20 in., and 6 ft. wheels, and again the biggest surface possible.' I was sick of it, and bolted for my Saturday afternoon.

We have now reached an event that is of peculiar interest, because it shows Joy's share in one of the most important developments of the steam locomotive – the production of the celebrated 'Jenny Lind' type of express engine. Joy describes the circumstances under which the design took shape.

Arrived at home, I thought over the engine to go for – and at once it struck me what a pretty engine it would make. So abandoned the Leeds and Dewsbury type, and all the feeling in favour of the old long boiler class. This was going back to the old engine, and my inoculation into Gray's ideas at once biassed me in favour of that type. I had studied very well for three weeks, and had ferreted among all the types of engines on the Brighton Railway, and had ridden on most of them, with the idea to get a definite opinion of my own which was best for big speeds.

So, having a sheet of double elephant mounted ready, as I mostly had now – as I spent all my evenings at drawing any engine I could ever get outside dimensions of – I set to work and drew out with Gray tendencies, a 10 ft. 6 in. boiler, as big in diameter as I could get it, and as low down as I could possibly get it – for the cry was one for low centres of gravity to secure steadiness, though Gray did not seem to care for it. Cylinders, 15 in. by 20 in.; drivers single,

and as far back as possible, 6 ft. diameter. Inside frames, which must be made to carry the cylinders, the frames stopped at the firebox, so that the firebox was got as wide as the wheels would allow it. This, of ordinary length, gave 80 sq. ft. of surface, and with 124 tubes 2 in. diameter gave 730 sq. ft., or a total of over 800 sq. ft. Then I put on Gray's outside frames for leading and trailing wheels, 4 ft. diameter, giving the bearings below, thus making a firm wheel-base, with no overhanging weight.

Cylinders and valves between, with ordinary (then approved) link, but slung only from one side. The steam dome on the middle of the boiler, and two safety valves under a cover on the firebox.

On the Monday morning Joy took his drawing to the Railway Foundry, where Wilson and Fenton approved it 'in almost every particular' and gave instructions to order boiler plates immediately and proceed with working out the details.

Having set the new engine design on its way for manufacture Joy spent considerable time at the end of 1846 making drawings for the new blacksmiths' and

The original Jenny Lind: London, Brighton & South Coast Railway No. 60, EBW 132/1847. An engraving from a drawing by David Joy. This is sometimes wrongly attributed to the Midland Railway on account of the engraving's having been used in the contemporary press to illustrate trials on the Midland Railway from Derby to Rotherham Masborough in May 1848. A Midland Railway 'Jenny' was compared with a Sharp Bros. engine to determine the best engine for fast trains. On like-for-like trains the Jenny averaged 51.9 m.p.h. and 36.2 lb. of coke per mile as against the Sharp's 48.1 m.p.h. and 44.5 lb. per mile.

INSTITUTION OF MECHANICAL ENGINEERS

erecting shops, which were to run alongside Cancel Street. In the new smiths' shop was one of the original Nasmyth's steam hammers. This fact is of interest because, in later years, Joy designed and constructed steam hammers extensively. Of this one he writes that 'it was under a general repair every Saturday but it did a lot of work.'

In May 1847 he records that the first Brighton engine, No. 60, was completed.

Boiler, etc. lagged with mahogany. Made first run with her to Wakefield via Normanton, and returning lost steam, and were nearly run into by the Manchester mail.

Opened new erecting and small tool shops with a big dinner and a ball to follow, when Wilson brothers – Charles and Arthur – and I fraternised. Charles is now head of the shipping firm Thos. Wilson & Sons, Hull.

First Brighton engine was called Jenny Lind *after the famous singer 'Jenny Lind', who was making a great excitement in London. I made a very highly finished drawing 1 in. to 1 ft. of her (the engine, not Jenny), which was lithographed, and sent about [see previous page].*

Got lots of orders here and there for this engine, and made at the rate of one per week – then a vast accomplishment. Now arranged same engine for a four-coupled [i.e. a 2–4–0].

Built the last Brighton engine, No. 69, with 6 ft. 3 in. drivers and 1,000 ft. surface.

With regard to the success of the 'Jenny Lind' class of engine, for many years they were to take the down express trains from London Bridge to Brighton – a distance of $50\frac{1}{2}$ miles – in 75 minutes. Joy writes:

Of course, it was the steam pressure that did it. But who was to blame, or to credit, for this bit of pressure from 80 or 90 to 120 lbs.? I have no note, but doubt not it was Jimmy Fenton. Thus the engine came out a mixture of the good points of the Gray, as first designed by Gray himself for the Brighton Railway, and the engine that had come of the long boiler, with its inside frame; from this type it got the elastic plate frame for leading and trailing wheels, but not rigid as in Gray's design. So these engines, at high speeds, always rolled softly, and did not jump and kick at a curve. Another thing I think came of my fancy was a very free exhaust. I always, from the first, saw the blast port cores made, and with my own hands passed over them, had them pared over, to get a free passage.

In the two years from 1846 to 1848 the combined entrepreneurial skills of Edward Brown Wilson and the common-sense draughtsmanship of David Joy had created at the Railway Foundry a locomotive works without equal. The 'Jennys' sold to the largest main-line railways – the Midland, the Great Northern, the London & North Western, the Manchester, Sheffield & Lincolnshire, the North Eastern, the London, Brighton & South Coast and the constituents of the Great Western – while standard 0–6–0 and 2–4–0 locomotives were equally widespread. Over five hundred main-line locomotives were built during the next ten years, mostly of standard types, covering all of the companies that later survived the Grouping to become the 'Big Four'. Remember that these locomotives were designed by the manufacturer, not the purchasing railway, available in some cases 'off the shelf', thereby anticipating an ethos not unlike the 1996 privatization plans for British Rail.

Seventy 'Jenny Linds' can be positively identified, including ten for the London, Brighton and South Coast Railway, eleven for the York & North Midland, and twenty-four for the Midland. Some historians credit the Midland with having fifty, but this is a misconception brought about by the fact that in addition to new construction Wilsons also undertook the repair and rebuilding of the engines belonging to the Midland Railway and others. Between 1849 and 1852 several old double-framed 2–2–2s and long-boiler 2–4–0s were comprehensively reconstructed or renewed in the 'Jenny Lind' style at the Railway Foundry, and in many cases emerged indistinguishable from the genuine article. The Midland's own works at Derby were still relatively small at that time.

Some measure of the size and capacity of the Leeds works can be gauged from the fact that, despite all this activity on standard types and on overhauls and renewals, there was time left to build at least eight of Thomas Crampton's patent rear-drive 4–2–0 express locomotives, six in 1848 and two in 1849. Of the first six, one went to the North British Railway as No. 55, and is recorded as having hauled the Royal train to Edinburgh after Queen Victoria had opened the Royal Border Bridge in 1850. For this occasion it is alleged to have been decked out in the Stuart dress tartan. The remaining five 1848 engines went to the Eastern Counties Railway, while the pair built in 1849, reputedly a cancelled export order, were purchased by the Aberdeen railway in 1850.

It was not all wine and roses, however. The 'railway mania' had burnt itself out, and by December 1848 many of the after-dinner 'gentlemen's agreements' to purchase locomotives were repudiated; and the Railway Foundry was left with about twenty engines well into completion.

Photographs of 'Jenny Linds' in original 'as built' condition are rare, but this view is a superb example. It shows Will Shakspere, *believed to be EBW 559 and certainly one of a pair (EBW 558 was the other), that Wilsons built to stock and sold to the Oxford, Worcester & Wolverhampton Railway in 1856. This one became OW&W No. 51, later Great Western Railway No. 208. Shown very early in life it was a 'large' Jenny with 6 ft. 3 in. driving wheels and 15 x 22 in. cylinders. It worked, mainly between Worcester and Oxford, until it was withdrawn from service in 1878 without ever having undergone any rebuilding.* NATIONAL RAILWAY MUSEUM

The Tartan Crampton. A painting by C. Hamilton Ellis purporting to show Queen Victoria's colourful chariot, North British Railway No. 55, after the Queen had opened the Royal Border Bridge in 1850. The E. B. Wilson fluted safety valve cover is much in evidence. MARTIN SMITH COLLECTION

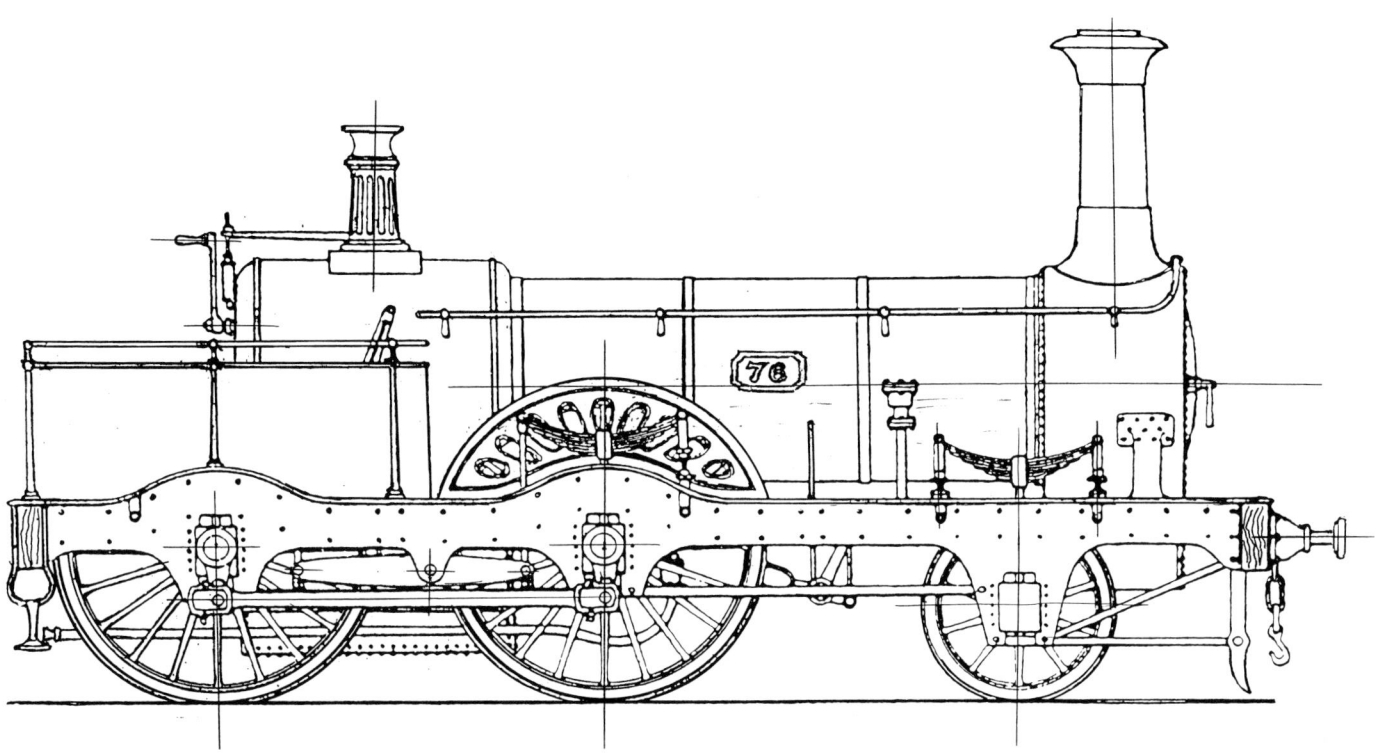

Archibald Sturrock's first design for the Great Northern Railway was a 16 x 22 in. 2–4–0 express engine with 6 ft. 0 in. diameter driving wheels. Five of them were built by R. & W. Hawthorn and fifteen by Wilsons, all in 1851. The first of the Leeds batch is shown in this drawing by G. F. Bird, which first appeared in the Locomotive Magazine in 1897. The classic Railway Foundry splasher and safety valve cover are noticeable.

1849 was a bad year, with little locomotive work, and Joy went to London as 'London Agent of the Railway Foundry Company', where he had ample opportunity to meet the 'swells' of the profession at the Civil Engineers Institute. The Foundry had entered into non-railway work in an attempt to balance the fall-off in locomotive orders, and Joy found himself involved in ironwork for 'baths and wash-houses' in Lambeth. 'Engineering work with a vengeance', as he confided to his journal.

By Easter 1850 the worst was over, and Joy was back in Leeds. Wilson negotiated an order with Archibald Sturrock, then Chief Mechanical Engineer of the Great Northern Railway, for ten 0–6–0 goods tender engines. Joy recalls that *Wilson came home with an order for ten Great Northern Railway goods engines, and telling Dickenson, the cashier, the price, Dickenson said, 'Why, Mr Wilson, that will only pay for the engines.' Wilson swore a big oath. 'D———. I never counted for the tenders at all.'*

There then follows a reference to Archibald Sturrock's first design of passenger locomotive, also for the Great Northern Railway – a 2–4–0 tender engine with domeless boiler:

Now we got another order from [the] Great Northern Railway for 10 four-coupled passenger engines, 6 ft. wheels. These became quite a type, and were the back stay for the running of the passenger trains, and all the specials of the Exhibition time the following year.

Joy is referring to the Great Exhibition of 1851 in Hyde Park, and the quantities mentioned must have been the subject of further discussion, for the Great Northern locomotive registers show sixteen 0–6–0 and fifteen 2–4–0 locomotives (GNR nos. 144–58, 167 and 76–90 respectively) supplied by E. B. Wilson in 1850–1, with many more to follow.

These were turbulent times, and the principal players exhibited the temperaments and characteristics of the true *artiste* in their business relationships, as witness a 'dust up' (Joy's expression) in June 1850:

Quarrelled with E. B. Wilson about forthcoming Exhibition for 1851, and went at a moment's notice, J. C. Wilson taking my place as shop foreman, and prigging all my books and papers in the shop foreman's office – all my John Gray's notes.

Summer, again a holiday.

True to form, by August, Wilson was begging him to come back.

E. B. Wilson fetched me on a Friday evening in a cab, took me to Arthington Hall to go next evening to open Nottingham and Grantham Railway on the Monday. He had taken it to work by contract at 2s. per mile run. No engines, nothing ready.

To Nottingham early Saturday. Midland Railway supplied us with two old Bury's singles to be at Grantham Sunday night. Saturday afternoon over the line with Underwood (engineer), Gough (secretary), and on the contractor's (G. Wythes) engine (ballast), went off the road, not very fast, but a jolly tumble about. Water tanks all to get ready by Monday.

Then came the opening.

Started at 9 a.m. with first train – five or six carriages – part second and third – and a lot of low-sided wagons.

Wilsons, through Joy, worked the line until October 1851, after which Joy went 'freelance' for a while until, again at Wilson's instigation, he was hired on 18 April 1852 as Locomotive Superintendent of the Oxford,

Worcester & Wolverhampton Railway. This was another line run on contract, this time by C. C. Williams, and sure enough it too was bereft of both locomotives and men. Joy persevered and collected what locomotives he could, and soldiered on until new locomotives were delivered, mainly 2–4–0 tender locomotives from R. & W. Hawthorn and E. B. Wilson.

On 1 February 1856 Williams's contract was terminated and the railway company took over its own locomotive department. Joy, now out of a job, returned to Leeds having transformed the 'Old Worse and Worse' into an efficient operation worthy of its incorporation first into the West Midland Railway in 1860 and, in 1863, into the Great Western Railway.

Meanwhile production at the Railway Foundry had returned to normal, the most prolific year being 1854, when over eighty locomotives were produced. These were

Although a small number of 2–4–0 locomotives based on the 'Jenny Lind' singles were built, having inside frames only for the driving wheels, the standard stock-built Wilson coupled passenger engine was an outside-framed 2–4–0 in three basic sizes having 5 ft. 8 in., 6 ft. 0 in. and 6 ft. 6 in. driving wheels respectively. Oxford, Worcester & Wolverhampton Railway Nos. 21–6 (EBW 330–5 but not in works number order) ultimately became Great Western Railway Nos. 182–7 and were of the small 5 ft. 8 in. wheel variety. All of them, though much rebuilt, lasted over thirty years, the last surviving until 1904. An unidentified member of the class is seen here, probably as near after its building in 1853 as 'sun painting,' or photography, would allow.

NATIONAL RAILWAY MUSEUM

Shown after its first rebuilding at Wolverhampton in 1871, GWR No. 184, formerly OW&W No. 23, has a Great Western Group 34b boiler but retains its semi-open splashers and rectangular fence plates, albeit with a rudimentary shelter for the driver.   NATIONAL RAILWAY MUSEUM

In this view GWR No. 184 is shown some time after its 1871 rebuilding, still with the GWR Group 34b boiler but with a GWR cab in place of the Wilson fence plates and weather-board. It was rebuilt yet again in 1893 with a larger R3 boiler with raised round-top firebox, and survived until October 1899.   NATIONAL RAILWAY MUSEUM

mainly 2–2–2, 2–4–0 and 0–6–0 tender engines for the Great Northern, the York, Newcastle & Berwick and the Oxford, Worcester & Wolverhampton, which were regular customers, though there were twenty-five 2–4–0 and ten 2–2–2 locomotives for the East Indian Railway. Four of the 2–4–0s were lost at sea – a common export hazard in the days of sail and, occasionally, steam. Six 2–4–0 tank engines went to Denmark.

At the time of Joy's return his diaries record that he 'got mixed up with Willis's road engine, from which all the agricultural engines have sprung'. Of this early road motor Joy says: 'I was at the road trials of this engine, with crowds all around – wondering.' Willis was the designer at Ransomes & May at Ipswich (later to become much better known as Ransomes, Sims & Jefferies, the Ransomes steam cranes and hydraulic buffer stops maintaining the railway connection). Sometimes known as the 'Farmer's Friend' this Railway Foundry 'one-off' appears to be in something of a time-warp; for Joy's diaries, if in sequence, suggest 1856 as the year of building, and some manu-facturing details support this, but Ronald H. Clark in his book *The Development of the English Traction Engine* says that it was exhibited at the Leeds and Norwich shows in 1849. Whichever date is correct the machine pre-empted all other makes of road locomotive – particularly those of the famed Steam Plough Works of John Fowler & Co., which was established after the Railway Foundry ceased trading.

Of the railways serving Leeds, the Great Northern, the London & North Western and the York & North Midland all purchased large numbers of Wilson engines, including 'Jenny Linds'. The Leeds Northern was not a Wilson customer, but the formation of the North Eastern Railway was in the air; and McLean in his *Locomotives of the North Eastern Railway* describes the event thus:

Ever since the Jenny Lind *attracted attention the directors (of the York, Newcastle & Berwick) had contemplated placing an order with E. B. Wilson & Co. of Leeds and they were only deterred from doing so in 1849 by reason of the success of no. 180 (the 'Plews') and no. 190 [built by Hawthorn and Stephenson respectively]. However in 1853, with the amalgamation [of the YN&B, Y&NMR and Leeds Northern] in sight, came the demand for additional engine power, and Messrs. E. B. Wilson were asked*

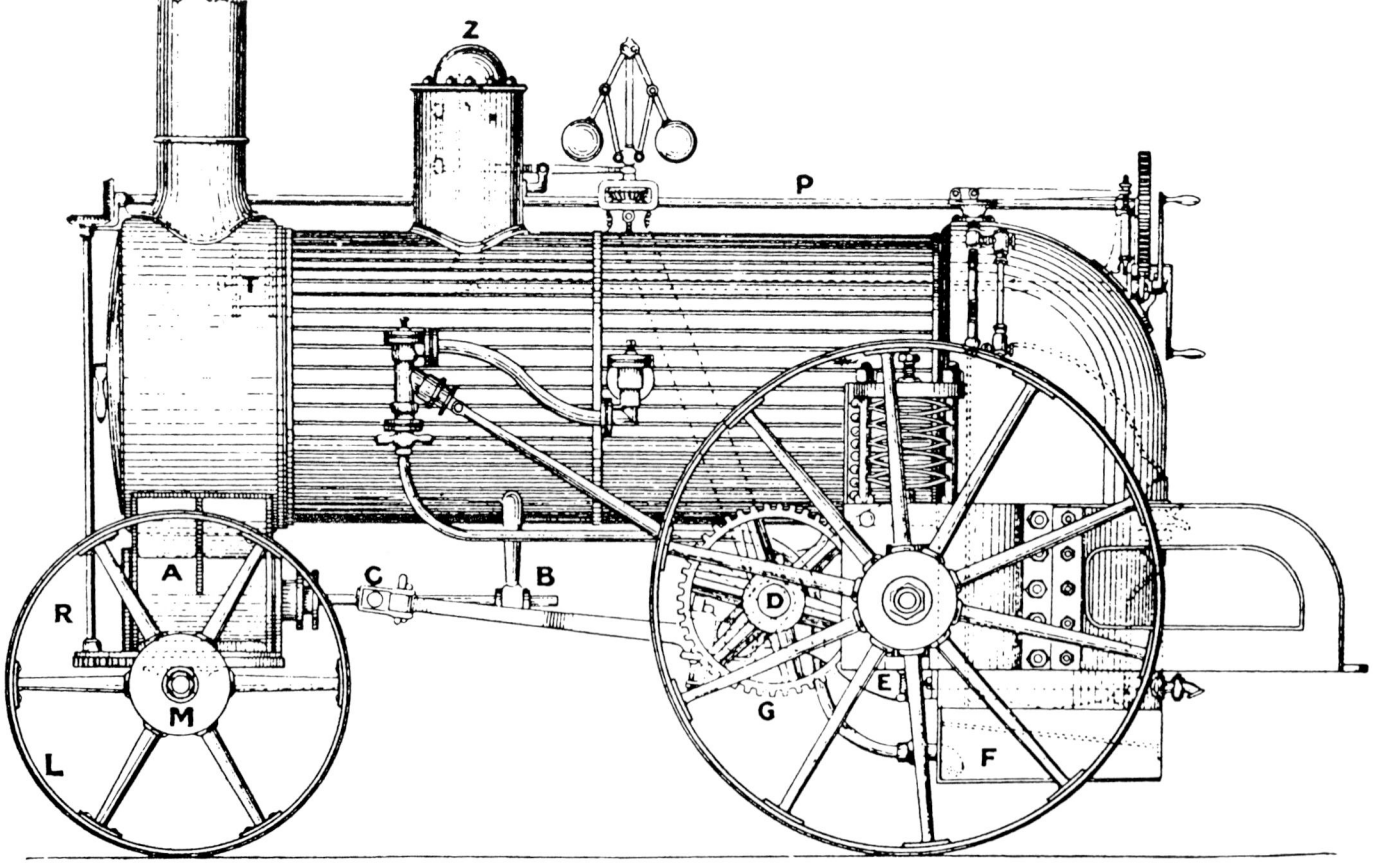

The progenitor of the traction engine was Willis's road locomotive, built at the Railway Foundry by E. B. Wilson & Co. Ltd. This drawing clearly demonstrates its locomotive works origin.

A. O. HUNT

to supply two large 'Jenny Linds' and three different sizes of coupled passenger engines as well, making a total of nine. The engines, the first lot of North Eastern Railway engines proper, all bore the well known classic features – fluted dome and safety valve covers in the style of a Corinthian column on a square base and radially slotted driving wheel splashers. This portion of the design, considered to be the most elaborate ever attempted, was the work of a lady, the wife of James Fenton, manager of the firm – a circumstance probably unique in the history of the locomotive. The two 'Jenny Linds' had 16 x 20 in. cylinders and 6 ft. 3 in. driving wheels. Then came seven 2–4–0 tender engines, one with 16 x 20 in. cylinders and 6 ft. 0 in. coupled wheels and finally two 16 x 22 in. cylinders with 6 ft. 6 in. coupled wheels. These last, known as the 'Big Wilsons', were the largest engines on the North Eastern at the time and later had the cylinders enlarged to 17 in. diameter.

The work done by Wilsons for the Midland Railway – new construction, repairs and rebuilding – has already been remarked upon. The Great Northern Railway business was of equal importance. Between 1851 and 1854 one hundred main-line engines, including seventy-four 0–6–0 goods and seventeen 2–4–0 express passenger, were supplied to the GNR. Additionally, an offer was made in 1851 by Wilson to overhaul and repair the entire fleet of some 200 engines at Leeds, and this received the most serious attention of Edmund Denison, the railway company's chairman. The GNR Board commissioned a panel of engineers to study the proposal, including, in addition to E. B. Wilson himself, several railway chief mechanical engineers, among them Joseph Beattie (LSWR), Thomas Cabery [Cabry?] (York & North Midland), J. C. Craven (LB&SCR), Matthew Kirtley (MR) and Archibald Sturrock (GNR).

No reason was given for rejection of the plan, but it is interesting to note that the Midland produced the first locomotive at its Derby Works in 1851, and that building of the GNR's own 'plant' works at Doncaster started in 1852. Battle lines had been drawn up for the division of British locomotive building into, broadly, 'railway company' and 'private' sectors, and so it would remain for a long, long time to come.

In the late autumn of 1854, British and French troops landed in the Crimea and the siege of Sebastopol commenced. By the winter, over thirty thousand soldiers were dug in, enduring weather of unprecedented severity and more likely to die from the cold, cholera or starvation than from enemy action. Florence Nightingale worked wonders with the organization of the nursing services, but the lack of an adequate supply line from the port of Balaclava was a matter of grave concern. Thomas Brassey, the millionaire contractor, offered to build a railway from the sea to the trenches, an offer that Lord Aberdeen, Prime Minister of the day, accepted. Brassey had probably twice as many men at his disposal as there were British troops in the Crimea, and his request of 2 December for volunteers from his work-force was equally over-subscribed. In a display of logistics that would not discredit – nay, they would embarrass – the 1990s, several hundred men with their material and equipment sailed in twenty-five ships, the first leaving Birkenhead on 21 December; and 29 miles of railway, main lines and branch lines, were complete by the first week of April 1855. The Illustrated London News wryly observed that the men 'who made England great by their skill, enterprise and powers of organization were of far different calibre from the officials the Government employs'.

On 9 December 1854 the same journal had described the proposed line thus:

The Railway will be a double line from Balaclava to near the batteries. From a certain point single lines will radiate to each of the latter and supply them with shot, shell, guns and stores from the fleet. Materials for fifteen miles are to be immediately despatched. The rails will resemble those of the Great Western. The line will be worked by means of stationary engines, four or five in number, and the trucks be drawn by wire ropes.

This may well have been the mode of operation for the first six months or so; and while the rails mentioned can be assumed to have been of Brunel's 'baulk road' pattern using longitudinal timber bearers and possibly even his characteristic bridge rail, the line must have been laid to standard 4 ft. 8$\frac{1}{2}$ in. gauge, for, on Saturday 8 September 1855 the Leeds Intelligencer reports:

We understand that a small locomotive called The Alliance was yesterday forwarded from the Railway Foundry, Leeds, to Balaclava. The engine is what is called a tank engine with 11 in. cylinders 17 in. stroke, six wheels of three feet diameter, all coupled. The engine was originally made for Messrs Leather, coal owners, for use on their tramway to and from their pits, and was purchased on Saturday last by a government agent. During the interval the words 'The Alliance' and the flags of England, France, Turkey and Sardinia have been painted on it.

On Tuesday 11 September the paper follows up with:

On Saturday morning a telegraphic message was received at the Railway Foundry Leeds ordering a second locomotive for the Crimean Railway, similar to the one purchased a few days previously. The engine was made for Sir John L. Kaye and at the time the order was received was working at Sir John's collieries. It

A 'Jenny' in later life: North Eastern Railway No. 326, originally York, Newcastle & Berwick Railway No. 95, as rebuilt by Edward Fletcher with North Eastern boiler and cab on lengthened and strengthened frames. It ran until 1884.

NATIONAL RAILWAY MUSEUM

Between 1850 and 1854 the Great Northern Railway purchased 90 locomotives new from E. B. Wilson and acquired a further ten Wilsons second-hand from absorbed local railways. Of the total of 100 (roughly thirty per cent of the GNR fleet at that time), seventy-four were standard 0–6–0 goods tender engines. This one is Great Northern No. 191 from a batch of thirty (Nos. 168–197) supplied in 1852. These and other pre-1852 0–6–0 engines had 5 ft. 0 in. diameter driving wheels; most later examples had 5 ft. 3 in. wheels. Many more were produced for the North Eastern, Scottish North Eastern, South Yorkshire, South Staffordshire and other railways. Twenty-eight from the Oxford, Worcester & Wolverhampton and the Newport, Abergavenny & Hereford Railways became Great Western Railway Nos. 248–259 and 264–279. The cut-down frames, motion and four of the coupled wheels of OW&W No. 34 (GWR No. 252) were set up as an instructional model at the GWR Stafford Road works, Wolverhampton, where they remained into the 1950s, and now reside in the Leeds Industrial Museum.

NATIONAL RAILWAY MUSEUM

The fluted safety valve cover and superb chimney affirm the E. B. Wilson design of Manning Wardle no. 16 built in 1860. Almost certainly identical to, and probably a replacement for, the locomotive Alliance *requisitioned for the Crimea, it would appear to be posed when quite new in the sylvan surroundings of Templenewsam, east Leeds, close by its colliery home.*

J. I. C. BOYD COLLECTION

was brought down to the Foundry on Saturday night and after being overhauled will be forwarded to its destination.

The *Leeds Mercury* adds:

It is one constructed a short time ago by Messrs E. B. Wilson & Co. and has been working a few months. It is a tank engine, namely, one which carries its own water on [in?] a tank placed on top of the boiler.

This episode is remarkable for a number of reasons, not least the positively factual reporting of the day without a hint of the media manipulation we know today. The speed at which action was taken and the enthusiasm shown are those of a nation good-humoured but not to be messed about. It is the first recorded mention of the true industrial locomotive on which Leeds's unparalled reputation of future years was built, and the first use of railway locomotives for military purposes.

Edward Brown Wilson was 'his own man', and in David Joy he had found a man of similar temperament. That they had several 'dust-ups' has already been recorded, but each man excelled in different areas, the one commercial and entrepreneurial, the other in engineering inventiveness. Consequently they were able to disagree and reconcile as the mood took them with no ill effect to the company. Call it a 'love–hate' relationship – not uncommon in successful businesses. But as a company grows so does the need for finance, and shareholders are less easy to contain than rebellious draughtsmen.

The shareholders eventually won, and Wilson departed from Leeds in 1856, later to die in obscurity in Tunbridge Wells at the relatively early age of fifty-six on 21 June 1874. Interestingly, volume 2 of the Railway Correspondence and Travel Society's *The Locomotives of the Great Western Railway*, in its coverage of the Oxford, Worcester & Wolverhampton Railway, states that 'from 1857 Edward Wilson was in charge of the locomotives'; that is, locomotives of the OW&W. It goes on to say that when the West Midland Railway was formed in 1860, 'Wilson became the Locomotive Superintendent of the new company and, after the amalgamation of 1863, he took charge of the Locomotive Department throughout the narrow gauge system of the Great Western Railway. He resigned in 1864.' Was this, as circumstantial evidence suggests, the same Edward *Brown* Wilson? for if it was he dropped nicely into the slot vacated by his talented former employee only a few months earlier.

Joy moved initially to the De Bergues bridge-building company, later into shipbuilding and finally into consulting engineering. He was a prolific producer of patents; he had been joint patentee with Wilson of the double-boiler locomotive exhibited in Hyde Park in 1851, and his other patents covered steam hammers, a steam riveting machine, a compound marine engine and of course his radial valve gear, which was patented in 1879. He died in 1903 at the age of 79. His misfortune was to have stayed in industry, and thus he missed the fame and glory attained by many of the railway company Chief Mechanical Engineers who, in a number of cases, achieved acclaim for being in the right place at the right time and not for what they actually did.

Some historians have said that the demise of the Railway Foundry came about as a result of building too many locomotives to stock, but this claim is difficult to reconcile with a rapidly-expanding British railway system hungry for locomotives and with a clear profit of £12,000 recorded in the final year of business. True, the growth of railway-owned workshops was bound to affect the private builders, but it was almost certainly board-room strife and meddling by absentee shareholders which, through an action in Chancery, brought about the winding-up of the company in 1858.

The eminent and reliable railway historian E. L. Ahrons suggested in an article published in *The Engineer* for 15 October 1920 that the demise of E. B. Wilson & Co. in 1858 was due in no small way to this policy of building standard locomotives to stock. To reinforce the argument he quoted from Joy's diaries that 'even for the alteration of the position of a clack-box £5 to £25 was demanded.' Several subsequent authors have rehearsed these comments down the years without really thinking things through, compounding the felony by suggesting also that the firm refused to build other than the standard types.

Building to stock enabled the company to keep its work-force in gainful employment at times of shallow demand, to retain their skills and to respond quickly to any railway company that was in urgent need of motive power – thereby outflanking the competition. Of course modifications in this case were expensive, especially if parts that had already been made needed to be modified.

Apart from the twenty 17¹/₄ x 23¹/₂ in. locomotives built for the 5 ft. 6 in. gauge Madrid, Zaragosa & Alicante Railway in 1857, inside-frame 0–6–0 tender engines of E. B. Wilson design were rare. This is Great Western Railway No. 279, formerly OW&W No. 33, one of two 16 x 24 in. 'ballast engines' with 4 ft. 6 in. driving wheels built in 1855. Two similar examples went to the Blyth & Tyne Railway in 1857. Manning Wardle perpetuated the design for a few years after the demise of Wilsons.

NATIONAL RAILWAY MUSEUM

Great Western Railway No. 189, formerly OW&W No. 41, was one of two 'big Wilsons' with 6 ft. 6 in. driving wheels supplied in 1855. North Eastern Railway Nos. 215 and 218 were identical, as were the South Staffordshire Railway's Esk and Justin. These last two went to the London & North Western's Southern Division in 1862 when the LNWR absorbed the South Staffordshire.
NATIONAL RAILWAY MUSEUM

North Eastern Railway 0–6–0 No. 2281 at Scarborough. A 16 x 24 in. engine with 5 ft. 0 in. driving wheels (EBW 592), it was originally NER No. 408, the fourth of a batch of ten supplied in 1857. Shown as rebuilt at York in 1882, it lasted until around 1920.
A. V. CHISHOLM/NRM

At the time of its closure Wilson's was in profit and only three locomotives remained unsold. Not exactly the scenario for catastrophe.

As for refusing other work, this argument is not supported by the facts. The standard 'Jenny Lind' and 2–4–0 passenger locomotives and the 0–6–0 goods engines sold well to most of the main-line railways, it is true, and monopolized the production line most of the time. However, production of the 2–2–2 passenger engines of McConnell's 'patent' design for the London & North Western's Southern Division, Cudworth's South Eastern railway 2–4–0s and the twenty 17¼ x 23½ in. 0–6–0 goods engines for the Madrid, Zaragosa and Alicante Railway, to name but three, gave the lie to any idea of a firm that was not prepared to build to suit a customer's requirements or, as one author deprecatingly put it, did not realize that specific types of engine were required for specific jobs. One should compare the ideals of Wilson and Joy with the rolling stock situation in Britain in 1997. One hundred and fifty years after *Jenny Lind* the wheel has just about come full circle.

To replace Wilson the shareholders had appointed one Alexander Campbell, from Scott's of Greenock, as manager in 1856, and under his management the Indian Ghat tanks were built and the first two locomotives to run in Argentina were despatched. The Newport, Abergavenny & Hereford Railway took two large 18 x 24 in. banking engines and four superb 2–4–0 passenger engines, but the most prestigious contract was for twenty 0–6–0 tender engines supplied in 1857 to the Madrid, Zaragosa & Alicante Railway, one of which survived under Spanish National Railways (RENFE) at Valencia until 1963. Six 0–6–0 tender locomotives went to the Newport, Abergavenny & Hereford Railway in 1858, two locomotives to New South Wales, an industrial locomotive to Derbyshire and a tank locomotive to the Scottish North Eastern Railway, and then the doors closed.

Alexander Campbell had seen the writing on the wall from the start and had 'plan B' in mind. By the time the Railway Foundry closed, Campbell had conspired, if that is the right word, with John Manning, C. W. Wardle and some Railway Foundry employees, backed by the Rt Hon. G. F. Russell (Viscount Boyne) to purchase part of Russell's estate immediately adjoining the western boundary of the Railway Foundry. They made a modest start as general engineers awaiting the crash they knew was shortly to occur next door.

And occur it did. On 20 July 1859 auctioneer Thomas Hardwicke called for bids for what he described as a works capable of employing 1200 hands who could turn out eighty locomotives a year to the value of 2500 guineas each. The 35,289 square yards (7.29 acres) was divided into seven lots, but no bid reached the reserve price and the property was withdrawn. For a period of 20 days from 15 August the drawings, tools and stock-in-trade were auctioned off in 4145 lots. This is where Alexander Campbell scored for Manning, Wardle & Co. Mannings acquired for £41.0s.0d. the whole of E. B. Wilson & Co.'s drawings, including of course the six-coupled industrial locomotive that was to be their staple product for many years, and the 'Jenny Lind' design. All the accumulated design experience of the country's largest locomotive works for less than the price of a set of six-coupled wheels and axles. A four-wheeled tank engine, probably the works shunter, was sold to a Doncaster contractor for £260, while two 2–4–0 tender locomotives with 16 x 22 in. cylinders and 5 ft. 6 in. driving wheels were shown as knocked down to the Leeds & Bradford Railway Company for £1900 each. This reference to the L&BR at such a late date is surprising, given that the company had been part of the Midland Railway since 1851. The author suggests that the entry was an auctioneers' shorthand for the Leeds, Bradford & Halifax Junction Railway, which was later to become part of the Great Northern Railway and which had recently entered Leeds at the newly-opened Central Station in 1857. Great Northern Railway records show two unidentified Wilson 2–4–0s as coming second-hand from the West Yorkshire Railway in 1866. The West Yorkshire was formerly the Bradford, Wakefield & Leeds Railway, which had inter-worked with the LB&HJR since its inception.

The Railway Foundry stood idle for a while until Kitsons purchased the old Shepherd & Todd buildings for use as a pattern store. Later this was used by C. A. Parsons for experimental work on steam turbines and naval torpedoes. These buildings survived almost intact for a further century, and part of them to the present day, as will be explained later.

Hudswell, Clarke & Co. was established in 1860 on one-and-a-half acres of the estate on the opposite side of Jack Lane from the main Railway Foundry works proper, where some small joiners' and paint shops had stood but which had been mainly unused land. Wilson's 'erecting hall', which on its opening in 1847 had entertained two thousand guests, now stood cold and empty. It was shortly to be demolished and the whole site cleared – to be painstakingly reassembled over the years by the Hunslet Engine Company.

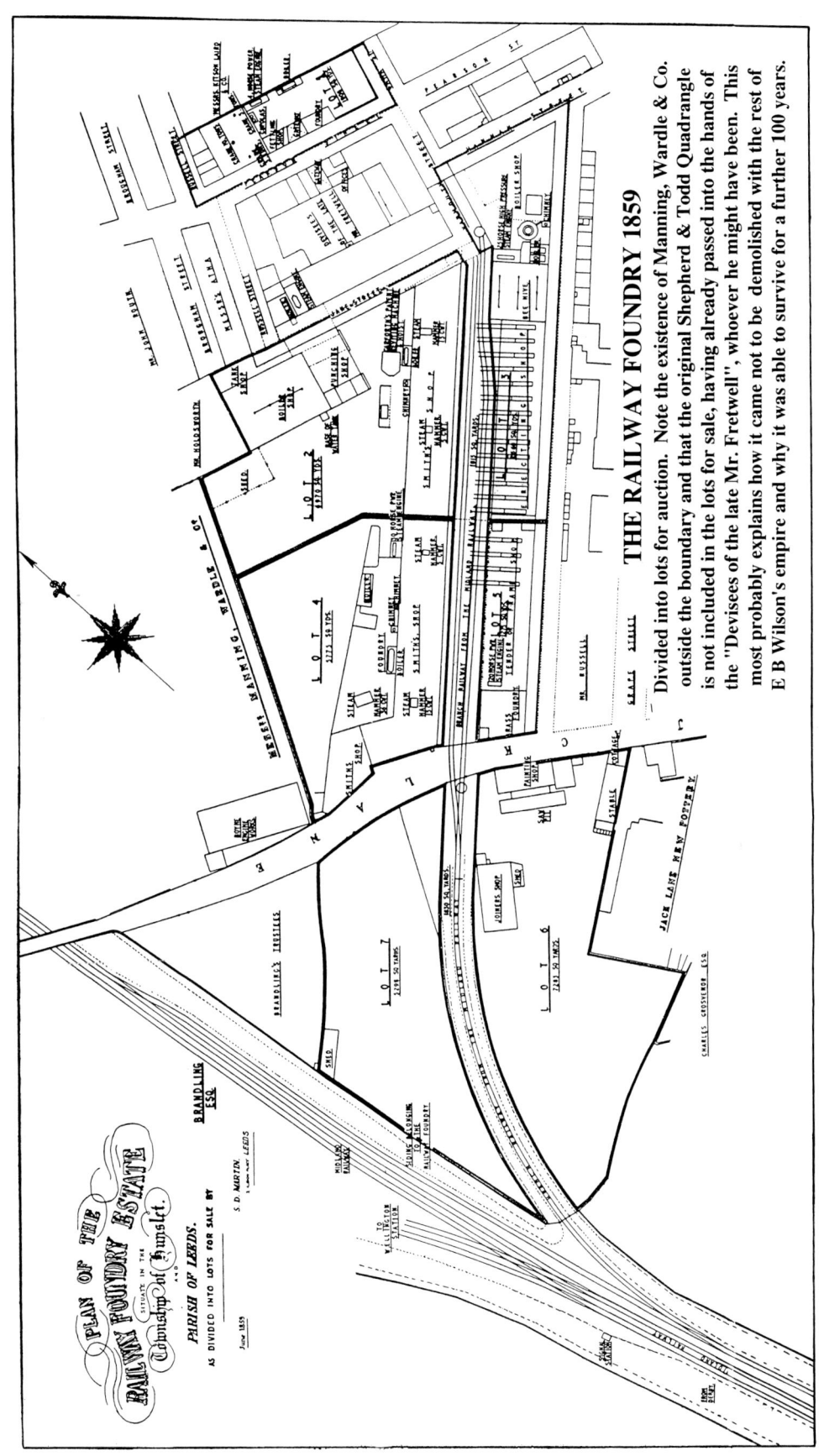

THE RAILWAY FOUNDRY 1859

Divided into lots for auction. Note the existence of Manning, Wardle & Co. outside the boundary and that the original Shepherd & Todd Quadrangle is not included in the lots for sale, having already passed into the hands of the "Devisees of the late Mr. Fretwell", whoever he might have been. This most probably explains how it came not to be demolished with the rest of E B Wilson's empire and why it was able to survive for a further 100 years.

CHAPTER 3

The Hunslet Engine Company – Victorian imperialism

In his otherwise excellent short history of E. B. Wilson & Company in *The Engineer* for 15 October 1920, already referred to, E. L. Ahrons unfortunately created another myth that has oft been repeated. He wrote of the happenings of 1858:

The old works then ceased operations as E. B. Wilson and Company, but the works are still in existence; a large part of them, together with the drawings and patterns, was taken over shortly afterwards by the present firm of Manning Wardle and Co. Limited and another portion of the works, situated on the opposite side of Jack Lane, Hunslet, is now in the hands of Messrs Hudswell, Clarke and Company. The latter firm still perpetuates the old name of the Railway Foundry.

As we have chronicled in chapter 2, this statement is only partially true. When John Towlerton Leather began the creation of the Hunslet Engine Company in 1864, nothing of E. B. Wilson's locomotive manufactory remained but for the site, bounded to the south by Jack Lane, to the east by Cancel Street, to the north by Shepherd & Todd's old Quadrangle and to the west by Manning, Wardle & Co. Ltd. Almost in the centre of this prime plot was placed the first small workshop of the new enterprise.

Leather's family were local colliery-owners, one of their pits being Waterloo Main, from where *The Alliance* was requisitioned for the Crimea as noted in chapter 2. The area south of Templenewsam House, where the colliery offices once stood, and which will soon resound to traffic on the M1–A1 link road, is still known to locals of a certain age as 't' back o' Leathers's'.

J. T. Leather himself was a civil engineer, a forerunner of Brassey, Ballard, T. A. Walker and the like. He was building the Erewash Valley section of the Midland Railway at the time (1847–50) when Joy was at E. B. Wilson presiding over the production of the 'Jenny Linds' for the same railway. Other works in which Leather was involved included the Portland breakwater, which had used broad-gauge 0-4-0 well tanks from Wilsons, and fortifications at Spithead. In partnership with George Smith his last major contract was the extension of the

Portsmouth Naval Dockyard, on completion of which, in 1877, he retired at the age of 73.

John Towlerton Leather had a son, Arthur, for whom the Hunslet Engine Company was intended as a vehicle for his future success. This was not to be, however, for while Arthur did enter the business he did not shine. James Campbell, the eldest son of Alexander next door at Manning Wardle, was brought in as Manager at the opening, and by the end of 1871 was sole owner in friendly competition with his father. Thus started the first dynasty at 125 Jack Lane.

Alexander Campbell (Alexander I we shall call him) had four sons. The said James was to be assisted in the new enterprise by the third son, George. The second son, Alexander II, was a disappointment. He ran a less-than-successful tweed mill in Glasgow, and was continually pressing James and George for financial help. Robert, the youngest, died at an early age.

Following the lead set by its two already-established neighbourhood competitors Hudswell Clarke and Manning Wardle, who had similarly risen Phoenix-like from the ashes and associated fall-out of Edward Brown Wilson's Railway Foundry, the fledgling Hunslet Engine Company modestly settled on a conventional six-wheeled contractors' locomotive as its first commercial offering. But in the Hunslet Engine Company's case, while the modesty was real it was intentional, and it was to be the hallmark that guaranteed independence and survival throughout the whole of the firm's 133 years in family ownership – first the Campbell family and then the Alcocks, each an autocracy of roughly equal tenure. The products were to be solid and reliable while at the same time innovative with minimum risk. And then there was the name. While Manning, Wardle & Company, Hudswell Clarke, John Fowler, Kitson and others were all *in* the district of Hunslet, the newcomer was *the* Hunslet Engine Company, operating from the *Hunslet* Engine Works, in time shortened colloquially to just 'Hunslet', thereby implying that the others only shared the territory by virtue of benign indulgence.

Hunslet no. I Linden *completed in June 1865, photographed prior to its despatch to Ampthill, Bedfordshire, for work on the construction of the London extension of the Midland Railway. The works plate did not identify it as No. I, merely carrying the legend 'THE HUNSLET ENGINE COY. LEEDS JUNE 1865'. Only nos. I, 2 and 7 were built to this design.*

This attitude was made flesh many years later in the person of John Alcock. The author well remembers introducing Mr Alcock to the Kenya Railways Board of Directors and Kenyan Government ministers at a tender meeting in Nairobi one afternoon in 1977. On being asked to address the meeting, 'Mr John', as he was known to employees and customers alike, remarked: 'I am not here to *represent* Hunslet: I *am* Hunslet.' The contract was won at that moment.

Locomotive no. 1 was a no-nonsense six-wheeled saddle tank with 14 x 18 in. cylinders – quite large by the standards of the period. Named *Linden* it was despatched on 18 July 1865 to the order of Thomas Brassey to Ampthill in Bedfordshire, one of several locomotives purchased by various contractors as maids of all work on the construction of the Midland Railway's new London extension into St Pancras.

That no. 1 had several generic features inherent in similar locomotives from Manning Wardle next door was understandable bearing in mind the Campbell family connection – indeed most of the castings, certainly the wheels, are believed to have come from Mannings – but the design generally was more chunky and masculine than that of the Boyne Works offerings. A Percheron to Mannings' Suffolk Punch, perhaps. The factory itself was modest, comprising in the main a blacksmiths' shop parallel to the reinstated Midland Railway branch, a very small detached erecting shop at right-angles to the track, and a small 'engine shed' astride the line itself. Enlargements to this arrangement were to take place at irregular intervals over the next century and a quarter.

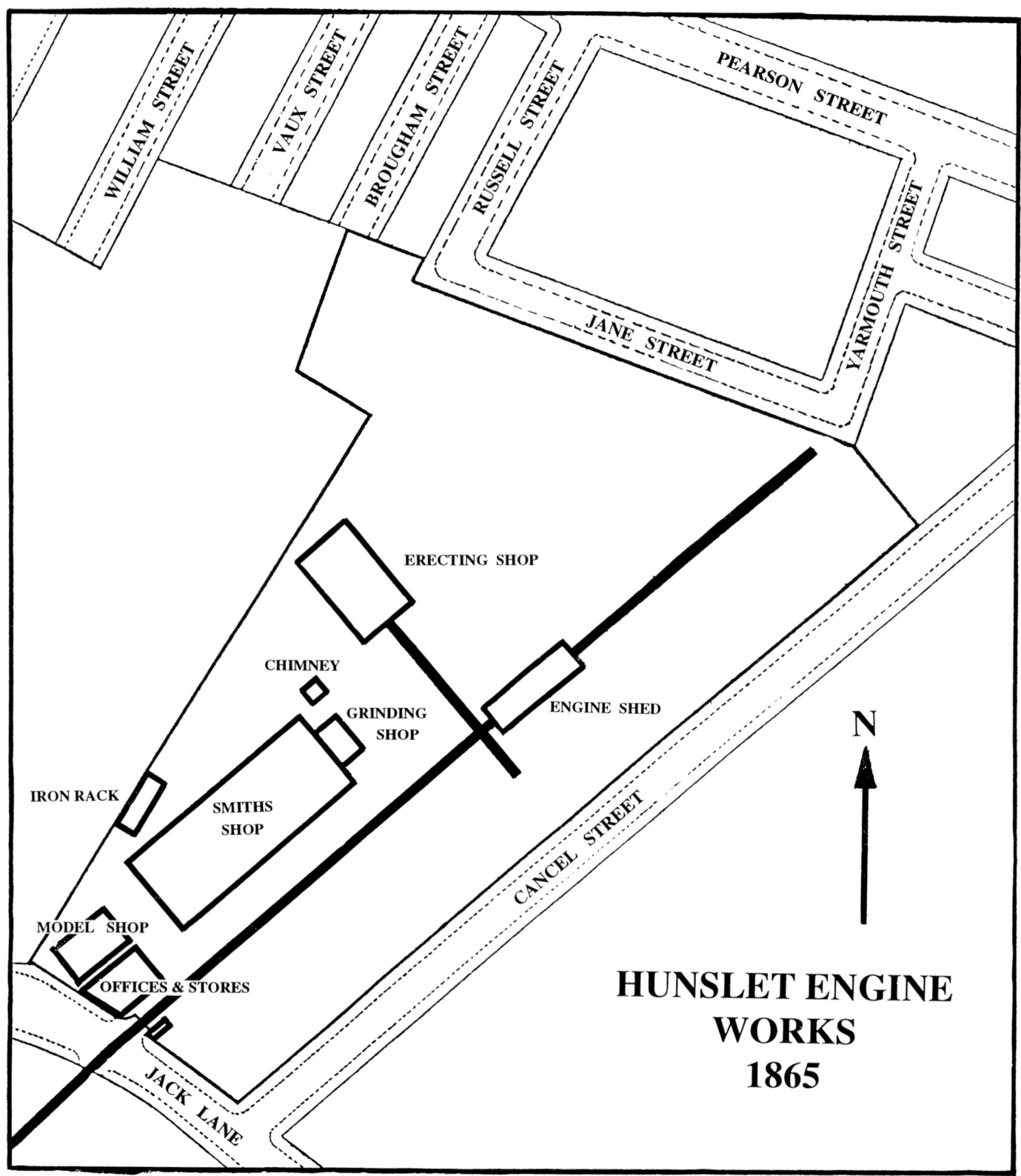

WILLIAM STREET

VAUX STREET

BROUGHAM STREET

RUSSELL STREET

PEARSON STREET

YARMOUTH STREET

JANE STREET

ERECTING SHOP

CHIMNEY

GRINDING SHOP

ENGINE SHED

IRON RACK

SMITHS SHOP

CANCEL STREET

N

MODEL SHOP

OFFICES & STORES

JACK LANE

HUNSLET ENGINE WORKS 1865

The contractors' locomotives of the late nineteenth century and their trains were very much the early equivalents of the modern load-haul dump trucks and passed from contract to contract through several owners before meeting their end in a breaker's yard. Many survived to a tremendous age on extremely arduous duties and appalling track. As a boy the author was fascinated by a locomotive at the nearby Sharlston colliery that was obviously of great age but carried no visible identity. The colliery records had it as purchased from Samuel Williams of Dagenham Dock on Thames-side, where it had carried their number 6 and was allegedly Hunslet no. 401. Years later, in the fifties, an order was received at Hunslet for a new saddle tank for this locomotive and no suitable drawing could be found, for no. 401 was recorded as having been built as a side tank. Still the penny did not drop, and a new tank was made by using the old rusting and leaking tank as a guide. Only after a letter from George Alliez, a well-known expert on industrial locomotives, did the author and a colleague, Geoff Horsman, visit Sharlston early in March 1966 to find that the locomotive had been cut up on site, just weeks before, by Wakefield Metal Traders, and only the 'new' saddle tank and a few small items remained. It was however established beyond our doubt that no. 1 *Linden*, Dagenham Dock no. 6 and the Sharlston locomotive were one and the same: 'old original' had survived to become a working centenarian but not long enough to be saved for posterity. In its peregrinations from contractor to contractor, and possibly at the Manchester Ship Canal, it had come alongside, and changed identities with, its much younger stable-mate no. 401. Historians should beware official records. Thus was a valuable historical item lost. Equally, one should be wary of uncritically accepting 'official' documents. Certainly there were several copies of the Hunslet engine record books, and these contained errors of the most elementary kind, particularly with regard to spelling and transcription. In some cases, the later copies of the books did not always show the original owner of the locomotives if a change of ownership had taken place meantime.

Of the 130 further locomotives built in the first ten years to the end of 1874, roughly twenty per cent were for contractors involved in building Britain's railways and railway-served docks and dockyards, thirty per cent went overseas, and the balance went to industrial users, mainly collieries and associated iron works.

The Midland Railway's London extension took nos. 1, 5, 7 and 8, while other railway construction contracts included the London & South Western at Barnstaple (locomotive no. 3), the Metropolitan District at Kensington (no. 6), and further Midland Railway work at Appleby (no. 72) and Cudworth (nos. 17 and 132). Notable among the dockyard contracts was Leather's own Portsmouth Dockyard Extension contract (as Leather, Smith & Company), which took no fewer than ten locomotives between 1867 and 1871. Geo. Wythes took three locomotives (nos. 22, 57 and 68) for contracts at East and West India Dock, Berkeley and Sharpness respectively, while Easton Gibb had nos. 20 and 24 at Cardiff. By this time the rate of building had grown to such an extent that undue individual treatment in a volume of this nature would become repetitive and tedious, and we shall therefore continue to highlight only the significant milestones in these early days.

No. 10 was the first export locomotive after only twelve months of manufacturing and went to Java in May 1866. A simple 4 ft. 8½ in. gauge 0–4–0 saddle tank, its destination was only a short distance from that of the last Hunslet steam locomotive, built in 1971. It is perhaps a little unkind to point out that in the intervening 105 years no other Hunslet steam locomotive set foot in what is now Indonesia.

In the main these early locomotives were pretty straightforward standard-gauge 0–4–0 and 0–6–0 saddle tanks, in almost equal numbers of each wheel arrangement; but there were some notable exceptions, particularly among the locomotives built for export.

The first narrow-gauge offering was no. 51. Tried in steam on 14 October 1870 and despatched from the works the following day, no. 51 was the harbinger of a cult following which from the mid-1960s has worshipped at the shrine of the 'quarry Hunslet'. It was in fact the first Hunslet locomotive to work in the slate quarries of North Wales, a breed of diminutive nominally 2 ft. 0 in. gauge 0–4–0 saddle tanks (actually 1 ft. 10¾ in. in the case of no. 51) which over a century later were numerically the second-largest class of locomotive in preservation in this country, possibly anywhere in the world. (The numerically largest preserved class is also a Hunslet design, but that must come later.) The purchaser of no. 51 was Charles Assheton Smith for his family's Dinorwic quarries in North Wales.

Locomotive no. 52, despatched on 8 March 1871, was also worthy of note, if only for its unusual appearance. Built to 5 ft. 6 in. gauge for the Oude & Rohilkund Railway

The unacknowledged centenarian. Scrapped in 1966 before its true identity was realized, Hunslet no. 1 rests at Sharlston Colliery, Wakefield, on 15 October 1961. But for a case of mistaken identity it might have survived for preservation. Several modifications, including cab, altered spring gear, the removal of compensating beams and new brake gear are apparent, as is the replacement welded saddle tank supplied in the fifties. D. TREVOR ROWE

John Leather owned Waterloo Main Colliery. It is therefore hardly surprising that his newly-established Hunslet Engine Company locomotives very quickly established a presence alongside the Manning Wardles already operating there. This one is Hunslet no. 2. The Hunslet–Waterloo Main business relationship lasted for over a century until the colliery was closed in 1968.

Far from the madding crowd. The graveyard on the right of this picture was the inspiration for Thomas Hardy's poem 'The Levelled Churchyard'. Hunslet 12 x 18 in. 0–6–0 saddle tank no. 8 Handy, new to Waring Bros. on 5 May 1866, disturbs the long-departed as Warings' navvies construct a bridge over the St Pancras burial ground on 2 July 1867 as part of the Agar Town to St Pancras stretch of the Midland Railway extension. The locomotive has a hinged 'falling chimney' for extra clearance when working in the restricted tunnel headings. NATIONAL RAILWAY MUSEUM

in India (the makers' engine book refers to India as the 'East Indies' in the early days), it was a 2–2–0 vertical-boiler locomotive supplied with a skeleton superstructure having the profile of a coach. The bodywork was completed in India to match the inspection coach that the locomotive was intended to haul. The same railway had previously taken four basically standard 10 x 15 in. 0–4–0 saddle tank locomotives in October and December 1870.

In 1871 the Huddersfield Corporation ordered *Deerhilt*, a 3 ft. 0 in. gauge 0–4–0 saddle tank with 8 x 14 in. cylinders and 2 ft. 4 in. diameter wheels for construction use on the eponymous reservoir. Carrying works number 71 it was despatched on 12 October 1871 and was followed by an identical machine, no. 74 *Huddersfield*, five months later.

Slightly larger than *Charlie*, the design became commonplace on reservoir and other construction contracts over a number of years, and variants were supplied overseas.

Hunslet's first bogie locomotives were nos. 84–89 and 95, seven 4–4–0 side and saddle tanks supplied to the 3 ft. 6 in. gauge Prince Edward Island Railway between September 1872 and June 1873. Victorian imperialism was now in full flow and British agents and contractors were buying locomotives in ever-increasing numbers for use in opening up and exploiting mineral deposits or for the construction and operation of new railways all over the world. The big household contracting names, Brassey, Leather, Oliver, Pearson, etc., were still there, but newcomers with special interests in certain countries now began to feature again and again in the order books of Hunslet and other British manufacturers.

J. Mason, later Mason & Barry Ltd, had laid claim to extensive reserves of copper, sulphur and manganese at São Domingos in the Alemtejo district of Portugal. Fifteen small 3 ft. 6 in. gauge 0–4–0 well tank locomotives had been purchased from Hawthorns of Leith over the ten years from 1864, but two conventional 3 ft. 6 in. gauge 0–6–0 saddle tanks, nos. 93 and 94, were provided by Hunslet in 1873. They were followed in 1874 and 1876 by nos. 127/8 and 149/50, four very stylish 0–6–0 side and well tanks which were unique at Hunslet in having the always-rare outside Stephenson's valve gear. It seems likely that in the present case this arrangement was requested by the customer to match the Hawthorns locomotives. The Hunslets were to carry out their duties for almost eighty years.

Clarke Punchard and Company were even more adventurous. They rounded off the first ten years of the Hunslet Engine Company by ordering eighteen locomotives of five different types for four widely-scattered railways. First came four 4 ft. 8½ in. gauge 14 x 22 in. 0–6–0 side tanks, nos. 107–110, which were delivered towards the end of 1873 to the North Western Railway of Monte Video. Next came a batch of eight very slender 3 ft. 6 in. gauge 14 x 20 in. 4–6–0 side tanks, which, apart from having six coupled wheels of 3 ft. 6 in. diameter instead of four of 5 ft. 4 in., looked for all the world like William Adams's 1868 design for the North London Railway. They bore Hunslet numbers 111–118 and originally took Nos. 1–8 in the stock list of the Tasmanian Main Line Railway. Nos. 1–6 and no. 8 duly went to Tasmania, but No. 7 was fitted with a number plate reading 'F. C. de R. T. No. 1' and went to the Rio Tinto Railway at Huelva in Spain to inaugurate the passenger service to the mines there. The eight locomotives were despatched between 2 February and 27 July 1874. They were closely followed by no. 122, which was another 0–6–0T for Monte Video. Nos. 123 and 124, both despatched on 16 November 1874, were 4–4–0

Serpa, no. 128/74, one of the four 3 ft. 6 in. gauge 0–6–0 locomotives produced for Mason and Barry's sulphur mines at São Domingos in Portugal. They pre-dated the widespread use of Walschaerts' valve gear; the outside Stephenson's gear both facilitated maintenance and allowed the small side tanks to be augmented by well tanks between the frames.

Tasmanian Main Line Railway 4–6–0T No. 8, Hunslet no. 118/74. The resemblance to the North London Railway Adams outside-cylinder 4–4–0 tank locomotives is remarkable. The polished brass bell behind the dome has fallen victim to a combination of the light and the retoucher's brush; it is a pity that manufacturers felt obliged to remove the background from most works photographs.

side tank engines with the unusual cylinder dimensions of 12 in. diameter x 19 in. stroke, and they also went to the Tasmanian Main Line Railway. Clarke Punchards' final purchases were three very large 15 x 22 in. 5 ft. 6 in. gauge 0–6–0 side tank engines (nos. 130–2) supplied at the turn of the year to the Buenos Aires and Campaña Railway (F. C. de B. A. C.), later the Buenos Ayres and Rosario.

The locomotives singled out for special attention above had been accompanied through the works by a veritable army of locomotives for construction and industrial use – mainly at home, but the occasional standard type found use overseas. A series of standard four- and six-wheeled designs developed, each in ascending order of cylinder size, wheel diameter and weight.

The standard industrial and contractors' locomotive sizes of this period ranged from the small 8 x 14 in. 0–4–0 saddle tank to the 15 x 20 in. 0–6–0 saddle tank, with outside cylinders exclusively on the four-wheelers, and mainly inside cylinders on the six-wheelers. A distinct house style had emerged. The shape of the tank, the chimney of Great Northern Railway appearance and the 'vinegar bottle' safety valve cover, and the general clean lines – all these things said 'Hunslet'. Locomotives were put in hand in batches, usually two or four in a batch, and two or three different sizes would be in hand to stock at any time to allow quick delivery when ordered. Out of the first hundred locomotives built at the Hunslet Engine Works the best-selling types were the 10 x 15 in. 0–4–0

Zambesian juxtaposition. This historic pair are imposters.

Stanley, *HE 137, was delivered on 1 July 1875 to the contractor T. A. Walker bearing the more down-to-earth name* Fred, *while the august doctor's name was taken in vain by HE 212, which went on 28 January 1879 to Wombwell, near Barnsley, labelled prosaically Cortonwood Colliery. No. 137 was a 10 x 15 in. engine; no. 212 had cylinders 13 x 18 in. Both were probably built to stock, painted photographic grey on one side and given fashionable names to boost both shop-floor morale and sales potential. It happened regularly – the camera could lie even 120 years ago.*

and the 12 x 18 in. 0–6–0, of which twenty-nine and twenty-one, respectively, were built – fifty per cent of the total output. This pattern was followed virtually throughout the construction of steam locomotives in Leeds, although the smaller types gradually fell out of favour as loads increased and larger types took their place.

Customers from more countries were added to the list when four 3 ft. 6 in. gauge 13 x 20 in. 0–6–0 side tank locomotives, nos. 141–4, were despatched in May and August 1875 to the Provincial Government of Otago in New Zealand, with nos. 145 and 146, two metre gauge 10 x 15 in. 0–4–0 saddle tanks going to the Ceara Railway Company in Brazil in July and September of the same year. Two 5 ft. 0 in. gauge 12 x 18 in. 0–4–0 saddle tanks, nos. 153 and 154, went to Russia two months later.

The early growth of railway company-owned workshops has already been remarked, and the private manufacturers increasingly became dependent on the export and industrial markets. To protect their interests the private builders in 1875 formed the Locomotive Manufacturers' Association, now the Railway Industry Association, and their concerted efforts resulted in a High Court injunction preventing the railway companies' workshops from manufacturing locomotives for sale or hire, limiting their construction to their own indigenous needs. Hunslet joined the Association in the first year of its formation and remained a member for 115 years.

A locomotive of particular note was despatched from Jack Lane on 26 July 1878. This was no. 206 *Beddgelert*, an elegant 1 ft. 11¼ in. gauge 0–6–4 saddle tank for working

Supplied new to Beckett & Bentley at Facit, near Rochdale, Lancashire, for use in the construction of the Lancashire & Yorkshire Railway branch line extension from Facit to Bacup, HE 222/79 Britannia later went to the Newport Alexandra Dock Co. From 1888 to 1894 it worked on the South Dock extension before going to Griff Colliery, Heath End, near Nuneaton, where it ended its days scrapped on site in September 1956. Very much of the old order, it was in superb original condition except for a new non-Hunslet chimney when photographed on 25 June 1937. GEORGE ALLIEZ COLLECTION/NRM

The North Wales Narrow Gauge Railway's Beddgelert, *Hunslet no. 206, when new in 1878, long before its graceful lines were broken when the boiler was inclined upwards at a slope of 1 in 40 in 1894 to give it a more functional physique for steep gradient work but a decidedly broken-backed 'kneeling cow' appearance.*

Titan, *Hunslet no. 324/83, was the fifth of nine locomotives supplied between 1880 and 1907 to the Carthagena & Herrerias Steam Tramway in south-west Spain. The first batches of these locomotives had a short wheelbase, with the trailing axle under the firebox, but later ones had the firebox between the third and fourth axles. The earlier ones were modified on site by splicing the frames to extend the wheelbase, so putting one axle behind the firebox. Photographed at Santa Lucia depot on 10 April 1956, the modern appearance of this septuagenarian is well displayed – the locomotive, that is, not the driver.*

D. TREVOR ROWE

passenger and freight trains on the North Wales Narrow Gauge Railway from Dinas Junction to Tryfan Junction and on to the Bryngwyn branch. The line was steeply graded for the whole of its length, averaging 1 in 40 all the way up to Bryngwyn, and for a century there has been much furious debate over the question whether *Beddgelert's* boiler was horizontal or not. The simple indisputable facts are that the locomotive was conventionally built with a horizontal boiler and put to work chimney-first up the incline from Dinas. It was an indifferent performer, probably owing to weight transfer on to the trailing Adams bogie; and in 1894 it came back to Leeds where it was modified to incline the boiler upwards at an angle of 1 in 40 from the cab front to the smokebox. It then returned to Dinas and operated the Bryngwyn service entirely satisfactorily, running bunker-first, until it was declared worn out and scrapped in 1908.

The boiler saga notwithstanding, *Beddgelert* was the standard-bearer for Hunslet's subsequent pre-eminence in the supply of narrow-gauge and light railway passenger and mixed-traffic locomotives. Hunslet became the largest single supplier of these types in Britain and Ireland, and arguably also throughout the developing world.

On 11 December 1880 two exceptional locomotives, nos. 253 and 254, were despatched to the Carthagena & Herrerias Steam Tramway in Spain. Named *Escombrera* and *Alumbres*, they were massive 3 ft. 6 in. gauge 13 x 20 in. eight-coupled saddle tanks. Another nine were supplied, at intervals, up to 1907, and thus the connection between Hunslet and the minor railways of Spain was established.

The following year no. 265, an 11 x 16 in. 0–6–0 side-tank condensing locomotive, was supplied to the Swansea and Mumbles Light Railway, where it became its no. 5, later renumbered 3 when Hunslet supplied two larger 13 x 18 in. locomotives, nos. 697 and 698, in 1899. These became S&MR nos. 4 and 5.

Hunslet's first tender locomotives were nos. 273 and 274, two metre-gauge 12 x 16 in. 4–4–0s supplied to the order of Megaw & Norton to Brazil on 30 January 1882. These were the forerunners of several 4–4–0 tender locomotives, mainly for South American countries but including some for the Tasmanian Main Line Railway and the Van Diemen's Land Company (also in Tasmania) in 1883.

It took some time for the slate quarries of North Wales to commit themselves fully to steam traction following the despatch of no. 51 *Charlie* in 1870. A second, similar locomotive, no. 154 *George*, went to Dinorwic in April 1877, and six months later came *Louisa*, no. 195. *Louisa* was a tiny little thing with cylinders only 5 inches in diameter, similar to *Little Egret*, which had been built earlier in the year for the Cransley Iron Company.

Apart from *Beddgelert* Hunslet did no further business in North Wales for five years (not that any other builder did any better), until in 1882 the major slate producers, Penrhyn quarries at Bethesda and Padarn quarries at Dinorwic, each bought a 'main-line' locomotive for transporting slate to Port Penrhyn and Port Dinorwic respectively.

The Penrhyn main-line locomotive was Hunslet no. 283 *Charles*, supplied on 27 May 1882. A 1 ft. 11½ in. gauge 0–4–0 saddle tank locomotive with 10 x 12 in. cylinders it was (and still is) a large locomotive by the standards of 'two foot' gauge four-wheelers. *Charles* was joined in June 1893 by sister locomotives nos. 589 *Blanche* and 590 *Linda*. These last two achieved fame as the mainstay for many years of the Festiniog Railway, where they are still active after over a century of arduous service. By coincidence (perhaps) *Charles* was built alongside four similar-sized 0–4–0 tank locomotives (Hunslet nos. 290–3) for the Darjeeling–Himalayan Railway, which were a subcontract from the Manchester firm of Sharp, Stewart & Company Ltd; perhaps one design influenced the other.

The Padarn Railway main-liner was a relatively conventional 0–6–0 side tank with 12½ x 20 in. outside cylinders and 3 ft. 6 in. wheels but running on the unusual track gauge of 4 ft. 0 in. It took Hunslet number 302 and the name *Dinorwic*; it was delivered on 15 December 1882 and was joined in 1886 and 1895 respectively by two similar locomotives, no. 410 *Pandora* (later renamed *Almathea*) and no. 631 *Velinheli*.

The year 1883 saw the beginning of large-scale purchase of narrow-gauge steam locomotives for the Welsh quarries. The two largest quarries, Penrhyn and Dinorwic, bought forty-two Hunslet locomotives between them over the years, and never bought a new steam locomotive from any other manufacturer after 1882. The prototype for the archetypal 'quarry Hunslet' was in fact no. 297, the 7 x 10 in. 0–4–0ST *Peep o' Day* which was supplied on 15 September 1882 to the Bold Venture Lime and Stone Quarry, Peak Forest.

The locomotive *Refael Nunez*, Hunslet no. 341, which was despatched on 21 May 1884 to the order of contractors Cotesworth & Powell, subsequently became no. 2 on the books of the Colombian Santa Marta Railway Company.

The first Penrhyn main-line Hunslet locomotive was Charles, HE 283, built in 1882. Blanche and Linda, HE 589 and 590, built in 1893, had slightly larger cylinders and firebox than Charles but were otherwise similar. Blanche is seen here in original Penrhyn condition before going, with her sister, to the Festiniog Railway, where both became famous on duties undreamt-of by their makers. The 10$\frac{1}{2}$ x 12 in. cylinders are steeply inclined to allow the connecting-rods to be inside the coupling rods, so bringing the cylinder centre-lines closer together and producing a powerful locomotive of minimum width.

R. E. VINCENT

Carrying the old-style name-plates and maker's plate, with recessed lettering and borders filled with beeswax, Winifred, HE 364/85, was the last to be built of the three Penrhyn 'Port' class locomotives used for shunting at Port Penrhyn. 7 x 10 in. diameter cylinders and 1 ft. 8 in. diameter wheels.

MAID MARIAN LOCOMOTIVE FUND

Typical of the many graceful 4–4–0 tank locomotives built by Hunslet (there was even a 2 ft. 0 in. gauge proposal for the Lynton & Barnstaple). No. 518/90 was the last of eight 12 x 18 in. cylinder metre gauge examples supplied through J. R. Banks for the Valencia Suburban Tramways (La Sociedad Valenciana de Tranvias) in Spain. It is shown here in the ownership of FESA (Ferrocarriles Económicos SA) at Tortosa on the Tortosa–La Cava Railway on 3 April 1961. Apart from the sandbox behind the dome, the locomotive is in 'mint' condition.   D. TREVOR ROWE

Carrying the name Valencia and numbered 109 at the Sagunto steelworks of Altos Hornos de Vizcayá on 5 April 1961, this neat 0–6–0 tank was originally SVT No. 9 of J. R. Banks. Hunslet no. 532 built in 1891.   D. TREVOR ROWE

This 3 ft. 0 in. gauge 4–4–0 tank locomotive with 9 x 16 in. cylinders must have served its new owners well, for they came back at intervals over the next 41 years to buy another twenty-four locomotives from Hunslet – a mixed fleet of 0–4–0, 0–4–2, 2–4–2, 0–6–0 and 4–4–0 tank locomotives and 2–6–0 tender locomotives.

Similar customer satisfaction appears to have been the case with J. R. Banks, who was responsible for a large number of light railways in Spain. Banks's first Hunslet locomotive had also been a 4–4–0 tank, no. 228, supplied on 9 September 1879, very similar to the Colombian machine but with slightly larger cylinders of 10 in. diameter and running on a track gauge of one metre. His further orders covered twenty-six locomotives, all for Spain and all but one of them for the metre-gauge light railways, over the period 1879–1910.

By 1877, work on the Manchester Ship Canal was well advanced, and T. H. Walker, the main contractor, was using a vast number of locomotives. Many of them had already passed through various hands and had been used on innumerable previous contracts. Included in their ranks was Hunslet no. 1, formerly *Linden* but now named *Patricroft*. Walker purchased no less than twenty-two new locomotives from Hunslet in 1887/8, these being a mixture of 0–4–0 and 0–6–0 saddle tank types.

This was boom time again, with Hunslet's versatility proving its worth. The bread-and-butter contractors' types went through almost without thought; then came the mid-range light railway passenger locomotives, and finally the 'exotica'.

M. Charles François Marie-Thérèse Lartigue had demonstrated his 'elevated single-line railway' in Tothill Fields, Westminster; and as a result of this demonstration the Listowel and Ballybunion Tramway Bill was promoted to provide a Lartigue railway between those two places in County Kerry, Ireland. The track, which was portable, consisted of a 27 lb. per yard running rail in 33 ft. lengths secured to the apex of a number of A-shaped iron trestles approximately 3 ft. apart and 3 ft. 3 in. high. This apparently odd choice of dimensions suggests that the components were originally conceived in metric modules, which would not be surprising, given the system's French origin. The cross-bar of the A-frames was 2 ft. 4 in. below the running rail, and at this point a pair of 11 lb. per yard guide rails in 20 ft. lengths was attached, one on each side. Hunslet supplied three locomotives, nos. 431–3, to the Lartigue Rail Construction Company in October 1887. The

locomotives, like the other rolling stock, straddled the trestle, and were of a twin-boiler type, one boiler each side of the running rail. Also embodying Mallet's patent in addition to that of M. Lartigue, they were in fact dual engines. Two 7 x 12 in. cylinders drove three 2 ft. 0 in. diameter wheels on the boiler (engine) portion giving a tractive effort of 3112 lb.; while two 5 x 7 in. cylinders drove two driving wheels on the tender portion to give another 926 lb. of tractive effort when required, very much in the manner of the Great Northern Railway steam tenders and the boosters on some later LNER engines. This auxiliary power facility was dispensed with after a time. Notwithstanding their peculiar design, the three Hunslet locomotives provided all the motive power on the eight-mile-long system until its closure in 1924.

Another example of the exotica was a series of large 15 x 22 in. 5 ft. 6 in. gauge 2–6–0 side tank engines for the Buenos Aires & Rosario Railway Company. In appearance they had much in common with the earlier Tasmanian 4–6–0 tanks, looking equally ancient. Three of them (nos. 446–8), were built in 1888, four more (nos. 486–9) in 1889, and another five (nos. 508–12) in 1890. Not for nothing did Argentina gain the sobriquet 'Britain's lost colony'.

Hunslet's first locomotive to go to Ireland had been a 5 ft. 3 in. gauge standard 12 x 18 in. 0–4–0ST, no. 178 *Faugh a Ballagh*, in 1878. No. 268, a 6 x 10 in. 2 ft. 6 in. gauge quarry type named *Fergus*, went to the County Clare reclamation scheme in June 1881. The contractor R. Worthington took two 5 ft. 3 in. gauge standard 9 x 14 in. 0–4–0 saddle tanks, nos. 315 and 319, in 1883.

It was Worthington who bought the first of those most attractive and much-photographed locomotives that were to epitomize the eccentricities of the Irish narrow gauge – the engines of the Tralee & Dingle Light Railway. Three 3 ft. 0 in. gauge 2–6–0 side tank locomotives, T&DLR Nos. 1–3 (Hunslet nos. 477–9), were supplied in 1889, with similar locomotives No. 6 (Hunslet no. 677) in 1898 and No. 8 (Hunslet no. 1051) following in 1910. T&DLR No. 4 (Hunslet no. 514), an 0–4–2T, was built in 1890 to work the Castlegregory branch; and No. 5 (Hunslet no. 555), a 2–6–2T, came out on 20 January 1892. No. 4 was alleged to be worn out by 1898 owing to neglect, but the remainder survived appallingly heavy work with minimum maintenance until the closure of the various Irish narrow-gauge lines in the 1950s. The Tralee & Dingle's 31 miles of roadside railways were the most westerly in Europe, and

Despite their unique design and 'oddball' appearance the three Lartigue monorail locomotives supplied to the Listowel & Ballybunion Railway in 1887 served the inhabitants and the livestock of County Kerry well for over thirty-five years; and stories, possibly apocryphal, of ingenious attempts to balance payloads – pianos with calves, passengers with manure, etc. – abound. This works photograph shows No. 1 (Hunslet no. 431/87) without the massive oil-burning headlamp, some two feet square, that dominates later views but obscures some of the mechanical details of the locomotive.

were a switchback of gradients, mainly at 1 in 31 to 1 in 29, with seventy level crossings.

Worthington purchased two other Hunslet locomotives for Ireland: no. 482, a 5 ft. 3 in. gauge standard 15 x 20 in. 0–6–0 tank supplied in 1889 which eventually found its way on to the books of the Great Northern Railway of Ireland; and no. 520 *St Molaga*, a special 5 ft. 3 in. gauge 0–4–2T supplied for the Timoleague & Courtmacsherry Light Railway a year later. The TCLR purchased a second Hunslet locomotive, no. 611 *Argadeen*, a 2–6–0 side tank, in 1894. This last was one of a pair, although its partner no. 610, which was otherwise identical in all respects, was built as a conventional 0–6–0, without the leading two-wheeled pony truck. No. 610 ended its days with the Irish

Great Southern and Western Railway at Inchicore, while no. 611 survived into CIE days.

The Crown Agents for the Colonies had purchased their first locomotive from Hunslet in 1874 – it was a 10 x 15 in. 0–4–0 saddle tank for the Colombo Harbour Authorities in Ceylon – and by 1894 they had taken twenty-four machines for clients throughout the colonies. Colombo Harbour, and indeed Ceylon generally, was to remain a Hunslet stronghold. Over the years thirteen Hunslet locomotives were supplied to the harbour and 104 to the island as a whole. The Crown Agents connection remained unbroken for the remainder of the Hunslet Engine Company's history.

A small $7^{1}/_{2}$ x 12 in. 2 ft. 6 in. gauge 0–6–0 tank locomotive, no. 650, also supplied through Crown Agents,

The sleek lines of Hunslet no. 478/89 belying its constitutional ability to survive 63 years on that most Irish of Irish railways, Tralee & Dingle No. 2 is shown here without the lower side aprons required and provided, but not necessarily used, to comply with regulations governing roadside running.

on 22 April 1896, was Sierra Leone Railway No. 1, and it began another long and almost exclusive connection, involving 0–6–0, 2–8–2 and, more famously, 2–6–2 tank locomotives, forty-seven of them in all. The relationship lasted until No. 85 (Hunslet no. 3815) was despatched to Freetown in 1954. This last returned to Britain in 1977 and has since worked continuously on the Welshpool and Llanfair Light Railway.

The Manchester Ship Canal opened on 1 January 1894. In the beginning, eight of the locomotives purchased new by T. A. Walker in 1884–8 were retained, augmented by two 0–4–0 and two 0–6–0 saddle tanks, purchased between 1890 and 1897. The 0–6–0s came one from Hunslet and one from Hudswell Clarke, while the four-wheelers were by Pecketts of Bristol. Once established, the new company

embarked on a standardization policy, and of the seventy-eight steam locomotives subsequently purchased only twelve were not side tanks of 0–6–0 wheel arrangement, and only three were not of Leeds origin. The first side tank was MSC No. 13 *Montreal*, Hunslet no. 685, despatched on 1 December 1898, a neat $15^1/_2$ x 20 in. inside-cylinder design of which ten were built for MSC up to 1902 with further examples for collieries and other users. From 1903 Hudswell Clarke produced its version of the design, as did Kitson some time later.

There was quite a lot happening in the closing years of the nineteenth century. The events in the Soudan were of concern again. In a rerun of the Crimea exercise Hunslet had supplied two 4–4–0 tank locomotives to the contractor Ralph Firbank, and a further four to the Directors of Army

Contracts in 1885 for extensions to a 3 ft. 6 in. gauge line being pushed forward from Wadi Halfa in Upper Egypt towards Khartoum. This had only reached Akaswa, 88 miles from Wadi Halfa and still hundreds of miles from Khartoum, by the time of the death of General Gordon at the hands of the Mahdi in January 1885. The embryonic railway was subsequently destroyed by Khalifa Abdullah's forces (the Khalifa having succeeded the Mahdi after the latter's death later in 1885), as was another British military railway heading west from the port of Suakin on the Red Sea.

In 1896, when events caused Britain again to move against the Khalifa, a campaign culminating in the Battle of Omdurman in September 1898, the railway from Wadi Halfa was rebuilt, this time taking a more direct route across the Nubian Desert and then southwards along the Nile to Atbara. This 385-mile line was built by General Kitchener's forces in 1897–8 and played an important role

in the success of the campaign. Among the locomotives supplied at the instruction of Lt-Col. Western were eight Hunslet 2–6–2 tender engines, nos. 662–8, between April 1897 and April 1898. All were originally intended, it would seem, to have been condensing engines with side tanks, separate condensing vehicles and additional eight-wheeled water tenders; but only three of them are recorded as having appeared in this form, the remainder being conventional tender engines.

A graphic contemporary account of the Soudan Military Railway, written by Col. A. O. Green, appeared in *The Royal Engineers Journal* during June and July 1898. Col. Green described a journey on the line during the first week of March 1898, by which time the railhead was near Sherik, 292 miles from Halfa, and construction was continuing at a rate of up to three miles a day towards Atbara. The account lists thirty-seven locomotives of eight different types. The six original 1885 Hunslet 4–4–0 tanks had

Manchester Ship Canal No. 19 (Hunslet no. 721/00) photographed at Mode Wheel locomotive shed. The design was a Hunslet standard, starting with no. 396 in 1886 and continuing to no. 1151 in 1914. Three of them operated the Admiralty railway conveying stone and convict labour for the construction of the Peterhead breakwater. Early examples had 15 x 20 in. cylinders; later ones, including ten for MSC, had cylinders bored out to 15$\frac{1}{2}$ in.

K. J. COOPER/IRS

obviously survived the Khalifa's destruction of the earlier line, since they are described as 'good, but light, still effective'. The Hunslet 2–6–2 tender locomotives are shown as four working and four 'not yet delivered in the country' and considered to be 'a good type of engine, Walschaerts gear. Design sacrificed for condensing gear, which is not now required. Single bogies troublesome.' The problems associated with building a railway across hundreds of miles of uncompromising desert while at the same time reinforcing and supplying an advancing army

can be imagined, and it is not surprising that the Colonel reports: 'The engines have to run exceptionally long distances with very heavy trains, and they hardly get the chance of being overhauled 'till they break down altogether, which is a not infrequent occurrence.' The need for condensing equipment was overcome by the fortunate discovery of water at No. IV station (77 miles from Halfa) at a depth of 55 feet below the surface, and again at No. VI (127 miles from Halfa) at 70 feet, in quantities sufficient for all the practical purposes of the line. Until this

An indifferent print but historically significant in that it is the earliest workshop view of Hunslet yet discovered and the only photograph that shows the City & South London Railway tube train locomotives under construction. Taken looking south towards the exit doorway of the still almost brand-new 1898 erecting shop, the date is almost certainly mid-September 1899. The order number 21890 painted on the boiler at centre-left identifies this as being for a 2 ft. 0 in. gauge 0–4–2 tank locomotive for Kalamia Sugar Mill, Australia, delivered on 31 October 1899; while the standard-gauge locomotive in the centre is 0–6–0 tank no. 700, which was despatched on 18 September 1899 as Manchester Ship Canal No. 16 Galveston. A 7 x 12 in. quarry locomotive is on the extreme left, and three City & South London superstructures are visible.

Hunslet nos. 695 and 696 were two 14 x 18 in. side tank locomotives built in 1899 for construction and maintenance work on the Central London Railway, now the Central Line of the London Underground. Built to restricted dimensions to suit the tube profile, they had condensing equipment to minimize smoke emission in the tunnels, and two sets of buffing and drawgear to allow operation with either tube or surface rolling stock. They lasted until 1925, ending their days at Lillie Bridge depot, Kensington. The term 'condensing' is used somewhat loosely in this context. In locomotives such as these the exhaust (smoke plus steam) was passed through the water in the side tanks. This did reduce the eventual emission of airborne solids somewhat, but nowadays we should call the process 'exhaust gas conditioning' or 'scrubbing'.

discovery of water, which was made by sinking wells through the soft, spongy sandstone rock, all water for the engines and the construction parties generally had to be brought out from Halfa itself.

Meanwhile, at home, the first of London's tube railways were coming on the scene. The pioneer was the City & South London Railway (now part of the Northern Line), which was opened in stages from 1891 using small four-wheeled electric locomotives with an appearance not unlike that of two upright pianos back-to-back. These were supplied by a number of electrical companies – Mather & Platt, Siemens, Crompton & Co. and the Electric Construction Company – but the manufacture of mechanical parts and the assembly of the locomotives were

subcontracted. Hunslet built a number of them, but the work was done to spares order numbers and consequently is not widely known or recorded.

For the Central London Railway (now the Central Line) two compact six-coupled condensing side tank locomotives, Hunslet nos. 695/6, were supplied in July and August 1899 to the order of the Electric Traction Company. Intended for works trains and for emergency rescue purposes, they had both normal and low-height buffing and drawgear, in much the same manner as the present-day London Underground Limited battery-electric locomotives.

Another major operation handled under a spares order number was the rebuilding, between 1897 and 1914, of a

With the ancient Midland Railway style of crossbar signal and the general air of timelessness, this photograph could have been taken at any time in the first half of the twentieth century. Bass, Ratcliff & Gretton No. 7 was in fact photographed in the early 1960s at Shobnall sidings, Burton-on-Trent. Rebuilt by Hunslet in 1899 from an 0–4–0 well tank that had originally been supplied by the local Burton firm of Thornewill & Warham in 1875, this was one of six similar rebuilds undertaken at Jack Lane to the instructions of Mr Couchman, the brewery's engineer. R. C. RILEY

number of Bass brewery locomotives. These had originally been built locally in Burton-on-Trent, but Hunslet turned them into very attractive 0–4–0 saddle tanks to the instructions of the very exacting brewery chief engineer, Mr Couchman. Painted a rich Turkey red and fully lined out in black and yellow they must rank among the prettiest industrial locomotives ever.

Hunslet had missed out on the Lynton & Barnstaple locomotives in 1897, the order going to Manning Wardle next door. Hunslet had put forward quotations for this railway, however, including a very neat 4–4–0 tank proposal, and one has visions of intense competition between the two firms. This was in the days before good long-distance roads, telephone and telex, and much use of rail, stage coach and horseback must have been made if the quick responses to the correspondence of the day are anything to go by. (There was of course the electric telegraph, but this was not much help with design sketches and specifications, brief though the latter were in those days.) Hunslet redressed the balance in 1900 by supplying three standard-gauge 12 x 18 in. 2–4–2 side tank locomotives, nos. 713–5, for the Bideford, Westward Ho!

Bideford, Westward Ho! & Appledore Railway No. 2 (Hunslet no. 714/00) complete with side screens for street tramway use. A rather effete 2–4–2 side tank, it incorporated the boiler and mechanical portion of a standard 12 x 18 in. outside-cylinder saddle tank industrial locomotive but with an extended frame and a pony truck at each end.

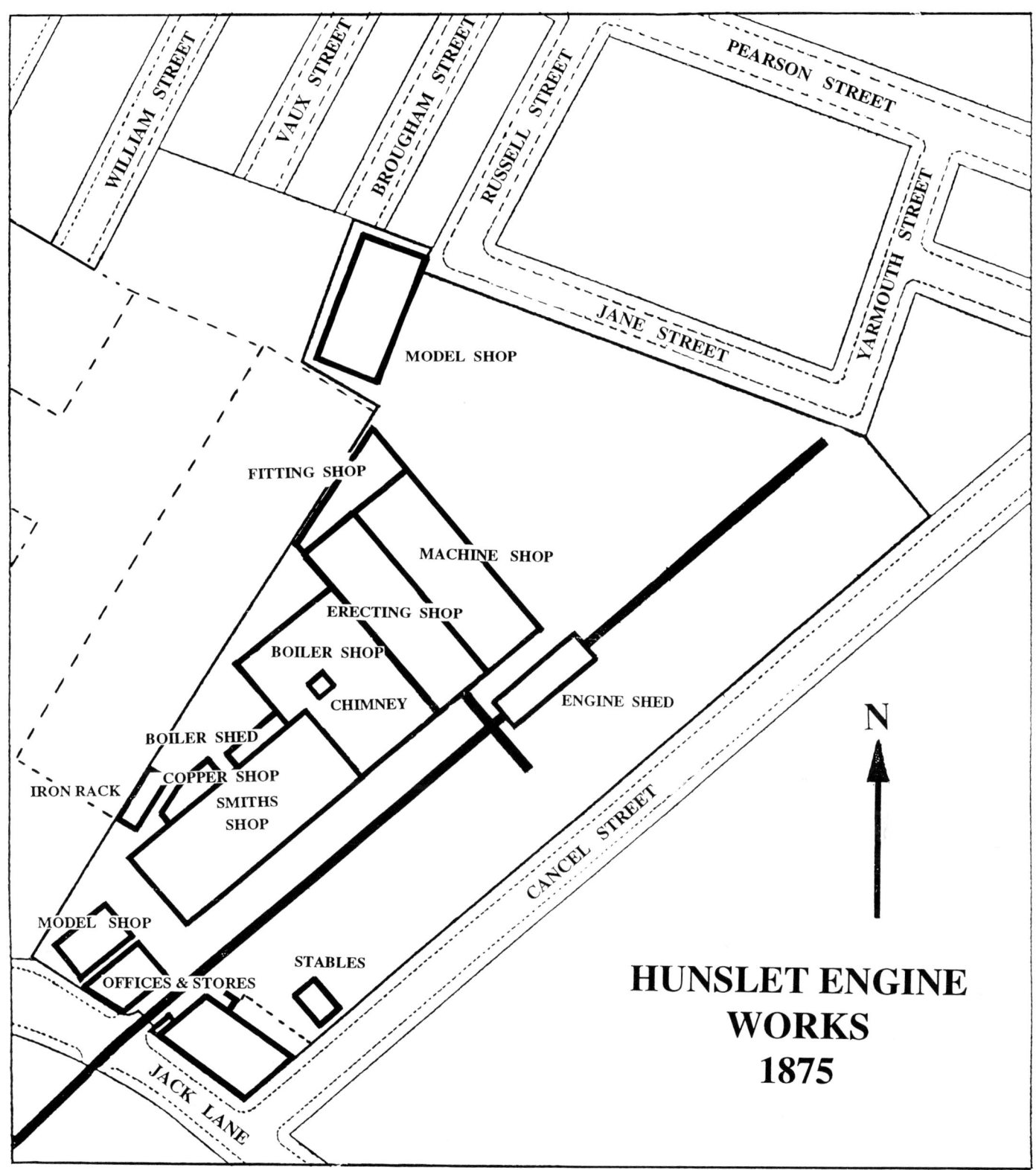

WILLIAM STREET

VAUX STREET

BROUGHAM STREET

RUSSELL STREET

PEARSON STREET

YARMOUTH STREET

JANE STREET

MODEL SHOP

FITTING SHOP

MACHINE SHOP

ERECTING SHOP

BOILER SHOP

CHIMNEY

ENGINE SHED

N

BOILER SHED

COPPER SHOP

IRON RACK

SMITHS SHOP

CANCEL STREET

MODEL SHOP

STABLES

OFFICES & STORES

JACK LANE

HUNSLET ENGINE WORKS 1875

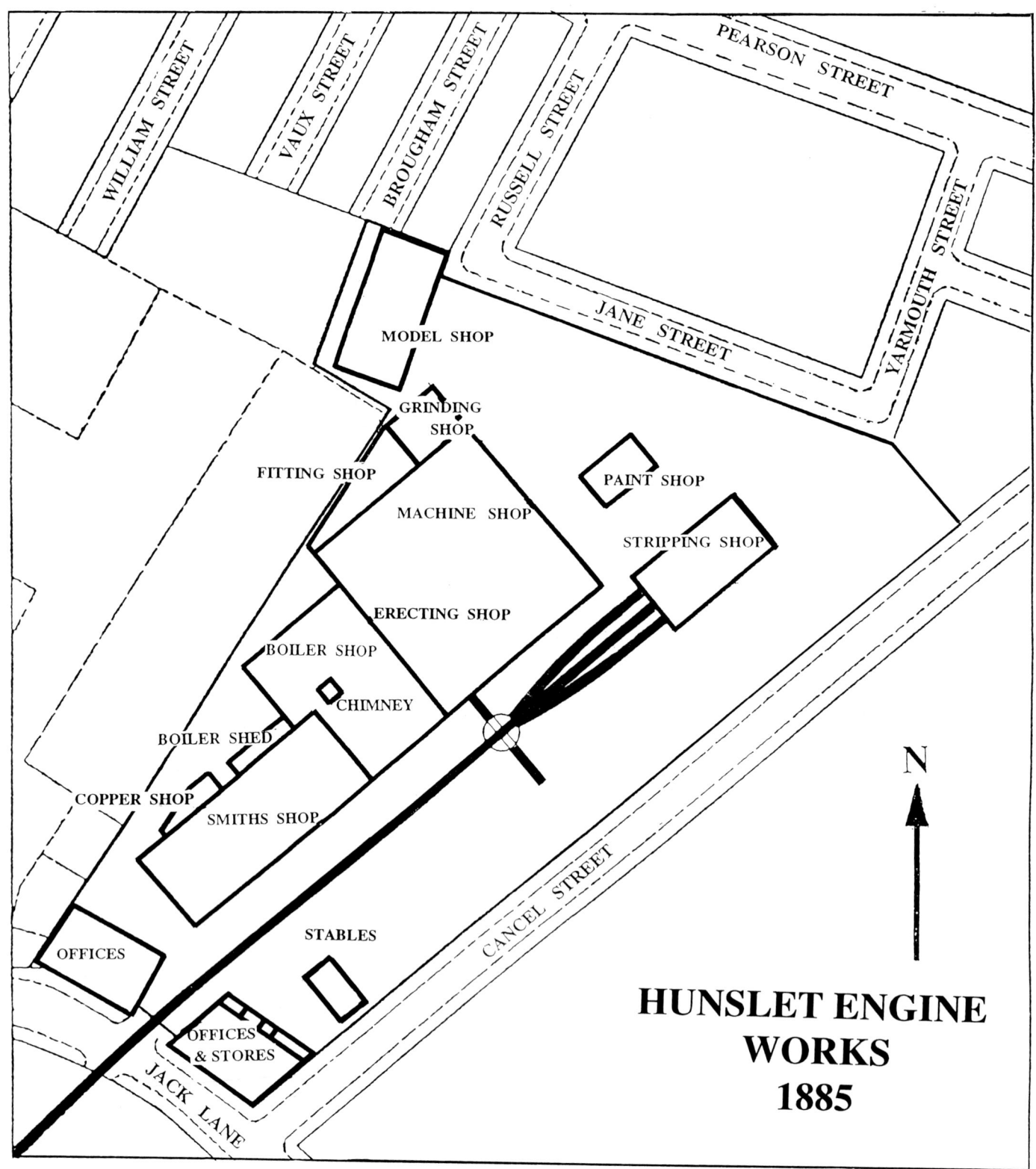

HUNSLET ENGINE
WORKS
1885

& Appledore Railway only 10 miles from Barnstaple. Named *Grenville*, *Kingsley* and *Torridge* these neat passenger engines gave the impression of being a brand-new, and indeed unique, design. Study a side elevation, however, and it quickly becomes apparent that the 'business' part of the locomotive is in fact that of a standard four-wheeled contractor's locomotive with side tanks substituted for the saddle tank and with the frame lengthened at each end to allow for a pony truck front and rear. A very simple way of providing an economical and steady-riding locomotive with adequate footplate area for the crew. Indeed, if one studies the multitude of Hunslet designs over the years it will be found that the various performance requirements were met in most cases by developments to, and different permutations and combinations of, well-established standard features. Evolution, not revolution.

By the end of the nineteenth century the fabric of the Hunslet Engine Works had changed dramatically since Wilson's Railway Foundry was demolished and Leather took it over as a 'brown-field' site. The branch railway from the Jack Lane crossing had swiftly been reinstated in 1864 and a large blacksmiths' shop was built on the west side of the branch. A small detached erecting shop extended northwards at right-angles from the smiths' shop. This latter necessitated locomotives being rolled out of the erecting shop and then manhandled through 90 degrees to be tested. A small engine shed sat astride the branch and a tiny office block fronted on to Jack Lane.

The sweet smell of success? This pristine 14 x 18 in. 0–6–0, HE 531/91 Mortomley, *is typical of the clean-lined 'second generation' of Hunslet industrial locomotives, usually manufactured two or three at a time to stock. Supplied new to the Whitwick Coal Co. Ltd, Coalville, Leicestershire, no. 531 had moved by 1910 to Newton Chambers, Sheffield, who were colliery and chemical works owners and makers of Izal disinfectant and lavatory paper, on whose premises at Thorncliffe it was photographed on 14 July 1937.*
GEORGE ALLIEZ COLLECTION/NRM

Lilla, *HE 554/91, was built for the Cilgwyn slate quarry but purchased by Penrhyn Quarries in 1928, the only Hunslet locomotive obtained second-hand by Penrhyn. Much larger than the other 'quarry Hunslets', but smaller than the Penrhyn main-line locomotives, it was mechanically similar to the Dinorwic 'Mills' class, although these last, Jerry M and Cackler, looked more powerful on account of their lower cab and shorter chimney. Lilla had 8^1/$_2$ x 14 in. cylinders and 2 ft. 2 in. wheels.* MAID MARIAN LOCOMOTIVE FUND

Tralee & Dingle No. 5, *HE 555/92, running into Ballinamore with the afternoon mixed train from Dromod on 19 July 1956. The only 2–6–2 tank on this uninhibited far west of Ireland institution. The locomotives had the constitution of a Roman god and just ran and ran. No. 5 survived to spend another quarter of a century in static preservation in the USA, returning to Ireland via Liverpool in 1986. On its return voyage it passed, in mid-Atlantic, Hunslet's last export to North America, a 603 h.p. diesel-hydraulic that is described in chapter 13.* D. W. WINKWORTH

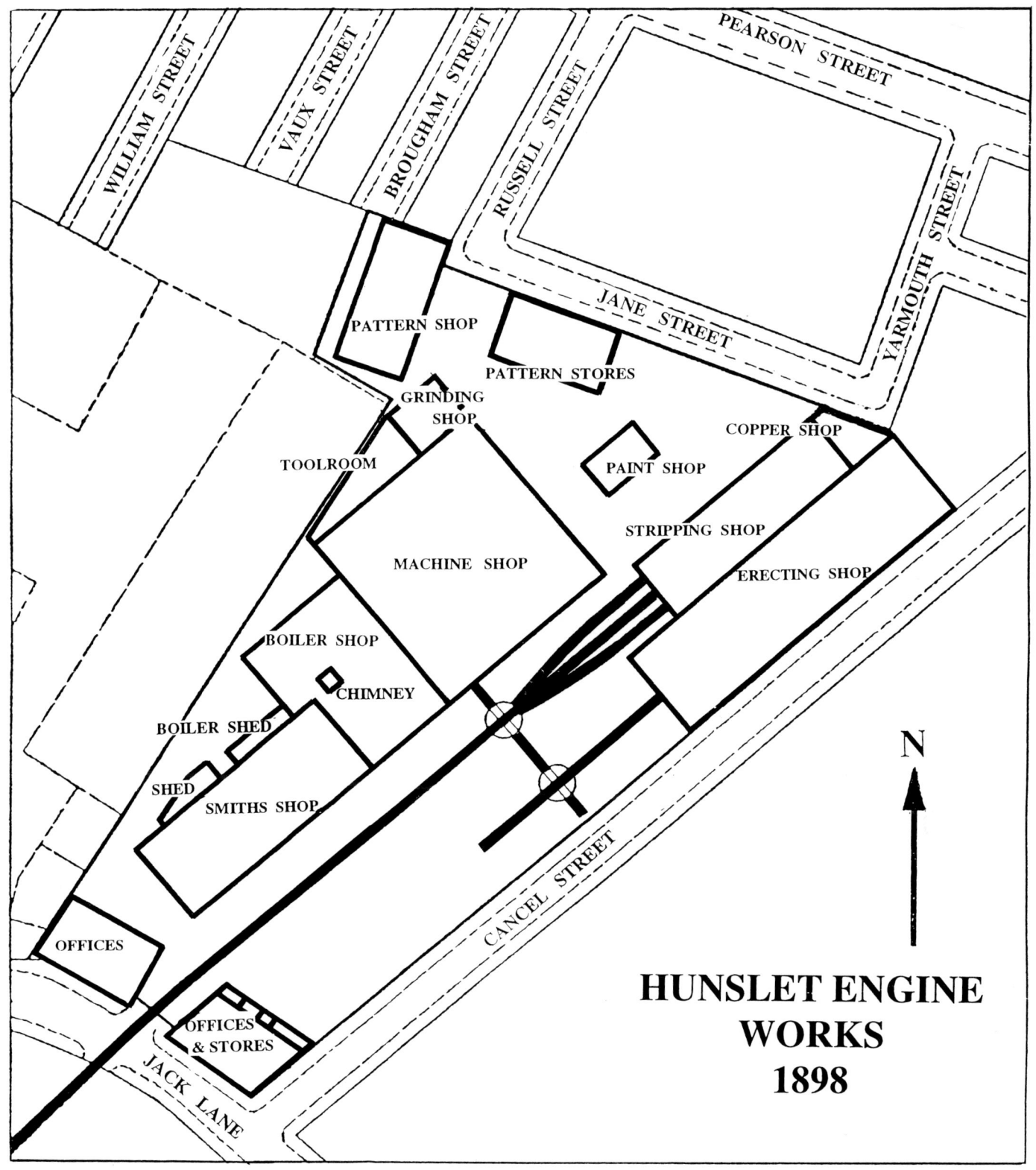

WILLIAM STREET

VAUX STREET

BROUGHAM STREET

RUSSELL STREET

PEARSON STREET

YARMOUTH STREET

JANE STREET

PATTERN SHOP

PATTERN STORES

GRINDING SHOP

TOOLROOM

COPPER SHOP

PAINT SHOP

STRIPPING SHOP

MACHINE SHOP

ERECTING SHOP

BOILER SHOP

CHIMNEY

BOILER SHED

SHED

SMITHS SHOP

CANCEL STREET

OFFICES

OFFICES & STORES

JACK LANE

N

HUNSLET ENGINE WORKS 1898

Francis and Bernard (HE 586/94 and 587/94) were two standard-gauge engines with collapsible chimney and cab canopy for use in tunnel headings. Supplied new to S. Pearson and Sons Ltd, contractors for the construction of the Lancashire, Derbyshire & East Coast Railway's Chesterfield to Warsop line they were later used by Pearsons on the Great Western Railway Patchway to Wootton Bassett contract. This 33-mile-long line employed over forty-five locomotives from 1897 to 1903, at least half of them from Hunslet. These two photographs show Bernard arranged for surface and underground duties.

Dick, Hunslet no. 628, supplied new in 1895 to S. Pearson and Son for the Port Talbot Docks construction contract and subsequently another of the locomotives used on the Patchway–Wootton Bassett main line construction from 1897 to 1903. Between 1903 and 1905, Dick helped in the construction of Seaham south dock, after which it soldiered on for the Seaham Harbour Dock Company until it was scrapped in December 1962. It was photographed, with typically north-eastern dumb-buffered wagons ('chaldrons') in the background, by its namesake Dick Riley on 8 June 1958.
R. C. RILEY

Nesta, HE 704/99, was the archetypal 'quarry Hunslet'. With 7 x 10 in. cylinders, 1 ft. 8 in. wheels and a domeless boiler, this example of the Penrhyn system's four-strong 'small' class was in all major respects the same as the thirteen examples of the Dinorwic system's 'Alice' class. The 'decanter' or 'vinegar bottle' style of safety valve cover denotes a Salter valve and therefore an earlier boiler; later examples had Ramsbottom safety valves. Boilers were exchanged frequently, however, as were other components.
MAID MARIAN LOCOMOTIVE FUND

By 1875 the erecting shop had been doubled in length and a machine shop added to its northern side, and a boiler shop filled the space between the erecting shop and the smiths' shop. Ten years later the new, and still existing, office block had replaced the old one, a further machine shop bay had been added and a large three-road stripping shop formed a termination of the branch line. This stripping shop was a necessity at a time of increasing exports to countries with primitive or non-existent harbour facilities, since most locomotives intended for these countries had to be shipped carefully dismantled with the parts packed and labelled for reassembly on site.

In 1898 a new, and much larger, erecting shop was built parallel to and between the railway line and Cancel Street, adjoining, but much longer than, the stripping shop. The old erecting shop became the heavy machine shop, and the line of rails out of it crossed both the branch line and the new erecting shop line at right-angles, with small turntables facilitating transfers from one line to the others. Thus the company entered the twentieth century.

The Victorian era came to an end with the death of Her Imperial Majesty on 22 January 1901. During her reign, locomotives from Hunslet had been exported to over thirty countries, and, at the time of transition from Victorian to Edwardian, the company was half-way through a batch of 2 ft. 6 in. gauge $11^{1}/_{2}$ x 18 in. 4–4–0 tanks for the Kelani Valley Railway in Ceylon. They carried KVR numbers 102–8 (Hunslet nos. 723–9). No. 104 was to have its brief moment of glory half a century later when it was 'destroyed' in a scene shot on the Kelani Valley Railway for the film *Bridge over the River Kwai*.

CHAPTER 4

Edwardian splendour and the Great War

Alongside the last of the Kelani Valley 4–4–0 tank locomotives, which spanned the Victorian–Edwardian transition, appeared three metre-gauge saddle tanks, one 0–6–0 and two 0–4–0s, ordered by Punchard Lowther & Co. for the constructional of the Hong Kong naval harbour.

Much of 1901's production at Jack Lane was directed at improving Britain's naval communications with her vast colonial acquisitions. Topham, Jones & Railton also took a metre-gauge locomotive for the Gibraltar Dockyard

extension; and Crown Agents had a further 5 ft. 6 in. gauge unit for the Colombo harbour extension, three metre-gauge 4–4–0 tanks for the Singapore Johore Railway and three quaint Emmett-like creations for the Lagos Steam Tramway. These last were small 2 ft. 6 in. gauge 0–4–0 well tank locomotives permanently articulated to a four-wheeled tender on which was mounted a small wooden coach body with slatted louvres. The purpose of this rear compartment is uncertain, but it could have been for the Governor, other VIPs, mail or objects of value.

In comparison with most other Hunslet locomotives, HE 776/02 was an ugly brute. New to the Featherstone Main Colliery Co. it never strayed out of the West Riding of Yorkshire, going in turn to Ackton Hall, Saville and Newmarket collieries as each pit closed, until its final demise in the wholesale slaughter of the sixties. When photographed at Ackton Hall on 6 June 1939, the original straight-backed cab had received a minute coal bunker reminiscent of a Peckett design. It was totally out of proportion – like a French maid's apron on a Russian bear. GEORGE ALLIEZ COLLECTION/NRM

Two batches of 3 ft. 6 in. gauge 2–6–0 saddle tank locomotives were built for Table Bay Harbour, Cape Zone – five in 1901/2 and six in 1903. With 13 x 20 in. cylinders and weighing 30 tons 5 cwts in full working order they were of similar size and capability to the Carthagena 0–8–0 tanks described in chapter 3 but with the leading coupled wheels replaced by a pony truck to provide more flexibility on curves. Hunslet no. 790 was identical but built as a one-off in 1902 for a colliery owner in Natal. When photographed in retirement at Rosterville it had successively been owned by the Victoria Falls & Transvaal Power Company and the Electricity Supply Commission, Johannesburg. Apart from the rather angular bunker tacked on the cab back, it is in original condition even down to the riveted cab, saddle tank and handrails.   D. TREVOR ROWE

The most numerous class of slate quarry locomotives was the Dinorwic system's thirteen-strong 'Alice' class, of which this one, Holy War, HE 779/02, was a typical example.   MAID MARIAN LOCOMOTIVE FUND

At the end of the year came the first two of a batch of five 3 ft. 6 in. gauge 2–6–0 saddle tanks, nos. 766–70, for Table Bay Harbour at the southern end of Cecil Rhodes's Cape-to-Cairo dream. The remaining three, plus an identical machine, no. 790, for a colliery in Natal, followed in 1902. Of the thirty-six locomotives built in this first Edwardian year, only ten were for customers at home. Six more of the unusual 2–6–0ST design (nos. 816–21) were supplied to Table Bay Harbour in 1903.

By 1902 James Campbell was well into his sixties and had run the Hunslet Engine Company virtually single-handed since the death of his brother George in 1890. The next generation was however ready to carry on the family business. Those were the days of large families: in addition to four daughters, James had four sons, Alexander Campbell III (b. 1870), who had stepped into George's shoes as Works Manager, William George ('Will', b. 1874) and Robert (b. 1877), who were also in the works, and finally Gordon (b. 1880), who was working in the drawing office.

A valuation of the works was made in preparation for the formation of a private limited liability company. Incorporated on 15 November 1902 the Hunslet Engine Company Limited had a capital of £80,000 in 2500 preference and 5500 ordinary shares of £10 each. James Campbell effectively sold the company for £38,200 and received his allotted 3820 ordinary shares. The remaining ordinary shares were all taken up by the Campbell family. James Campbell's wife Helen, their four sons, their eldest daughter Christina and George's widow Eliza all had holdings to different degrees.

The Articles of Association of the new company were a masterpiece of draftsmanship, seeing much further into the future than mere locomotive building, and although they have been published before they are worth repeating in full:

To carry on all or any of the businesses of electrical and light, heat, motive power, water supply, and sanitary and general engineers, manufacturers of and dealers in motor carriages, traction engines, locomotives, railway carriages, tramway cars, cycles, bicycles, tricycles, velocipedes and any other kinds of carriages, engines and vehicles whether drawn, propelled or worked by electricity, steam, gas, oil, animals, human beings or otherwise; and plant and machinery of all kinds and also miners, smelters, colliery proprietors, coal and coke merchants, machinists, fitters, metal founders, workers, converters, and merchants, metallurgists, *boilermakers, millwrights, smiths, carpenters, and joiners, woodworkers, builders, painters, enamellers, annealers, gas makers, printers, box and packing case makers, proprietors of vans, wagons and other vehicles, carriers and warehousemen.*

Also in 1902 the building of a new boiler shop was begun, taking up much of the long-vacant land disused since the demolition of the E. B. Wilson tender shop towards the Jack Lane end of the site. This boiler shop, which was opened in 1903, was in line with, but separated from, the 1898 erecting shop, and facilitated the direct transfer of boilers by a straight line of rails between the two shops. The vacated boiler shop, like the old erecting shop before it, became an extension to the machine shop, in which augmented facility was provided what is believed to have been the first large-scale application of electric power to drive machinery to be introduced in Leeds.

The line of rails connecting the boiler and erecting shops lay parallel to the branch line from the Midland Railway, which, at its extremity, entered the old stripping shop. Two turntables connected both tracks, and also provided a further track at right-angles, which entered the old erecting shop, now the heavy machine shop, thereby making possible the convenient movement of all heavy components throughout the works.

James Campbell was Chairman and Managing Director of the new limited company. Alexander continued as Works Manager and Will became 'secretary and traveller'. James's health was failing, however, and on his death at the age of 68 on 14 October 1905 Alexander took his place, with Robert stepping up to works manager. Not many years later, Robert was seriously injured in a locomotive accident during tests at a customer's premises and was left permanently disabled, while Will, presumably at odds with his brothers, resigned at the end of 1911. Robert took up the less strenuous duties of secretary on Will's departure, until in 1923 he resigned this post also because of continued ill health. He remained a director until his death in 1926 at the early age of 49. The youngest of the four brothers, Gordon, went into partnership around 1905 with two Germans, Rothert and Mayer, and the trio dabbled with a petrol-engined road lorry with a patented friction-drive transmission. Gordon prevailed upon his brothers to the extent that one, or perhaps two, prototypes were built at the works. The project was a failure – hindsight suggesting that friction drive was perhaps not such a good idea, given the scientific knowledge of the day – and Gordon emigrated in 1907 to Canada, where he died in 1926.

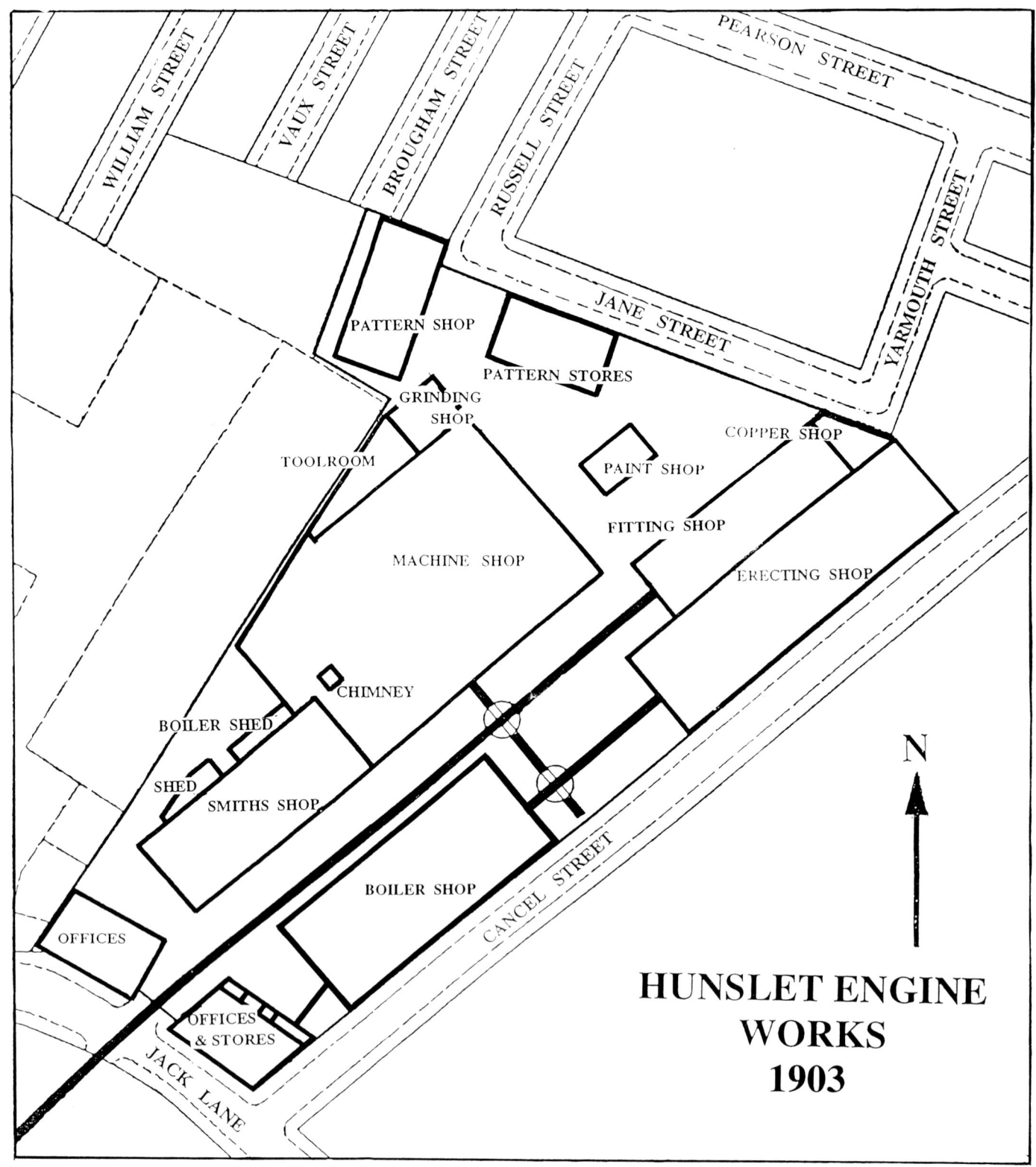

WILLIAM STREET

VAUX STREET

BROUGHAM STREET

RUSSELL STREET

PEARSON STREET

YARMOUTH STREET

JANE STREET

PATTERN SHOP

PATTERN STORES

GRINDING SHOP

TOOLROOM

COPPER SHOP

PAINT SHOP

FITTING SHOP

MACHINE SHOP

ERECTING SHOP

CHIMNEY

BOILER SHED

SHED

SMITHS SHOP

BOILER SHOP

CANCEL STREET

OFFICES

OFFICES & STORES

JACK LANE

N

**HUNSLET ENGINE
WORKS
1903**

The Hunslet erecting shop (northern end) in mid-September 1902. In the foreground, right, is no. 794, a 3 ft. 6 in. gauge 12 x 16 in. 0–6–0 saddle tank for the East London Harbour Board, South Africa. Centre, no. 783, ordered by J. R. Banks for the metre gauge Vasco Asturiana Railway in northern Spain as their No. 21 Nalon. Far left, the half-built no. 790 The Collier, a quaint 3 ft. 6 in. gauge 0–6–2 saddle tank, now preserved, for Natal. In the back corner, right, is Lomax no. 799, a standard-gauge 14 x 18 in. 0–6–0 saddle tank for Sutton Heath and Lea Green Collieries, St Helens.

No. 21 Nalon, Hunslet no. 783/02, shown in the previous photograph, in less pristine condition ninety-one years later. Sold by the Vasco Asturiana Railway in 1961 for industrial use it is seen here at Ujo on 7 December 1993 in store for a proposed transfer to the railway museum at Gijon.
          D. TREVOR ROWE

Hunslet 3 ft. 0 in. gauge 4–4–0T no. 842/02 carrying second-hand plates as Valencia Suburban Tramways (La Sociedad Valenciana de Tranvias) No. 8 at Villanueva de Castellón on 3 May 1953.   D. TREVOR ROWE

The Edwardian period had started with the Kelani Valley Railway 4–4–0 tank locomotives, and the KVR, like Sierra Leone and others yet to come, was to provide a steady trickle of repeat bread-and-butter business for very many years.

The first seven Sierra Leone 2–6–2 tanks, Nos. 21–7 (Hunslet nos. 673–5, 709–10 and 744–5), built between 1898 and 1901, had a relatively small boiler with a raised round-topped firebox and 10 x 15 in. cylinders. Commencing with No. 28 (Hunslet no. 800) in January 1903, and continuing right through to No. 85 (Hunslet no. 3815), which was supplied half a century later in 1954, the boiler diameter was increased by just short of 9 inches and the cylinder diameter by three-quarters of an inch. The series thus ran from 1898 to 1954, with the increased dimensions from 1903 onward. In the new form the round-top firebox was flush with the top of the boiler. The elements of this larger Sierra Leone design reappeared in 1905 in a 1 ft. 11½ in. gauge 0–6–2 tank for the Leeds City

Council's Masham Reservoir project – echoing the comments towards the end of the last chapter regarding standardization, specifically with respect to the Bideford, Westward Ho! and Appledore 2–4–2 tanks.

Leeds No. 1, Hunslet no. 865, was in effect a Sierra Leone locomotive with fundamentally the same boiler, wheels and cylinders but with shorter side tanks, a shorter smokebox and no leading pony truck. Spacing the frame plates closer together meant that the firebox had to be lengthened slightly to maintain the same firegrate area. More significantly for the railway enthusiast, the Masham locomotive design was used as the basis for no. 901 *Russell*, built in 1906, and on this locomotive the longer tanks, smokebox and 2–6–2 wheel arrangement returned. The lengthened firebox meant a two-inch increase in the distance between the trailing driving wheels and the rear pony truck as compared with the Sierra Leone locomotives, and the pony trucks had inside bearings. *Russell* was supplied to the North Wales Narrow Gauge Railway, where it replaced the 0–6–4ST *Beddgelert* and was later joined by Hunslet's only Fairlie locomotive, no. 979 *Gowrie*, also an 0–6–4, in 1908. *Russell*, after a very chequered career, has survived to be restored on the revived Welsh Highland Railway, while Sierra Leone Railway No. 85 is on the Welshpool and Llanfair Railway. It is therefore possible in the space of one afternoon to compare two locomotives built 48 years apart and virtually identical but for minor details occasioned by the difference in gauge and the modifications made during refurbishment by successive owners.

For almost thirty years *Beddgelert* had been the only Hunslet example of the 0–6–4 wheel arrangement – never

a really common configuration, in Britain in any event. Hunslet nos. 878–80 revived the type in November 1905, and in doing so created a 'niche market' in compact, heavy-haul narrow-gauge locomotives for difficult terrain. These three 0–6–4 side tank locomotives, together with six further identical units, nos. 907–12, the following year, were quite large locomotives – 36 tons 2 cwt, with $15\frac{1}{2}$ x 18 in. cylinders – and were built for the 2 ft. 6 in. gauge Antofagasta (Chile) and Bolivia Railway. Six more, 3 tons heavier but with cylinders reduced to 15 x 18 in., nos. 945–50, were built for the same customer and appeared in 1907. In addition to these fifteen 0–6–4 tanks, the FCAB (Ferrocarriles Antofagasta Bolivia) also took fourteen

Leeds No. 1, HE 865/02 on a Leighton reservoir construction train between Masham and Leighton Yard. The presence of only a few marks on the front buffer suggests that the locomotive was virtually brand-new, a suggestion reinforced by the viaduct timbers loaded on the first wagon. Imagine the side tanks extended forward to the front of the smokebox and the frames lengthened to accommodate a front pony truck, and you have a fair picture of the North Wales Narrow Gauge Railway Russell in its original condition.
R. N. REDMAN

Hunslet no. 901/06 Russell was a pretty engine when built. It was cruelly modified in a vain attempt to make it negotiate the confines of the Festiniog Railway's Moelwyn tunnel in 1923 when the Welsh Highland Railway initiated through trains on to the Festiniog system. The Welsh Highland closed in 1937, and Russell was stored until 1941, when the Ministry of Supply acquired it for industrial use. It is seen here on 9 July 1945, running as a 0–6–2 tank at the Brymbo Steel Co. Ltd's Hook Norton Ironstone Mines in Oxfordshire. Although a fair representation of the original style, Russell as now running on the Welsh Highland is brand-new above the frames. D. W. WINKWORTH

A proud work-force stands in front of Hunslet no. 874/05 Dom Carlos, having reputedly built this 14 x 18 in. 3 ft. 6 in. gauge 0–6–0 side tank in 21 days from scratch. The customer was Griffiths & Co., contractors building the Benguela Railway from Angola to the Belgian Congo. The locomotive stayed with the railway for at least seventy years.

The archetypal Hunslet 15 x 18 in. 0–6–4 tank locomotive. This one is Kyshtim Corporation No. 3 (Hunslet no. 1027/10), built to 3 ft. 0 in. gauge for the Karabash copper railway, although the metre gauge examples built for Chile were virtually identical. The same basic design was also supplied to Bolivia and Brazil. The Adams bogie is reminiscent of Beddgelert and Gowrie.

Pre-Revolutionary Russia – just. Kyshtim–Karabash Railway (Kyshtim Corporation) No. 7 (Hunslet no. 1073/11) photographed some time around 1915. Some minor modifications to the as-built condition are apparent; a cab side window has been added and larger sandboxes fitted. The additional bunker rails and spark-arresting chimney indicate a change from coal- to wood-burning, while the typically Russian headlamps add to the 'Doctor Zhivago' image.

ALEKSANDER KOLESOV, EKATERINBURG, VIA KEITH CHESTER

2–8–2 and ten 2–8–0 tender engines (nos. 888–91, 922–31 and 958–67) over roughly the same period. The largest of these tender engines weighed over 74 tons in working order – very big for a rigid-frame 2 ft. 6 in. gauge locomotive.

The 0–6–4 side tank design sold well to a number of customers. The FCAB 2 ft. 6 in. gauge examples were outside-framed, but an inside-framed version to metre gauge sold to the Chile Longitudinal Railway and the Brazil North Eastern railway, while eight 3 ft. 0 in. gauge examples went to Russia. In all, thirty-five of this type were produced between 1905 and 1916.

Although virtually identical in appearance to the South American examples, the Russian engines are perhaps the more interesting. Four of them, nos. 1027, 1029–30 and 1045, were supplied in 1910; they were followed by no. 1073 in 1911, nos. 1093 and 1104 in 1912 and no. 1162 in 1916. All of these, plus a little 0–4–0 saddle tank, went to the Kyshtim–Karabash Copper Railway in the Ural Mountains. They are all believed to have survived until 1960, some indeed until the railway closed in 1978. The Urals region produced 50 per cent of all Russia's copper in 1913, of which half came from the Kyshtim Corporation. To put it another way, a quarter of all Russia's copper was hauled by Hunslet locomotives.

Coincidental with the building of *Russell* and the FCAB tank and tender locomotives in 1906 was the emergence of two further noteworthy narrow-gauge designs. One was a batch of five 12 x 16 in. 2–8–2 tanks with Belpaire fireboxes for Sierra Leone, nos. 883–7, built in March of that year; the other was a sleek and very nippy $8^1/_2$ x 14 in. 0–4–2 side tank for the Howrah–Amta Railway in Calcutta. This initial batch of Howrah–Amta locomotives, nos. 902–5, were shipped in pairs on 29 August and 30 September, and became known as the 'Eva' class after the name carried by no. 904. This was another design that was to stand the test of time: nineteen in all were built for the Howrah commuter line over the years, the last one, no. 3867, being despatched to Calcutta on 28 March 1955.

It is appropriate to comment here on the proportion of the Hunslet Engine Works business that was now going overseas. Exports had always been healthy, averaging between 40 and 60 per cent of the work-load for the last three decades of the nineteenth century. From 1901 to the start of the Great War in August 1914 the balance shifted to 80 per cent export. Of the remaining 20 per cent, half was devoted to narrow-gauge locomotives, typified by

those destined for the Welsh slate quarries and the stone quarries of the Midlands. Out of almost 450 locomotives built during this 14-year period, only 36 were conventional standard-gauge industrial and contractors' types. That the Company could cope with the almost insatiable appetite of the Colonial railway builders was probably due only to the large numbers of standard-gauge contractors' locomotives rendered surplus by the completion of the home railways. The Great Central Railway's extension to London Marylebone, for example, had just been finished at the end of the century; so had the Manchester Ship Canal and the many harbour projects, all of them releasing quantities of locomotives. But as one door closes, another opens – 'twas ever thus – and there was a good repair business as these surplus units were dispersed.

The year 1907 saw thirty-three locomotives go overseas out of a total production of thirty-nine. Sixteen of these were for the Antofagasta (Chile) and Bolivia Railway (ten 2–8–0 tender engines and six 0–6–4 tanks) already mentioned, while a further eight were stylish 2–6–2 tanks for the 2 ft. 6 in. gauge Shahdara–Saharanpur Railway near Delhi. This railway, like the Howrah–Amta, was one of several light railways operated in many parts of India by T. A. Martin of Calcutta, who might be thought of as the 'Colonel Stephens' of India. Another was the Baraset–Bashirat, which had taken a 2 ft. 6 in. gauge Hunslet 0–4–2 tank engine, no. 918, in 1906. The Shahdara locomotives, nos. 913–6 and 932–5, had a similar overall style to that of *Russell* and the Sierra Leone engines; but, with $12^3/_4$ x 16 in. cylinders and weighing just under thirty tons in working order, they were much larger. The design later developed into a 2–6–4 tank, the trailing bogie allowing for a rear coal bunker; the last of this design was built as late as 1955. In all, eight 2–6–2 and eleven 2–6–4 locomotives were supplied by Hunslet to the Shahdara–Saharanpur over the 48 years, including two assembled by the railway itself during the World War II from parts supplied prior to 1939. All were still in service on the $92^1/_2$-mile-long line in 1969.

To complete the export picture for 1907, there were two more 3 ft. 6 in. gauge 0–8–0 saddle tanks for the Carthagena and Herrerias Steam Tramway in Spain; three 0–6–0 side tanks for the 3 ft. 6 in. gauge Lagos Government Railway; a metre gauge 4–4–0 tank for the Singapore Government Railway; a 2–4–2 tank for Martin's 2 ft. 6 in. gauge Bukhitapur Railway; an 0–6–0 side tank for the metre-gauge Cordoba & Rosario Railway in Argentina;

Pope and Pearson Ltd owned the appropriately-named West Riding Colliery at Altofts, north of Normanton, right through to nationalization in 1947. One of the many victims of widespread closures in the 1960s, the colliery was situated in the fork of the junction where the line to York diverged from the Midland main line to Leeds – a site now occupied by the Port Wakefield freight terminal for Channel Tunnel traffic. Apart from one Manning Wardle 0–6–0 saddle tank purchased in 1868, it remained a steadfast Hunslet customer, buying ten new four-wheelers over the years between 1866 and 1925: no second-hand cast-offs here. Except for once using the name Diamond on HE829/03, only four names – Altofts, Haigh Moor, Silkstone and West Riding – were ever used, the names being handed down as each locomotive was retired. All were 0–4–0 saddle tanks except for the third West Riding which was a side tank, purchased in 1925. The example pictured here is the second Altofts, HE 940/07, a 15 x 20 in. machine in superb mechanical condition but in need of a repaint when photographed on 17 June 1938.   GEORGE ALLIEZ COLLECTION/NRM

an 0–4–0 saddle tank for the metre-gauge Valencia Suburban Tramways in Spain; and finally a small 2 ft. 0 in. gauge 7 x 12 in. 0–4–0 saddle tank for Sarawak in the Straits Settlements. The mix was typical of the period, and said much for the quality of the draughtsmen employed.

The large Antofagasta (Chile) and Bolivia Railway 2–8–0 tender engines nos. 958–67 dominated the major part of 1908, but there was still time for three 3 ft. 0 in. gauge 0–6–0 tanks for British Honduras, two 5 ft. 6 in. gauge 4–6–0 tender engines for Ceylon Government Railways and three more 0–6–0 tanks for the Lagos Government

Railways. The Kelani Valley Railway took a large 4–6–4 side tank, the first of many, which with 14 x 20 in. cylinders and a weight of 46¾ tons, was the equivalent of two of the line's 4–4–0 tanks of seven years earlier. Taking KVR no. 142 (Hunslet no. 977), it was the first of the railway's 'J2' class. There was *Gowrie*, of course, for North Wales, another 'E' class for the Howrah–Amta, a small metre gauge 0–6–0 saddle tank for Argentina and, just before New Year's Eve, the first two of a batch of large 5 ft. 6 in. gauge 0–6–0 saddle tanks for the Buenos Ayres and Pacific Railway. The year's production was completed by a

Pictured at the Sagunto steelworks of Altos Hornos de Vizcaya, Gilet 105 was originally Valencia Suburban Tramways (La Sociedad Valenciana de Tranvias) No. 20, when supplied new through J. R. Banks. Hunslet no. 957/07, metre gauge 7$^{1}/_{2}$ x 12 in. cylinders. The original nameplate has been used, the previous lettering 'SVT 20' remaining just visible after being ground off prior to the new name and number being superimposed.   D. TREVOR ROWE

Hunslets' only Fairlie: the North Wales Narrow Gauge Railway's fabled Gowrie (Hunslet no. 979/08), photographed at Dinas, presumably when quite new. Although, at 9$^{1}/_{2}$ x 14 in., much smaller than the Russian locomotives, the beginnings of the later Kyshtim design were starting to emerge. Beddgelert—Gowrie—Kyshtim is a logical progression. The Adams bogie still reigns supreme.

Gertrude, *HE 995/09, was one of Penrhyn's six 'large' class locomotives. These had a domed boiler and $7^1/_2$ x 10 in. cylinders.*

Bland, *Hunslet no. 1003/09, was a 15 x 20 in. locomotive for the Silkstone and Haigh Moor Colliery Company nine miles away to the south-east of the Hunslet works at Allerton Bywater, Castleford. The collieries of the Castleford area played a large part in the development of Hunslet industrial and mining locomotives through the years, providing both a faithful clientele and a useful testing-ground.*

standard-gauge six-wheeler for the Groby Granite Co. and a 2 ft. 6½ in. gauge four-wheeler for Jee's Hartshill Granite Company.

There were six of the Buenos Ayres and Pacific Railway 0–6–0 saddle tanks, nos. 2501–6 (Hunslet nos. 982–7). Two were built in 1908 and the rest the following year. They were very British in outline, somewhat similar to the Kitson-built Taff Vale Railway 'V' class, and were fitted with standard British-style side buffers. The official photograph shows American-style cow-catchers, or 'pilots', at each end, extending some eighteen inches beyond the buffers; this may have worked when coupling to wagons, but would have been a mite problematical when trying to couple two locomotives together.

The year 1909 saw the production of Hunslet no. 1000: *Taki*, a 2 ft. 6 in. gauge 2–4–2 tank for Martin's Baraset–Bashirat Railway, was outshopped on 14 June.

The rest of 1909 was very much taken up with repeat orders for existing customers – T. A. Martin's various railways in India, Crown Agents for Sierra Leone and Nigeria, and an 0–4–2 tank for Santa Marta. There were also three more 'large' class 7½ x 10 in. locomotives for Penrhryn Quarries, four small 5 x 8 in. 0–4–0s for Indian steelworks, and *Nonus*, a 2 ft. 0 in. gauge 0–6–0 saddle tank for Groby Granite. Two standard-gauge locomotives for the Appleby Iron Co. and Allerton Bywater colliery contrasted with a steam inspection car for Buenos Ayres & Pacific, and with a sugar estate locomotive for Queensland completed the year's tally of thirty-three locomotives.

1910 was a busy year. Only three locomotives were built for the home market: these were no. 1028, a 600 mm gauge 0–4–0 wing tank named *Microbe* for the Leeds Corporation Sewage Works; another, no. 1039, of similar size for Jee's quarries at Hartshill; and *Sextus*, a duplicate of *Nonus*. There were forty-one locomotives for overseas, though, including fifteen almost identical 0–6–4 tank locomotives. Four of these were the Kyshtim 3 ft. 0 in. gauge examples mentioned earlier. The other eleven were the same but to metre gauge: two, nos. 1053/4, went to the South American Construction Company for the North Eastern Railway of Brazil, and nine were supplied to Griffiths & Company Contractors Ltd for the southern

Hunslet no. 921/05 Sybil Mary, one of the larger 7½ x 10 in. cylinder 1 ft. 10 ¾ in. gauge engines at Penrhyn Slate Quarries. Penrhyn purchased 17 Hunslet locomotives, all but one of them new, between 1882 and 1909. All went into preservation, Sybil Mary being latterly with the Lynton and Barnstaple Railway Association.
MAID MARIAN LOCOMOTIVE FUND

Close by the Waterloo Main Colliery locomotive shed was the Leeds Corporation Knostrop (sic, a corruption of Knowsthorpe) Sewage Works where Hunslet no. 1028/10 Microbe worked on the 600 mm gauge tramway serving the filter beds. It shared duties with Bacillus, a similar-sized (7 x 10 in.) locomotive, purchased at the same time from the neighbouring works of Hudswell, Clarke & Co. Ltd to avoid any hint of favouritism on the part of the City Council.

section of the Chile Longitudinal Railway. Griffiths had worked on the Benguela Railway in Angola five years previously. The Griffiths locomotives were Hunslet nos. 1040, 1044, 1047–50 and 1057–59, and were despatched between 19 July and 16 December 1910.

The railways of West Africa were becoming deeply committed to Hunslet locomotives. Sierra Leone had its fleets of 0–6–0, 2–6–2 and 2–8–2 tank locomotives; Lagos, Southern Nigeria and Northern Nigeria had a variety of small locomotives; and in April 1910 came four elegant and workmanlike 2–6–2 tank passenger locomotives for the Gold Coast Railway, nos. 1022–5 (GCR Nos. 31–4). Again the locomotives had the unmistakable family likeness that had been apparent in and had slowly developed from *Beddgelert* through *Russell*, the Shahdara's

and other locomotives. By now the outside cylinders and Walschaerts valve gear had been developed to provide almost standard engines in a range of sizes. The smallest was a 5 x 8 in. machine; 14 x 20 in. had been reached by 1910; and the range eventually reached 18 x 24 in. cylinders. The shape of the components – crossheads, valve rods, connecting-rods, etc. – had become identifiable just as surely as the shape of the chimney, the long side tanks, the cab cut-outs and the bunker with its sloping back. At the risk of pointing out the obvious, a set of steam locomotive cylinders complete with attendant valve gear, connecting rods and crank axle or driving wheels was equivalent to the whole assembly of petrol or diesel engine, transmission and final drive yet to come – or to the 'traction package' of today's electric trains. At Hunslet, developing what might appear to be a brand-new special design for a customer was more often than not an exercise in experience, whereby the draughtsman 'mixed and matched' the most suitable existing components to provide the optimum result. Once that 'special' design had been produced and proved itself it became yet another 'standard', to be progressively developed if or when a suitable opportunity arose. Such a process of common-sense development has much in its favour. Modern jargon would call it 'management of risk'.

This 'pick and mix' practice was noticeable in seven 9½ x 14 in. 2–4–2 side tank locomotives supplied to T. A. Martin. They were similar in overall dimensions and appearance to *Russell* but had inside frames to the driving wheels, as on the Sierra Leone 2–8–2 tanks, while the boiler and cylinders were developed from *Gowrie*. The driving wheels were 2 ft. 6 in. in diameter, as on the 'Eva' class. Seven of them were supplied between 10 June and 7 October 1910. The first five of the series, nos. 1033–7, went to the Arrah–Sasaram Railway. The other two, nos. 1038 and 1041, were for the Baraset–Bashirat Railway. Despite their later numbers, these last two were actually despatched first, probably with maker's plates switched to keep the records straight, since the ones for the Arrah–Sasaram Railway would appear to have been running a little behind schedule – a small six-wheeled contractor's locomotive named *Constructor III* (Hunslet no. 1031) had gone to the Arrah–Sasaram only at the beginning of May – perhaps construction of this railway was running into some difficulty.

The Tralee and Dingle Railway in Ireland had taken delivery of six Hunslet tank locomotives, four 2–6–0, one 0–4–2 and a 2–6–2, between 1889 and 1892; but by 1902 the railway was in a dreadful state owing to bad maintenance and general penury. In response to an offer from Kerr Stuart, the beleaguered railway purchased a 'bargain' 2–6–0 tank that had been 'built for another overseas customer', and was followed up with a second, similar machine in 1903. The Kerr Stuarts did not have the stamina or the ability to withstand abuse that the older locomotives had shown, and the ninth and last Tralee and Dingle locomotive was supplied by Hunslet (no. 1051) on 22 November 1910.

Also of 2–6–0 wheel arrangement, but this time with eight-wheeled double-bogie tenders, were Santa Marta Railway Nos. 10, 11 and 12, Hunslet nos. 1015, 1019 and 1046. These identical 15 x 20 in. locomotives, against

three individual orders, were supplied on 1 January, 2 February and 19 December 1910 respectively. The ordering procedure for all the Santa Marta engines down through the years was unusual, with no apparent system, and the Hunslet approach seems to have been equally cavalier when one looks at the variety of types supplied. Either this or they were simply happy to acquiesce in the whims of a capricious customer.

The remaining eight 1910 locomotives included two more Lagos Tramway 0–4–0 tender engines, two 0–4–2 tender engines for sugar estates in Queensland and Natal, an 0–6–2 side tank for Rangoon and three small narrow-gauge four-wheelers for Trinidad, Spain and South Africa. Quite a spread!

1911 very nearly became remarkable for being the only year in the locomotive-building history of the Hunslet Engine Company in which not one new locomotive was built for use in England. Not until 20 December did the overseas monopoly break, when two 0–4–0ST locomotives went to the South Durham Steel and Iron Company. The remaining twenty-six locomotives went overseas. Many were repeat orders – two more 2–6–0s for Santa Marta (Hunslet nos. 1071–2, Santa Marta Nos. 13 and 14) in May were followed in July by Nos. 15 and 16 (Hunslet nos. 1074–5); but characteristically the former was a 12 x 18 in. 2–4–2 tank and the latter a 9 x 16 in. 4–4–0 tank. Presumably they knew what they wanted. South America in fact took over half the year's production: two large 53 ton standard-gauge 4–6–0 tank locomotives with 16½ x 22 in. cylinders (nos. 1052 and 1070) for the North West of Uruguay Railway Company; two more 2–8–0 tender loco-motives (nos. 1066–7) for FCAB; and another five 0–6–4 tanks for the Chile Longitudinal Railway. All the earlier Longitudinal Railway locomotives had been for use by Griffiths on the southern section of the line. Of the 1911 deliveries, no. 1064 also went to Griffiths, but nos. 1062–3, 1065 and 1078 were purchased by MacDonald, Gibbs, McDougall for the northern section, and are recorded as having been sent out with additional tenders, presumably on account of the desert conditions in the Atacama region.

Kyshtim also purchased a further 0–6–4 tank in 1911 (no. 1073), together with a diminutive 7 x 10 in. 0–4–0 saddle tank locomotive. T. A. Martin took five locomotives, comprising another 'Eva' class 0–4–2T for the Howrah–Amta, two 7 x 12 in. 0–4–0 side tanks for the 2 ft. 0 in. gauge Jagadhi Light Railway in Calcutta, and a pair of tiny 5 x 8 in. 0–4–0 tanks, also 2 ft. 0 in. gauge, for the

Larger than its delicate lines would suggest, this 3 ft. 6 in. gauge 4–8–0 tender engine (no. 1060/11) had 17 x 21 in. cylinders and weighed 68 tons with tender, in full working order. It was one of thirty-nine steam locomotives of several types supplied by Hunslet to the Gold Coast Government (later Ghana) Railways.

Indian Iron and Steel Company. These last were similar to four that had been taken by the same customer in 1909/10.

There were four locomotives for West Africa: one 0–6–0 tank and one 2–6–2 tank for Sierra Leone, and two 17 x 21 in., 3 ft. 6 in. gauge 4–8–0 tender engines for the Gold Coast Railway. A 2 ft. 6 in. gauge 0–4–2 side tank was supplied through Matheson & Co. for Peru.

The emphasis on exports was to continue for many years. Only four locomotives were built for the home market in 1912, these being no. 1094, a 15 x 20 in. 0–6–0 saddle tank for the local Silkstone and Haigh Moor Colliery Company, no. 1101, a small 7$^{1}/_{2}$ x 10 in. saddle tank for Penmaenmawr, and two 16 x 22 in. standard-gauge 0–4–0 saddle tanks for Pope and Pearson Ltd at Altofts Colliery

(no. 1107) and for the South Durham Steel & Iron Company (no. 1108). In 1913 there was only one, no. 1146 *Violet*, a 14 x 20 in. 0–4–0 saddle tank for the Mountsorrel Granite Company in Leicestershire. The first eight months of 1914, before war broke out, produced two 0–6–0s, a side tank named *King George V* for James Roscoe & Co. of Little Hulton, Lancashire, and no. 1147 *Strafford*, which was identical to no. 1094 mentioned above. Over the same period, however, these seven home-market locomotives had seen the production of no less than sixty-eight for overseas.

Before a description of these last pre-war export locomotives, many of which were of large and impressive designs, let us look at the dynastic change that was taking place at this time. The Campbell family had ruled the roost in both of the locomotive-building firms on the north side of Jack Lane (Manning Wardle and the Hunslet Engine Company) for 56 years since the end of Edward Brown Wilson's concern. The departure of brothers Will and Gordon Campbell to Canada and the debilitating effects of Robert's injuries were putting increasing pressure on Alexander Campbell III, who had combined the duties of managing director and works manager since 1909. On 5 January 1912, Alexander, who was then 42 years old, advised the Board that he intended to advertise for applicants for the post of Works Manager.

The successful applicant was the 35-year-old Edgar Alcock, who left his post of Assistant Works Manager at the Gorton, Manchester works of Beyer, Peacock & Co. to become works manager at Hunslet.

Born in Macclesfield, Edgar Alcock had been apprenticed at the Lancashire & Yorkshire Railway works in Horwich. A restless man, he spent a short time in the army and then several short spells in various engineering establishments, including the 'Lanky' running sheds at Bolton, before returning in 1898 to Horwich in the Mechanical Department under Sir John Aspinall, then Chief Mechanical Engineer of the LYR. When Sir John became General Manager of the railway in 1899 he was succeeded as CME by H. A. Hoy, and Edgar Alcock worked directly under Hoy as one of his personal assistants.

In 1904 Hoy left Horwich – he could be said to have been head-hunted – to become general manager of Beyer Peacock, taking Edgar Alcock and a man named Rogerson with him. At that time Beyer Peacock appeared to be a firm in terminal decline. Cosy restrictive practices seem to have been rife at Gorton, and Edgar Alcock is said to have dismissed several leading hands before he had created working arrangements that were to his liking.

Edgar Alcock's notes of the period dwell at length on the one aspect of Beyer Peacock's history that possibly saved the firm and certainly had a profound effect on world steam locomotive history – the Garratt locomotive. Alcock recalls Garratt as a tall, bearded, intemperate man making short visits to Gorton two or three times a week. The inventor's design was only a rough sketch giving a bare idea based on seeing a long bogie wagon on a reverse curve. Edgar Alcock was the link man between Garratt and Beyers; and, helped by the enthusiastic efforts of a young and forthright draughtsman, Sam Jackson, the first two Beyer–Garratt locomotives were built in 1910 for

Tasmanian Government Railways and the Darjeeling Hymalaya Railway. Sam Jackson was the real driving force behind the Beyer–Garratt, and went on to become arguably the best steam locomotive designer in real terms (as opposed to those who took the credit) in the country.

In this tale there were two links with Edgar Alcock's future destiny. First, Garratt had originally offered his patented idea of an articulated locomotive design to the neighbouring firm of Kitson & Co., who had declined in favour of the Meyer principle; and, second, the most difficult design problem on the Garratt was the flexible steam joints. This was solved by recourse to investigation of the Fairlie locomotives on the Festiniog and Welsh Highland Railways, for which latter railway Hunslet was building *Gowrie* at the time when Garratt was knocking on the various doors.

When Hoy died, Rogerson succeeded him as General Manager of Beyers. As is so often the case, Rogerson did not honour Hoy's known intention of promoting Edgar Alcock to Works Manager in his place, and therefore Alexander Campbell's advertisement was well timed.

Edgar Alcock brought from his Beyer Peacock and earlier LYR days considerable experience of large locomotives, and this was to be of tremendous advantage in the difficult years ahead. What he probably did not realize fully at the time, however, was that he was starting a second dynasty at the Hunslet Engine Works that was to endure for seventy-five years.

The export designs of these immediate pre-war years were impressive in their size and variety. Among them was the first superheated locomotive built by the company, no. 1084, a 4–8–0 tender engine with 18 x 21 in. cylinders for the 3 ft. 6 in. gauge Shire (pronounced 'Shirrie') Highlands Railway in Nyasaland – now part of Malawi State Railways. The use of superheated steam was synonymous with piston valves, on account of lubrication difficulties if a combination of slide valves and superheat is attempted, and the machinery at Jack Lane was updated to provide the greater machining accuracy demanded by the more advanced designs.

Locomotives for the metre gauge featured prominently in the Jack Lane order book at the time. The most noteworthy were three 4–6–0 tender locomotives (nos. 1091–3) for the Southern São Paulo Railway; four 4–6–4 tanks (nos. 1091–2 and 1109–10) and two 0–6–4 tanks (nos. 1138–9) for the British North Borneo Railway; five 2–6–4 superheated tank engines (nos. 1140–4) for the

Buenos Aires Midland Railway, and two 2–8–4 tanks (nos. 1102–3) for the Antofagasta and Bolivia Railway. All were large machines, but the last two, with 18 x 24 in. cylinders and 3 ft. 8 in. driving wheels, weighed 86 tons and had a tractive effort of 23,835 lb. – large indeed for the metre gauge over eighty years ago.

An interesting design from 1912/13 was a batch of eight standard-gauge 2–6–0 tender locomotives with 18 x 24 in. cylinders for the New South Wales Railway. Oddities were three engine portions of 9 x 14 in. cylinder 0–4–0 side tank pattern (nos. 1095–7) for standard-gauge steam rail motors being built by the Gloucester Railway Carriage and Wagon Company for the Egyptian State Railways, a keen advocate of this form of steam propulsion.

Ceylon took eight more 'J2' class 4–6–4 tanks for the Kelani Valley and seven large 0–6–0 tanks for the 5 ft. 6 in. gauge Government Railway. Six 0–6–0 side tanks were built for the 3 ft. 6 in. gauge Nigerian Railways. The Santa Marta Railway, the Kyshtim Corporation and others took repeat orders, and there was a 3 ft. 0 in. gauge 4–6–0 tank for the West Clare Railway in Ireland, no. 1098 *Kilrush No. 1*, in August 1912.

The Great War of 1914–18 saw none of the tension and build-up that characterized the 'phoney war' in the years before 1939. So naïve were British industrialists that, during June 1914, Alexander Campbell and Edgar Alcock toured locomotive factories in Germany and Italy without any realization on their part that Germany was preparing for war. Edgar Alcock did however observe of the German engineers that they were 'inspired copyists and born adaptors' and worked longer hours for less money than their British counterparts. The Italians were found to be even lower down the wages ladder but produced better hand-work than the Germans. In summary, the two innocents abroad considered that American manufacturers posed no threat to British exports because of their 'indifferent' standards, but that the Germans, with their passable imitations and low prices, were 'a menace of the greatest seriousness'.

The most serious menace, the threat of war, went unnoticed. But scarcely had the pair returned to Jack Lane before all Europe became embroiled in the aftermath of the assassination, on 28 June, of Archduke Franz Ferdinand in Sarajevo. By 3 August Germany had invaded Belgium, and Britain in retaliation had declared war first on Germany and then on Austria–Hungary. Factories in Britain turned to war production.

From the death of Queen Victoria in 1901 to the declaration of war, 421 locomotives had been built by the Hunslet Engine Company, of which only fifty were for home consumption. At the start of the war there were overseas orders on the books for another forty locomotives, plus a very small handful for home use. Some of the export orders were delayed until after the war, or cancelled altogether – these will be mentioned later – but, after a very slow start, 28 were completed at intervals in wartime. There were various reasons for the slow start. An immediate call-up of younger men for the armed forces, coupled with preparation for machining and testing howitzer shells, the manufacture of shell-turning lathes, production of gun carriages, etc., was countered by the employment of women on the factory floor. This use of female labour on a large scale was hitherto unknown, and the Ministry of Munitions used the Hunslet initiative as a model for the rest of industry to follow.

The older workmen trained the women on the munitions work and then returned to locomotive-building. As a result, no overseas locomotives left the works between August and November 1914; but two of the large Gold Coast Railway 4–6–4 tanks (nos. 1155–6) were despatched during November; and nos. 1168–9, two 0–6–0 saddle tanks for Trinidad Railways, departed one each side of Christmas. Also completed on 23 December 1914 was no. 1162, the eighth and last 0–6–4 tank for Kyshtim, but it was not despatched until eighteen months later, on 8 June 1916. Germany had declared war against Russia on 1 August 1914, Britain declared war against Turkey on 5 November, and this effectively put paid to any shipment through the Black Sea ports until after the trauma of Gallipoli. Then, in 1917, came the Russian Revolution, and Kyshtim never took another British locomotive.

Eighteen locomotives were shipped abroad in 1915: ten 14 x 20 in. 0–6–0 tanks to Nigerian Railways; three sugar estate locomotives for Australia and one for South Africa; two 4–4–0 tanks for the Santa Marta Railway; another 0–6–0 saddle tank for Trinidad; and a large 2–6–2 tank for the Puerto Cabello and Valencia Railway in Venezuela, which, although completed on 21 July 1914, was held at the works until 8 February 1915. Five 'F' class Indian Railways standard 0–6–0 tender engines went to the South India Railway in April and June 1917, and two Gold Coast Railway 4–6–4 tanks on 29 December 1914 and 16 January 1918 finished the quantity of twenty-eight 'civilian' exports.

THE HUNSLET ENGINE CO. LTD *Engineers* LEEDS ENGLAND

2-8-4 TYPE

SIDE TANK ENGINE

Gauge of Railway	3 ft. 3⅜ in.
Size of Cylinders	18 in. dia. × 24 in. stroke
Dia. of Coupled Wheels	3 ft. 8 in.
,, Bogie Wheels	2 ,, 4 ,,
Rigid Wheelbase (Engine)	11 ,, 9 ,,
Total Wheelbase (Engine)	32 ,, 1 ,,
Height from Rail to Top of Chimney	12 ,, 8⅞ ,,
Extreme Width	9 ,, 9 ,,
Heating Surface—Small Tubes ... ... 1825 sq. ft.	
,, ,, Firebox ... ... 112 ,,	
Total ... ... 1937 ,, ...	1937 sq. ft.
Grate Area	26·4 ,,
Working Pressure	180 lbs. per sq. in.
Tank Capacity	2850 gallons
Fuel Space (Coal)	4 tons 0 cwts.
Weight Empty (Engine)	64 ,, 4 ,,
,, in Working Order (Engine)	85 ,, 18 ,,
Total Weight on Coupled Wheels	58 ,, 4 ,,
Maximum Axle Load	15 ,, 13 ,,
Tractive Effort at 75 per cent. of Boiler Pressure	23835 lbs.
Ratio Adhesive Weight ÷ Tractive Effort	5·46
Minimum Radius of Curve Engine will traverse with ease	250 ft.
Weight per Yard of Lightest Rail advisable	80 lbs.
Load Engine will haul on Level	1235 tons
,, ,, ,, up Incline of 1 in 100	600 ,,
,, ,, ,, ,, ,, 1 in 50	330 ,,

*Code Word—*ANGASTA

C P 25 400—3/31

Order **34280**

Individual catalogue sheets were produced for the majority of new designs and significant variations. Order no. 34280 covered Hunslet nos. 1102/3, a pair of 2–8–4 side tank locomotives built in 1912 for the Antofagasta (Chile) & Bolivia Railway. The weight and tractive effort of these locomotives was not much different from those of the L1, V1 and V3 passenger tanks of the London & North Eastern Railway, or indeed the Southern Railway Maunsell 'W' class heavy freight locomotives, yet the track gauge was only one metre.

Home market civilian production during the war comprised two 0–6–0 and six 0–4–0 saddle tanks, all for essential industries. The six-wheelers were no. 1164 *Ravine*, a large 17 x 22 in. engine for the West Leigh Colliery Co. and no. 1167, a 13 x 18 in. machine for Charles Baker of Rossett, near Wrexham. Both were supplied in 1915. Of the six four-wheelers, five had 15 x 20 in. cylinders. No. 1163 *Silkstone* was supplied to Altofts Colliery near Normanton in 1914. No. 1184 went to Thos. Firth in Sheffield and no. 1191 to Groby Granite in 1915; no. 1292 to the Schoen Steel Wheel Company at Newlay in 1917; and no. 1294 *Roseberry* to the Tees Furnace Company Ltd in 1918. A solitary 10 x 15 in. engine, no. 1293 *Viking*, was taken by Wigan Rolling Mills, also in 1918. Newlay, it will be remembered, was where the Leeds Forge had built its 1896 demonstration railway and Schoen had absorbed Fox's American works by 1899.

The first locomotive supplied against a miliary contract after the start of the war was no. 1189. This was a 2 ft. 6 in. gauge 5 x 8 in. 0–4–0 saddle tank to the order of the 'Inspector of Iron, War Office'. It was despatched, painted khaki, to France on 17 May 1915.

That day also saw the despatch of the first of twelve little 18 in. gauge 0–4–0 well tank locomotives to the Royal Army Service Corps Depot at Deptford, which was formerly the City of London's Foreign Cattle Market and bordered on the Thames wharf of the Royal Naval Victualling Yard. The site had been requisitioned as a ration depot shortly after the declaration of war. The twelve locomotives were ordered in three batches – Hunslet nos. 1196–8, built in 1915, nos. 1207–11 in 1916, and nos. 1288–91 in 1917. Again painted khaki, they were similar in appearance to a much earlier version of the type, no. 684/98 *Jack*, which had been supplied to the clay processing works of John Knowles at Wooden Box near Woodville in Leicestershire. *Jack*, a coal burner, had 6 x 8 in. cylinders, a wheelbase of 3 ft. 0 in. and a tractive effort of 1518 lb. The Deptford engines, by comparison, were oil burners, with 6$\frac{1}{2}$ x 8 in. cylinders, a wheelbase of 3 ft. 6 in. and a tractive effort of 2192 lb. A water tank (48 gallons on *Jack*, 58 gallons on the others) filled the space between the frames forward of the leading wheels, and the Deptford engines had a 25 gallon oil fuel tank in the forward left-hand corner of the cab.

It is a sobering fact that nothing unites people more strongly than large-scale warfare. Unpalatable as it may sound to modern liberalized, equality-conscious minds, the threat of a common foe focuses the mind tremendously, and produces a sense of purpose and industry of prodigious proportions. There were no half measures; you were either for it or agin it and took the consequences accordingly. The later years after the Second World War have produced a welcome semblance of 'peace in our time', but it is a fragile and uneasy peace too often punctuated by the effects of easy access to destructive technology and a multiplicity of misdirected and conflicting 'causes'. Terrorism knows no conscience.

Equally ironic is the fact that manufacturing industry benefits from and thrives in situations of total war, and this is nowhere more apparent than in the railway industry. In both world wars, and in minor skirmishes before and since, the railway workshops not only produced locomotives, rolling stock, trackwork, etc. but turned out ships, tanks, guns, aircraft, jerry cans, tin hats and more at a rate as never before. Thus were fortunes made and industrial empires formed.

The Crimea had seen the first use of railway locomotives for military purposes. Of the eight thousand or so locomotives produced by the Hunslet Engine Company in its lifetime, no less than fourteen hundred had been sold direct to the Army or its agents and the Allies; and the sum total for military service from all constituents of the subsequent Hunslet Group (Hunslet, Hudswell Clarke, Avonside, Kerr Stuart, Manning Wardle and Andrew Barclay) was in excess of 1800. Fortunes of War indeed!

The standard-gauge locomotives produced in both wars for service at home and abroad are well known to most railway enthusiasts, but of equal significance were the narrow-gauge units intended for front-line support in the 1914–1918 trench-bound stalemate. Constant shelling turned the ground around the front lines into muddy swamps virtually impassable by the primitive road transport of the time, and in which the early lorries, with their thin tyres, were helpless.

The Germans had foreseen this eventuality and with typical thoroughness had provided light railway equipment even before war broke out. The French followed suit, but it was not until the end of 1915 that the British started using light railways on a very large scale, perhaps more efficiently than their opponents.

The British adopted a track gauge of 60 cm in conformity with the French Decauville system, and contemporary figures suggest that no less than 725 locomotives were operating by the summer of 1917. The majority of these

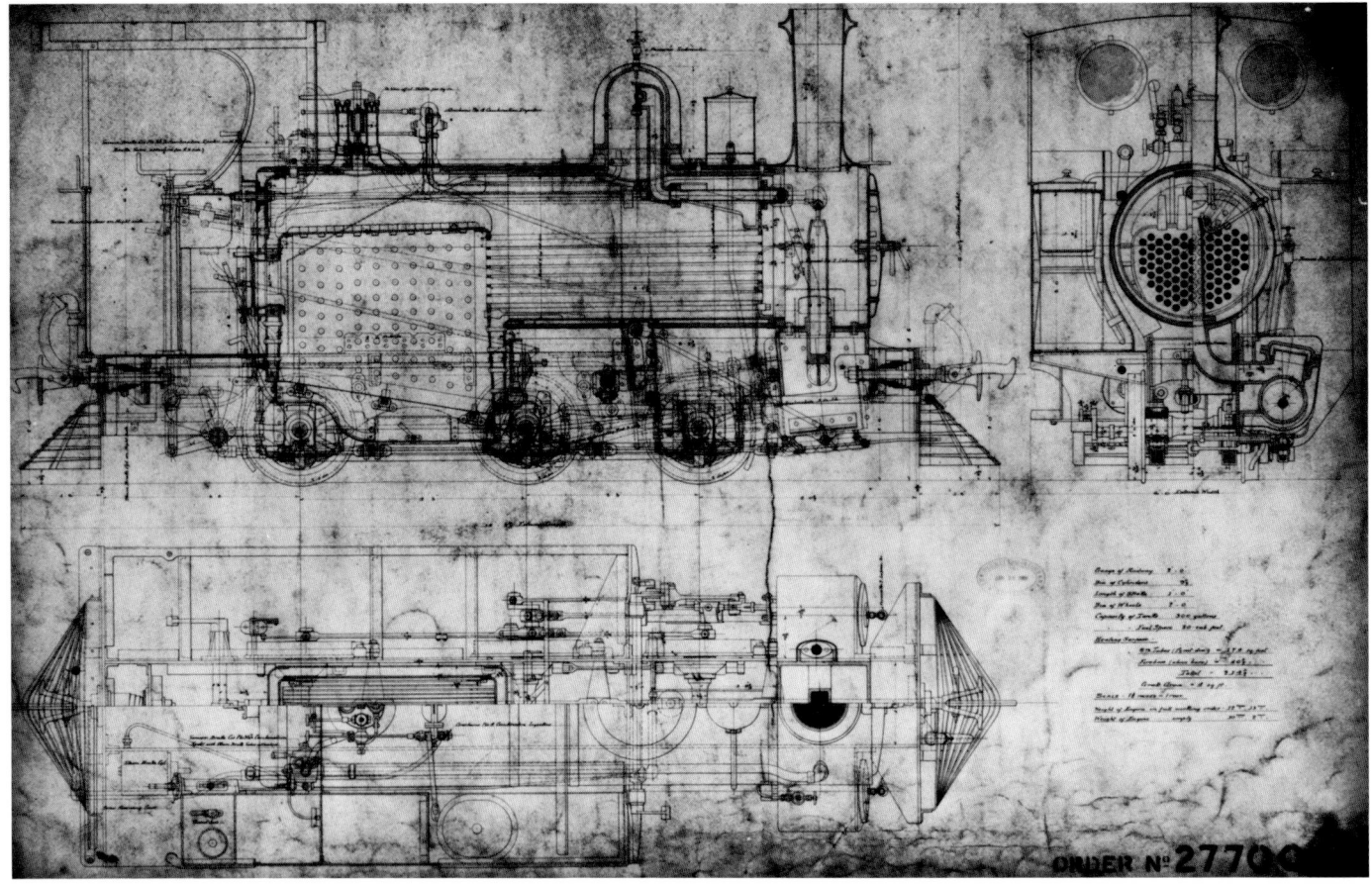

This typical general arrangement drawing is for HE 867/05 Hans Sauer, a 9¹/₂ x 15 in. 2 ft. 0 in. gauge 0–6–0 side tank locomotive. This was supplied to the Lomaguada Railway Company serving the Ayrshire Gold Mines in Southern Rhodesia. The War Office 4–6–0 tank engines used in the Great War were developed from this design.

were of the familiar Alco and Baldwin types – due no doubt to the greater building capacity available in America, coupled with the desperately late time of ordering. The British-built Hunslet 4–6–0 tank locomotives are less familiar, and it is these that we must now consider.

With cylinders, motion and wheels based on *Hans Sauer*, an 0–6–0 side tank locomotive built for a gold mining concern in Rhodesia in 1905, a total of 155 locomotives were built to the order of the War Office and a further nine built for private users. The War Office locomotives were ordered in five batches and may be subdivided under the maker's order numbers as described below.

Order no. 37400
Works nos. 1213–1222 (WD nos. 301–310)
This initial order for ten locomotives of 60 cm gauge was placed on 22 March 1916 and requested delivery of two locomotives by 1 June 1916. Delivery took place between 10 August and 15 September 1916, a remarkable

achievement that one feels would be impossible even to contemplate nowadays.

Order no. 37460
Works nos. 1223–1257 (WD nos. 311–345)
A further order for thirty-five locomotives identical to those of order no. 37400, also placed on 22 March 1916. They were completed between 29 September 1916 and 5 April 1917.

Order no. 37930
Works nos. 1258–1287 (WD nos. 346–375)
A batch of thirty locomotives with minor detail additions ordered on 6 October 1916 and delivered between 24 March and 22 September 1917.

Order no. 38810
Works nos. 1295–1334 (WD nos. 2323–2362)
With the possible exception of one locomotive which spent a short time at Longmoor, the previous 75 locomotives

went direct to the western front in France. On 22 November 1917 an order was placed for a further forty locomotives, also of 60 cm gauge. Completion dates ranged from 21 June 1918 to 30 January 1919, and owing to the cessation of hostilities in November 1918 very few of them saw active service. Fifteen or so reached Italy and at least two went to Egypt, but the remainder went into store at Purfleet or Barnbow (Leeds).

Order no. 39170
Works nos. 1336–1375 (WD nos. 3220–3259)
This final batch of forty locomotives was ordered on 11 July 1918 and they were completed between February and November 1919. Consequently none saw active service, and while the first twenty were built to 60 cm gauge, the order was amended to provide for the final twenty to suit a track gauge of 2 ft 6 in, since this was considered to be advantageous from a selling point of view. Delivery was made to either Purfleet or Barnbow, with the exception of locomotive no. 1356, which was modified in building to suit 2 ft 6½ in gauge track and delivered direct

to Jee's Hartshill Granite and Brick Company Limited, near Nuneaton, on 28 June 1919.

Older drawing office staff of my acquaintance in the fifties quite often referred to the War Office 4–6–0 tank as 'George McArd's' or 'Mac's' engine, and perhaps a few words on personalities may be appropriate here, if only to emphasize that while social habits may change with time, the basic animal instincts of survival and pride remain unaltered.

George McArd, born in Whitehaven, served an apprenticeship at Kitsons, and after two years at the North British Locomotive Company was appointed by Alexander Campbell in 1908 to understudy the Hunslet Chief Draughtsman, Arthur Hird. The possibility of his new appointee's succession to the Chief's post in due course must have been in Campbell's mind, for the rapid growth of the company and the heavy work-load was putting increasing pressure on Mr Hird, who, in Mac's own words, was then 'approaching retiring age [he was in fact 54] and feeling the strain'.

W. D. L. R. 4-6-0T No. 302, HE 1214/16, in service in France during the spring of 1918. These locomotives were well liked in military service and consequently almost all quickly found new owners after the cessation of hostilities. The same could be said of the Hunslet designed Austerities produced in the second world war.

THE LATE W. E. DUNNING, ANDREW NEALE COLLECTION

The War Office locomotives saw service in many guises after the war. This example (makers' number unknown) was running as a 4–6–0 tender engine, with makeshift tender, as Ferrocarril de Correntino No. 658 when photographed at Corrientes depot, Argentina, on 23 August 1968. D. TREVOR ROWE

Hird had served an apprenticeship at Hunslet from 1869 to 1875 and had been in the drawing office ever since, and he had imparted his style to the later nineteenth-century Hunslet locomotives just as surely as did the Chief Mechanical Engineers of the main-line railway companies.

The First World War delayed Arthur Hird's retirement, and he soldiered on until 1922, then aged sixty-seven. In the mean time, in 1912, Edgar Alcock had come from Beyer Peacock and brought several people with him. One of these was H. E. Dean, originally from Nasmyth Wilson. Dean worked for some time on jig and tool design until he, not McArd, became Chief Draughtsman on Hird's retirement. Willie Morrell, who had also started at Hunslet in 1912 and was chief of the gearbox design office during the author's time in the drawing office in the fifties, said that when the appointment of a successor to Mr Hird was announced, Mac 'up and left the same night'! The events of 1910 at Beyer Peacock, when Edgar Alcock was the loser, have a very similar ring. Changes at the top have most effect on those in the upper middle echelons of management; it was always so and always will be.

After a very short spell at Fowlers just around the corner, George McArd went back to North British and eventually to Sir W. G. Armstrong-Whitworth and Company, where he was much involved with that company's diesel locomotives in the early 1930s in competition with Hunslet. He retired from Whitworth's in 1949.

In his fourteen years at Jack Lane McArd contributed much to the Hunslet design process – not only in the locomotives themselves, for he was the catalyst for the subsequent 1920s standard locomotives, but also in overhauling the drawing system and instigating the production of a full-size valve gear model for design purposes. This model was to grace the drawing office wall for some fifty years.

Edgar Alcock had no intention of being the loser a second time around. Subsequent to the war he was appointed Managing Director of the Hunslet Engine Company, and Alexander Campbell became Chairman. The inter-war years consolidated the change-over from the Campbell dynasty to that of the Alcocks.

CHAPTER 5

Between the Wars

Inevitably a period at war brings difficulties in its wake, and the Armistice of 11 November 1918 was sudden and unexpected, unlike the end of World War II in 1945. It had been a war of attrition with very little movement, industry remaining on a war footing up to the last. Overseas markets needed to be re-established, and while there was some build-up of neglected repair work, no home market orders of any consequence were on offer. Herein lies the contrast between the two major conflicts of the twentieth century. The Great War of 1914–18 was tragically expensive in terms of human life, but the material damage was local to the front line or within the range of naval bombardment. World War II, on the other hand, compounded human suffering with widespread airborne destruction, necessitating the replacement of both infrastructure and equipment on a large scale.

In Hunslet's case the overrun of the War Office 4–6–0 tank locomotive contracts provided some degree of continuity, and after the war, many of the locomotives were returned to the makers for overhaul and regauging with a view to resale. This applied both to locomotives that had seen active service and to locomotives that were still being held as new units in store. To detail all the alterations and destinations would require a complete volume, since examples have been reported working in Argentina, Australia, Brazil, Burma, Chile, France, Hungary, India, Italy, Malaya, Mauritius, Nepal, Palestine, Spain and the Sudan in addition to the United Kingdom.

One of the War Office type 4–6–0s turned out after the war. HE 1537/26 Guhyeshwari stands at Khajuri works in Nepal on 28 February 1982. By this time the side tanks had been extended to the front of the smokebox and the bunker had been enlarged. A brand-new set of wheels for use as spares for this locomotive and its partner HE 1536/26 Pashupati was supplied three years later.   D. W. WINKWORTH

After the war, nine further new locomotives were built, as shown in table 5.1. How many of these locomotives were completely new is difficult to ascertain; at least two of them took boilers from earlier locomotives that had been cannibalized for components.

Table 5.1

'War Office' type 4–6–0 tank locomotives built for private users after the Great War			
Works No.	Date	Gauge	Customer and Destination
1457	9.1.23	2 ft. 0 in.	Robt. Hudson, Calcutta
1454	16.1.23	2 ft. 0 in.	Robt. Hudson, Calcutta
1453	17.1.24	2 ft. 6 in.	London Nitrate Co., Chile
1498	31.7.25	2 ft. 0 in.	Sir Douglas Fox, Australia
1536	29.6.26	2 ft. 0 in.	Robt. Hudson for Nepal
1537	29.6.26	2 ft. 0 in.	Robt. Hudson for Nepal
1538	14.6.26	2 ft. 0 in.	A. & W. Smith for Zululand
1539	25.10.26	2 ft. 0 in.	Robt. Hudson for Nepal
1556	9.9.27	2 ft. 0 in.	Robt. Hudson for Nepal

Early in 1919 came the first large post-war order, an order that was to initiate a major reorganization of the Hunslet Engine Works. Twenty-seven supremely stylish superheated 4–6–0 tender locomotives (nos. 1376–1401 and 1416) were ordered for the metre-gauge railways of India, and all were despatched between 27 August 1920 and 6 June 1921. Twenty-one were of the Indian Railways standard 'B' class and the remaining six were of the standard 'M' class. The 'B' class had 16 x 22 in. cylinders and 4 ft. 0 in. driving wheels, while the 'M' class were 16½ x 22 in. with 4 ft. 9 in. drivers. All had Walschaerts valve gear.

To handle the 4–6–0 contract alongside the bread-and-butter industrial and narrow-gauge locomotives that were now beginning to re-emerge, the erecting shop and boiler shop were joined together by building a new central section between them. At this central section the works siding was curved round to enter the erecting shop at an angle, thereby doing away with the previous pair of totally inadequate turntables and creating a feature familiar right up to the end of locomotive building on this spot. At right-angles to the new central link was formed a roofed connection to the machine shop – a *porte-cochère*, so to speak – with a loading bay and high-capacity overhead crane allowing large components such as cylinders and, later, frame castings for mining locomotives, to be transported to and from the erecting shop and down the full length of the machine shop centre bay.

Several classes of 4–6–0 tender locomotive were built for various Indian railways over the years. The metre gauge 'M' class with 4 ft. 9 in. diameter driving wheels first appeared from Jack Lane in 1920. This is a later example from a batch supplied to the Jaipur State Railway. No. 31092 of the by then redesignated Indian Railways 'MJ' class is seen pulling out of Ajmer Junction on 12 December 1979. The first vehicle in the train is the Research, Design and Standards Organization's 'Oscillograph Car'. Lucknow and the RDSO are the Indian equivalent of Derby and the now-dismembered Railway Technical Centre.

D. W. WINKWORTH

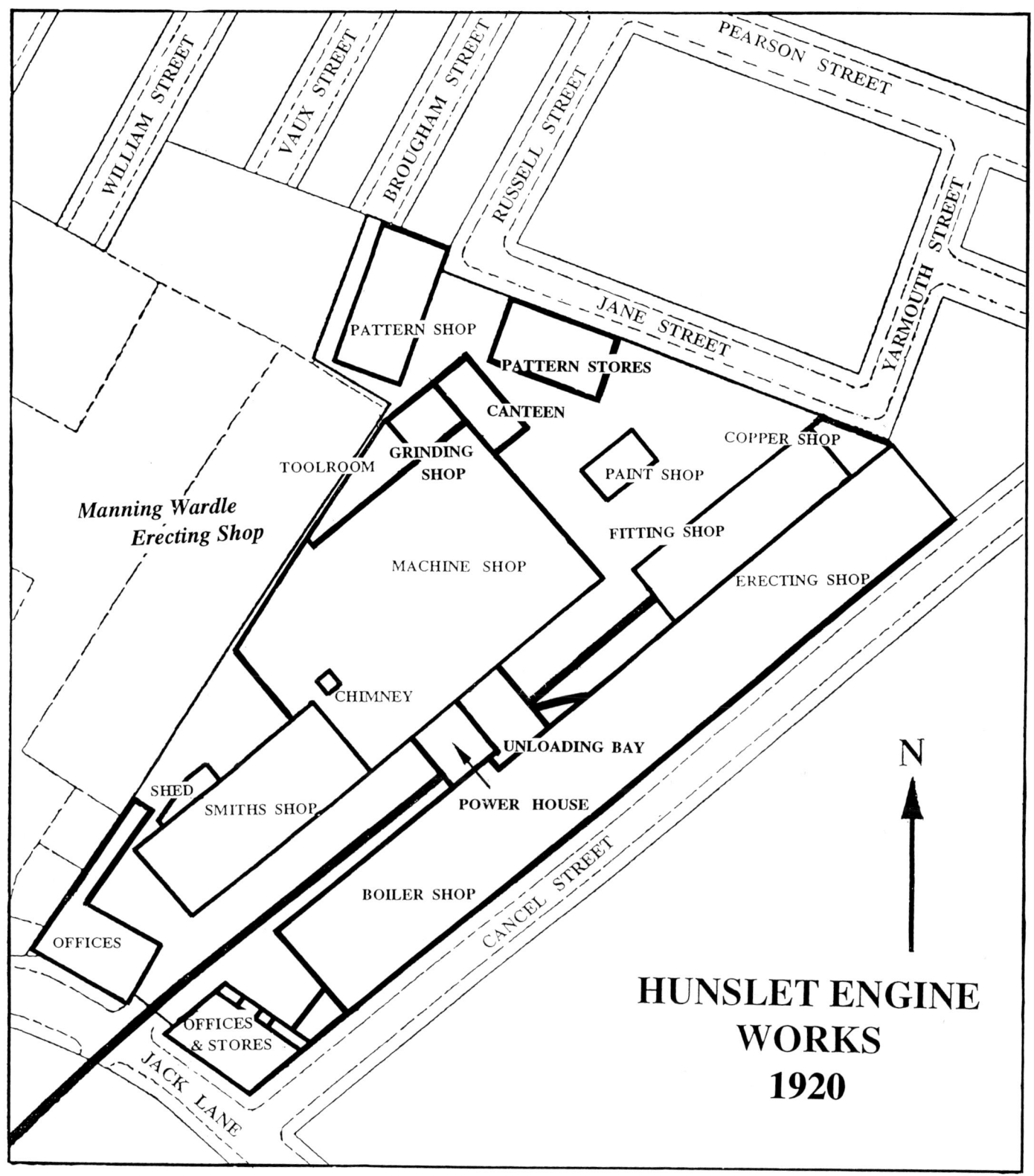

WILLIAM STREET

VAUX STREET

BROUGHAM STREET

RUSSELL STREET

PEARSON STREET

YARMOUTH STREET

JANE STREET

PATTERN SHOP

PATTERN STORES

CANTEEN

COPPER SHOP

TOOLROOM

GRINDING SHOP

PAINT SHOP

Manning Wardle Erecting Shop

FITTING SHOP

MACHINE SHOP

ERECTING SHOP

CHIMNEY

SHED

UNLOADING BAY

SMITHS SHOP

POWER HOUSE

OFFICES

BOILER SHOP

CANCEL STREET

N

OFFICES & STORES

JACK LANE

HUNSLET ENGINE WORKS 1920

In 1922 two 'Port' class locomotives, HE 1429 and 1430, were supplied to the Dinorwic system for shunting at Port Dinorwic. A third locomotive, HE 1704, followed in 1932. These were the last steam locomotives built for the Welsh slate industry.   MAID MARIAN LOCOMOTIVE FUND

Two 3 ft. 0 in. gauge 4–6–0 tanks, HE 1432 and 1433, were supplied in 1922 to the West Clare Railway, becoming WCR No. 3 Ennistymon and No. 4 Malbay. HE 1433 is seen here as CIE class BN3 No. 7C at Kilrush on 6 June 1954.

D. W. WINKWORTH

Some £22,000 was spent on expanding and improving the facilities. This was a lot of money in 1919, but it went a long way towards increasing efficiency and capacity and therefore offsetting the near-doubling of labour costs that was another consequence of the war years. The improved production flow increased capacity without adding to work force levels or to administration overheads, and thereby enabled unit costs to be kept down to a reasonable level. This in turn provided the necessary competitive edge in a depressed market. Nine locomotives ordered prior to the outbreak of war had been put 'on ice' pending the cessation of hostilities. These appeared but slowly, some of them as long as five years after the order had been placed. No. 1170, an 0–6–0 tanks for Nigeria, emerged on 9 July 1919, closely followed by *Blarney*, a 4–4–0 tank for Ireland's Cork and Muskerry Railway. Two 4–6–4 tanks for the Kelani Valley Railway (nos. 1185/6) had been the first detainees to escape, on 15 April 1919, but all orders for three 2–6–2Ts for Sierra Leone and two Shahdara–Saranpur 2–6–4Ts were not completed until 8 January 1920 and 27 January 1921 respectively. These last were developments of the 2–6–2 tanks mentioned in chapter 4.

Adding to the activity surrounding the building of the Indian 4–6–0 tender engines and the wartime backlog, the previously mentioned 'run of the mill' output during 1920 and 1921 was mainly for a variety of long-established Hunslet customers who regularly came back either for repeats of what they had purchased before or for something a little different to cope with changing operational requirements. Thus we see five miscellaneous standard-gauge 0–4–0 saddle tanks – one each for Robeys of Lincoln, John Knowles, the South Durham Steel and Iron Co., Mountsorrell Granite and Barrow Lime Works; two 2 ft. 6 in. gauge 2–6–2 side tanks for Sierra Leone; a 2 ft. 6 in. gauge 0–4–2 tank for Ceylon and a 2–4–2 tank of the same gauge for India; two 5 ft. 6 in. gauge 0–6–0 tanks for Ceylon; three 3 ft. 6 in. gauge 0–6–0 tanks for Nigerian Railways; a 2 ft. 0 in. gauge 0–4–0 saddle tank for Groby Granite; and, smallest of all, an 18 in. gauge well tank, *Gwen*, for John Knowles. A remarkable mixed bag in two years, covering the narrowest to the widest commercial track gauges. Forty-three locomotives of fifteen different types for seven track gauges in five different countries over a period of 16 months.

1922 was a much quieter year; the depression was deepening and locomotive production was down to

nineteen locomotives. Seven of these were 3 ft. 0 in. gauge, comprising a 2–6–0 tender locomotive and a 4–4–0 tank locomotive (Hunslet nos. 1434/5) for the Santa Marta Railway in Colombia (their numbers 21 and 22); two 4–6–0 tanks, *Ennystymon* and *Malbay* for the Irish West Clare Railway (nos. 1431/2); three 0–4–0STs for Preston Water Works (*Hamilton*) and Fylde Water Board (*Stocks* and *Hollins*). There were four 5 ft. 6 in. gauge 4–6–0 tender locomotives of class B8 for Ceylon Government Railways; three impressive 4–6–4 passenger tank locomotives for South Indian Railways; a 3 ft. 6 in. gauge 0–6–0 tank for the Nigerian Eastern Railway; and finally a pair of 2 ft. 0 in. gauge 0–4–0 saddle tanks of the 'Port' class for Dinorwic Quarries at Dinorwic on the North Wales coast.

The flattest period was 1923, probably induced by the grouping of the British railways into the 'Big Four' – change and uncertainty always seriously affect industry despite the protestations of the politicians, and a meagre twelve locomotives saw the light of day. Three of these were neat 3 ft. 6 in. gauge 0–6–0 side tanks for the Public Works Department of New Zealand in Wellington, and two were standard-gauge 4–6–4 tanks for British Guiana. A solitary reconstructed War Office 4–6–0 tank ventured off to Nepal and another one to Calcutta.

It was the remaining four 1923 locomotives that were of the greatest historic value. Despite the very heavy work-load of the previous two years the drawing office had been developing an integrated range of rugged, cheap inside-cylinder standard-gauge industrial locomotives that were easy to build and simple to operate, to cope with any foreseeable client requirement. All were 0–6–0 saddle tanks (with one notable side tank exception) and progressed from the smallest, 12 x 14 in., through 14 x 20 in., 15 x 20 in. and 16 x 22 in. to 18 x 26 in. sizes.

The first 'standard' to appear was 16-inch no. 1438 *Fitzwilliam* for Hemsworth Colliery south of Wakefield on 28 February, closely followed by no. 1439 *Leigh* as the first 15-inch. Nos. 1440–2 (*Airedale*, *Diana* and *Dora*) appeared later in the year for three local collieries. These had a shorter wheelbase than *Leigh* and became the future standard 15-inch design.

Along with the inside-cylinder standard classes was also developed a range of standard-gauge locomotives with outside cylinders and Walschaerts valve gear. The range covered 12, 14 and 16 in. 0–4–0 side tanks; 14, 15 and 16 in. 0–6–0 saddle tanks; and 0–6–0 side tanks with 16 and 18 in. cylinders.

The 16 x 22 in. standard class gave the impression of having been 'stretched' from its smaller 15 in. brother. The cab was the same, and the boiler the same diameter but longer in the barrel. Water and coal capacity were increased, the frame and wheelbase lengthened, and there were slightly larger wheels to allow increased bearing surfaces in the axleboxes. This is HE 1438/23.

The first Hunslet standard 15 x 20 in. locomotive, no. 1440/23, for Airedale Collieries, Castleford. Simple and rugged – handsome maybe – they had all the characteristics of the eponymous terrier and got everywhere.

Most of the inside-cylinder locomotives were built to stock in pairs, while their outside-cylinder brethren, being more expensive to build but cheaper to maintain, were usually the result of firm orders. This led to some interesting combinations of works numbers and building dates as attempts were made to keep the work-force busy in times of recession. A similar situation was to arise in the 1960s.

Collectively these new standard locomotives firmly established a readily identifiable family image, with clean lines neatly executed but without unnecessary embellishment. The sloping flat back to the bunker, the shape of the cab openings and the proportions of the water tanks identified a Hunslet locomotive just as clearly as, for example, the shape of the cab identified a Gresley locomotive on the LNER.

The full list of 4 ft. 8½ in. gauge standard locomotives, the largest 18 x 26 in. version of which led directly to the Austerity 0–6–0 saddle tank of World War II, is given in tabular form at the end of the chapter, in order of appearance of the first of each class.

All of the above designs of locomotive incorporated a high degree of interchangeable components, and were essentially intended for the home market. Only twenty-two of the total left mainland Britain, and the majority of these were to return, as we shall see later.

The 14 in. classes of 0–6–0 saddle tank, both inside- and outside-cylinder versions, throw an interesting light on the perfidiousness of building to stock. Four outside-cylinder and twelve inside-cylinder examples went to Palestine to transport stone from the quarries to extensive harbour and breakwater works being carried out at Haifa.

It is worth dwelling in some detail on this, since straightforward logic does not explain why two different classes of locomotive, with consequent variation in some spare parts, should be purchased new to work on the same contract.

The inside-cylinder locomotives, as already explained, were cheap but rugged and were not usually offered for use overseas. Also they were usually laid down to stock in small quantities to provide a balancing load on the shop floor and give a competitive edge to cope with urgent requirements.

The Hunslet standard inside-cylinder locomotives were about as simple as you could get. There was nothing that was not needed but the style and proportions were right. 16 x 22 in. HE 1590/32 Stella stands at South Leicester Colliery on 29 April 1963.   R. C. Riley

Haifa Harbour Works Dept. No. 3, HE 1645/30, repatriated and seen at the Oxfordshire Ironstone Co. on 9 April 1958. It has been rebuilt with a buckeye coupler for handling the ore-tipping wagons (one of them is seen behind the locomotive) at the calcining banks, but the original double-roof cab has not been replaced. A non-standard cab backplate has been fitted; compare Bramley No. 4 on page 160. R. C. RILEY

By comparison, the outside-cylinder locomotives were more expensive, and, as previously mentioned, they were usually built only against firm orders. However, the Walschaerts valve gear was more acceptable to overseas clients than the Stephenson gear of the inside-cylinder machines, and was more accessible for maintenance.

Crown Agents were responsible for acquisition of the Haifa locomotives, and it seems reasonable to speculate that their colonial railway background caused them to specify outside cylinders. It is also reasonable to assume that the advance estimate for the construction contract was twelve locomotives only, and not the sixteen that were ultimately involved. This theory is reinforced by the fact that only twelve of the outside-cylinder 14 in. class were ever built, these being Hunslet nos. 1643–8 and 1685–90. Nos. 1643–6 went to Palestine in October and November

1929. I think that all 12 were ordered by Crown Agents, but it proved impossible for Hunslet to deliver the first locomotive in time for the commencement of the contract in Haifa. To ease the situation Hunslet substituted the two stock inside-cylinder machines nos. 1585/6, presumably at a lower price, and continued with nos. 1643–6 as soon as possible.

This, I think, is where things started to go wrong. The two stock units obviously did the job; they were cheaper; and perhaps the order was modified to cancel the remaining eight outside-cylinder machines and replace them with a larger quantity of inside-cylinder units for the same total price.

The displaced remaining eight outside-cylinder locomotives, nos. 1647/8 and 1685–90, were subsequently taken by John Mowlem for the Southern Railway's

Visit of H.R.H. PRINCE GEORGE to the works of the HUNSLET ENGINE Co. LTD., Leeds, May 1931. In background one of eight engines built for the Southampton Graving Dock to the order of Messrs John Mowlem & Co. Ltd., London.

PARTICULARS OF ENGINE

TYPE O.6.O.
SADDLE TANK.

OUTSIDE CYLINDERS,
14" x 20".

HRH Prince George (later King George VI) talks to Alexander Campbell during a visit to Hunslet in May 1931. John Alcock, looks on at left. In the background is one of the eight 14 x 20 in. outside-cylinder locomotives for John Mowlem & Co. Ltd's Southampton Graving Dock contract.

A party of South African students photographed on a visit to the Hunslet Engine Works on 8 January 1932. The fashionable dress of the day, plus-fours and trilby hats, is to the fore. The two stock 14 x 20 in. locomotives in the background are HE 1691/35 and 1692/36, eventually despatched to Shap Granite and the Austin Motor Company respectively.

LEEDS MERCURY

Southampton Graving Dock contract, and were despatched between 15 June and 27 July 1931. The speed of delivery suggests that some of them had been completed earlier, and this is corroborated by the fact that one of them (believed to have been no. 1648) was on display at the Liverpool & Manchester Railway Centenary Exhibition in Wavertree Park in September 1930, in the company of Hunslet no. 1671, a metre gauge 2–6–2 side tank for Tanganyika Railways which was shipped on 4 October 1930.

The sixteen locomotives shipped to Palestine were fitted with an open-backed cab and double-skinned roof with polished mahogany inner lining. Any displaced cab components from stock batches were put on temporary stock and used as and when required on future locomotives. This fact explains anomalies on a photograph dated 8 January 1932 showing two 14 x 20 in. inside-cylinder locomotives almost complete in the Hunslet erecting shop. It was the practice to mark components with the order number in whitewash, and the leading locomotive in the picture clearly has '43500' painted on the cab fence plates and bunker side. Now 43500 was the internal order number for locomotives nos. 1585/6, which had gone to Haifa in August 1929. No. 1585 was photographed at Haifa in October 1929 with an open-backed cab and riveted saddle tank. The locomotives in the works picture had closed cabs and flush-welded saddle tanks.

A partially-obscured number on the smokebox reading '**510' gives the explanation. 44510 was the internal order number for nos. 1691/2 which, although virtually finished in January 1932, eventually became Shap Granite Co. Ltd *Haweswater* and Austin Motor Co. *Austin II*, despatched on 28 October 1935 and 16 March 1936 respectively. The two locomotives feature in the background of several photographs from 1932 onwards, and give the impression of having been 'put on one side' while the diesel revolution took place.

Three of the outside-cylinder and seven of the inside-cylinder Haifa locomotives came back to Hunslet after completion of the harbour works in 1938 to be overhauled for other uses, both civilian and military. No. 1643 had a particularly varied career, first with the Army at Bramley in Hampshire, ultimately going to the Peruvian Corporation in 1953, and serving in the interim as Hunslet's hire locomotive. While locomotive hire was not a regular 'core' business it could sometimes be handy to have a good general-purpose locomotive around to help customers out in times of difficulty.

The solitary 14 in. 0–4–0 no. 1509, supplied to the Bradford Corporation's Laisterdyke gas works, looked deceptively small, with its cut-down rounded cab to clear the retort house entrance. Nevertheless it had the same boiler, cylinders and valve gear as its six-wheeled counterpart, with the same overall weight and performance. Just one fewer set of wheels and an increase in axle load from 10 tons 13½ cwt to 14 tons 17 cwt.

By the standards of the day the two outside-cylinder 18 in. side tanks nos. 1456 and 1475 were large indeed. Ordered in November 1923 by Bridgewater Collieries Ltd (later Manchester Collieries Ltd) they were used on the various colliery lines radiating from Walkden yard, Brackley, Sandhole and others, and gained a reputation as prodigious load-shifters. Unusually among the 147 standard locomotives tabulated, they had Belpaire fireboxes, a distinction shared only with no. 1506, with a boiler not very dissimilar to that of the LMS 'Jinty' 0–6–0 tank.

No. 1506 has always puzzled the author. The number suggests that it was first put in hand for a date of building around February 1926, but it was not despatched until 16 June 1930. Either its building was delayed by construction of the ninety LMS 'Jinties' between October 1924 and February 1929, a reasonably enough assumption, or it hung around waiting for a buyer – or a combination of both. It was a significant machine none the less, despite being a 'one-off'. A boiler almost identical to those of *Bridgewater* and *Joseph*, complete with Belpaire firebox but with a slightly smaller grate area owing to the smaller wheelbase, was mounted on frames and running gear that were destined to become the prototype for the standard '48150' class of steelworks shunter, of which works number 1849 was the first in February 1937. With modifications to suit wartime conditions this was to develop into the Austerity of the World War II. No. 1506 was eventually purchased by John Bowes and Partners to operate on the Pontop and Jarrow Railway, later known as the Bowes Railway, at Marley Hill locomotive shed in north-west Durham. The railway eventually became National Coal Board property and carried coal from a number of connecting collieries to staithes on the River Tyne over a succession of locomotive-hauled lines and rope-worked inclines. Some historians suggest that no. 1506 was specifically designed for the very steep Hobson

The two Manchester Collieries 18 in. engines Bridgewater and Joseph were big for their day. HE 1456/24 Bridgewater is seen here in the company of an Austerity at Walkden shed on 26 May 1951. Two tons heavier than the Austerity, Bridgewater was slightly lower on tractive effort (20,736 lb. as against 21,060 lb.) only by virtue of the lower boiler pressure of 160 p.s.i. which it shared with the LMS 'Jinty' and the Bowes Railway No. 15 (HE 1506). The Austerity was pressed to 170 p.s.i.

C. A. APPLETON/J. A. PEDEN COLLECTION

bank between Marley Hill and Burnopfield. Although it was ideally suited to this duty, it is unlikely that the customer would have waited effectively $5\frac{1}{2}$ years for his engine. What is more likely is that this was an opportune requirement for a locomotive already built to stock.

A similar situation surrounds the solitary 15 in. outside-cylinder locomotive no. 1446, which the makers' records show as having been 'tried in steam' on 10 January 1927 but was not despatched to the Oxfordshire Ironstone Company at Banbury until 16 July 1929.

The remaining outside-cylinder classes shown in the table appear to have had buyers from the start, and were despatched from the works more or less at the times one would have expected. The smallest (12 x 18 in.) and largest (18 x 26 in.) inside-cylinder 'standards' were still to appear, as we shall see later.

But the last five years of the 1920s were not by any means devoted to standard locomotives. Ninety of the London, Midland and Scottish Railway's popular 'Jinty'

class 3F 0-6-0 side tank shunting locomotives were produced, in four batches, as follows:

Hunslet works nos.	LMS numbers	Date	Qty
1460–1474	7135–1749	7.10.24–9.2.25	15
1511–1535	16510–16534	16.8.26–12.7.27	25
1557–1582	16625–16649	25.11.27–8.5.28	25
1591–1615	16650–16674	4.9.28–6.2.29	25

There were export orders also, and, despite the 1920 expansion of the works, space was at a premium, The whole of the available former E. B. Wilson land on the north side of Jack Lane had been taken up by this time. Fate took a hand, for while 'Hunslet' were working flat-out, their neighbours, Manning Wardle, found life difficult, and closed down in 1927 – an inevitable result of having failed to modernize their facilities to meet an increasingly demanding business environment. Manning Wardle goodwill passed meantime to Kitson and Company (not for long, as it turned out), while the Hunslet Engine Company

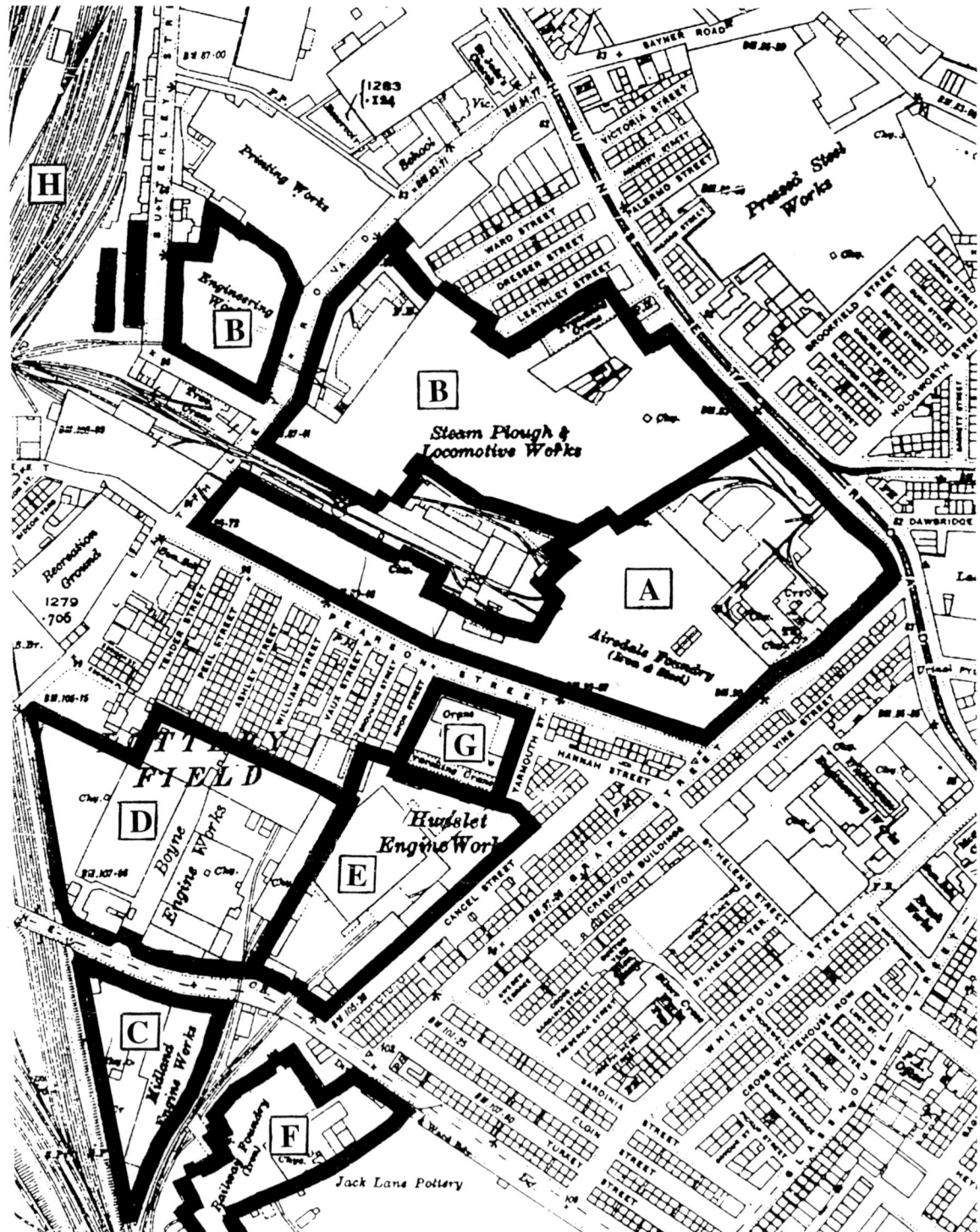

**THE DISTRICT OF HUNSLET AT ITS LOCOMOTIVE & TRACTION ENGINE
BUILDING PEAK IN THE LATE 1920's**
Within half a square mile can be seen Kitson's Airedale Foundry (A), John Fowler &
Sons Steam Plough Works (B), J & H McLaren's Midland Engine Works (C), Manning
Wardle's Boyne Engine Works (D), Hunslet Engine Company's Hunslet Engine Works
(E) and Hudswell Clarke''s new Railway Foundry (F).   Shepherd & Todd's old
Quadrangle, at that time leased to Kitson's is shown at (G) and Hunslet Lane Goods
Yard at (H). Manning's was the first to go, closing before the end of the decade.

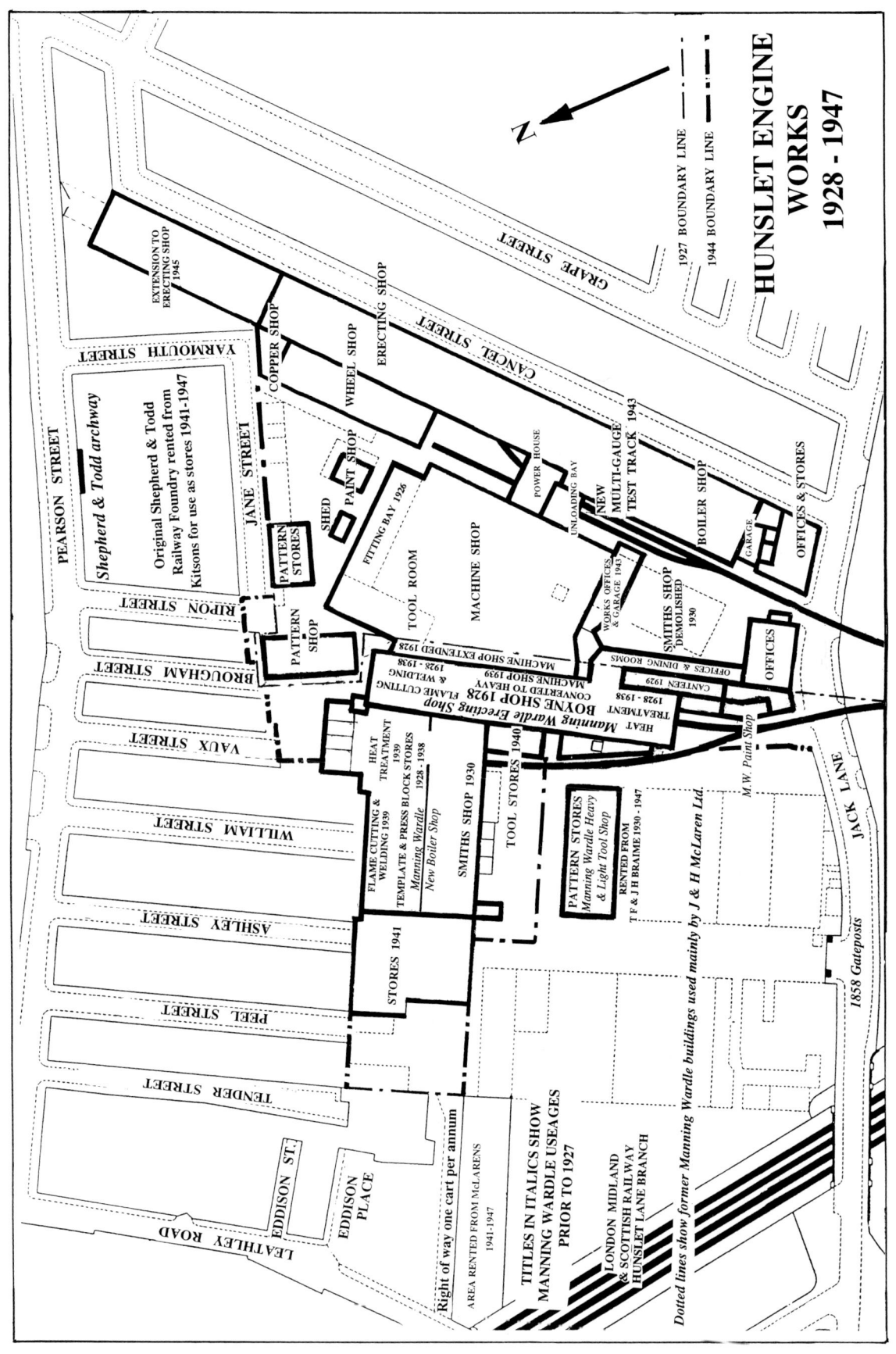

HUNSLET ENGINE
WORKS
1928 - 1947

N

1927 BOUNDARY LINE ————
1944 BOUNDARY LINE —·—·—

GRAPE STREET

CANCEL STREET

ERECTING SHOP

EXTENSION TO
ERECTING SHOP
1945

COPPER SHOP

WHEEL SHOP

PAINT SHOP

SHED

PATTERN
STORES

FITTING BAY 1926

POWER HOUSE

UNLOADING BAY

NEW
MULTI-GAUGE
TEST TRACK 1943

BOILER SHOP

GARAGE

OFFICES & STORES

PEARSON STREET

YARMOUTH STREET

JANE STREET

Shepherd & Todd archway

Original Shepherd & Todd
Railway Foundry rented from
Kitsons for use as stores 1941-1947

RIPON STREET

PATTERN
SHOP

TOOL ROOM

MACHINE SHOP

WORKS OFFICES
& GARAGE 1943

MACHINE SHOP EXTENDED 1928

SMITHS SHOP
DEMOLISHED
1930

OFFICES & DINING ROOMS

OFFICES

BROUGHAM STREET

BOYNE SHOP 1928 FLAME CUTTING
& WELDING
CONVERTED TO HEAVY
MACHINE SHOP 1939

Manning Wardle Erecting Shop

1928 - 1938

HEAT
TREATMENT
1928 - 1938

CANTEEN 1929

VAUX STREET

HEAT
TREATMENT
1939

FLAME CUTTING &
WELDING 1939

TEMPLATE & PRESS BLOCK STORES
1928 - 1938

Manning Wardle
New Boiler Shop

SMITHS SHOP 1930

TOOL STORES 1940

M.W. Paint Shop

WILLIAM STREET

PATTERN STORES
Manning Wardle Heavy
& Light Tool Shop
RENTED FROM
TF & JH BRAIME 1930 -1947

JACK LANE

ASHLEY STREET

STORES 1941

1858 Gateposts

PEEL STREET

TENDER STREET

EDDISON ST.

EDDISON
PLACE

LEATHLEY ROAD

Right of way one cart per annum

AREA RENTED FROM McLARENS
1941-1947

TITLES IN ITALICS SHOW
MANNING WARDLE USEAGES
PRIOR TO 1927

LONDON MIDLAND
& SCOTTISH RAILWAY
HUNSLET LANE BRANCH

Dotted lines show former Manning Wardle buildings used mainly by J & H McLaren Ltd.

and two other local companies, J. & H. McLaren and T. F. & J. H. Braime, each took a part of the works. The Hunslet portion comprised, in the main, Mannings' erecting shop and a boiler shop that had been built as recently as 1923. A passageway was provided to link the erecting shop to the Hunslet machine shop, the paint shop was renovated for additional office space, and the boiler shop initially gave much-needed storage for patterns and press blocks. Over the next few years the relocation of the Hunslet blacksmiths' facilities into the Manning Wardle boiler shop would allow the demolition of the old smiths' shop, which had restricted operations in the front yard. New hammers, together with fabrication and flame-cutting equipment and new machinery, notably a number of large planing machines, were to provide the means whereby production could be sustained and the further acquisition of the designs and goodwill of Kerr, Stuart & Co. and the Avonside Engine Company exploited to the full. By 1928 the Hunslet Engine Works had reached a size that exceeded that of the old Railway Foundry by a considerable margin, and it was to remain this size until the end of the World War II.

Between 1923, when the first 'standard', *Fitzwilliam*, appeared, and the end of the decade, just short of three hundred locomotives were produced in total. In addition to the standard designs there were some eighty locomotives of miscellaneous designs, all but one of them for export, and one can only marvel at the quality of organization that was able to manage such diversity using purely manual design, costing and recording systems.

The one locomotive for home use that was not from the new standard range was a low-height but otherwise traditional 0–4–0 saddle tank with outside cylinders: no. 1493, supplied on 1 January 1925 to James Oakes at Riddings Colliery. The export units ranged in size from the nine War Office 4–6–0 tank locomotives already mentioned to very large tender engines, 4–8–0s for Ceylon Railways and 4–6–0s for Paiana Plantations and for the Rohilkund and Kumeon Railway. Particularly noteworthy were the six 'J1' class 4–6–4 tank locomotives for the 2 ft. 6 in. gauge Kelani Valley Railway in Ceylon. These large passenger engines, with both side and back frame tanks and 14 x 20 in. cylinders, first appeared with Hunslet numbers 1478 in 1924 and 1497 in 1925, with nos. 1583/4 following in 1928 and finally nos. 1635/6 in 1929. They were larger versions of, and supplemented, the 'J2' class, of which nine had been supplied between 1908 and 1913. The 'J2s' had slide valves; the 'J1s' were superheated and in consequence had piston valves.

Kelani Valley Railway class J1 A 4–6–4 tank No. 263, built in 1928, on a freight train at Padukka on 28 January 1970. This locomotive has 14 x 20 in. cylinders with piston valves, and carries a superheated boiler; the earlier J2B class had slide valves and were unsuperheated. The wagons echo the Leeds Forge demonstration railway of 1898.
D. TREVOR ROWE

The Santa Marta Railway continued its random procurement with three 2–6–0 tender locomotives, two 0–4–2 tanks, a 4–4–0 tank and an 0–6–0 tank before being nationalized by the Colombian government in 1926. Two more 2–6–0s were purchased by the state railway in 1926/7; thereafter there were no more Hunslet steam locomotives for this long-established customer.

Run-of-the-mill but useful orders were nos. 1622–34, a batch of thirteen 14 x 20 in. metre gauge 0–6–0 side tanks for contractor Sir John Jackson in the construction of the Singapore Naval Base – locomotives with mechanical parts similar to those of the HHWD and Southampton 0–6–0 saddle tanks but with a longer wheelbase and a slightly larger boiler – and five 5 ft. 6 in. gauge 16 x 20 in. outside-cylinder 0–6–0 saddle tanks for the Madras Port Trust, nos. 1507/8 on 12 February 1926 and nos. 1616–18 on 29 August 1928. These last were very much the design behind the British outside-cylinder 16 in. 0–6–0 saddle tanks for Birch Coppice colliery and Tir John power station, and had mechanical components similar to those of *West Riding* and the London Underground pair.

The year 1930 dawned at Hunslet with very little cause for rejoicing. The economic depression was at its worst. The LMS class 3F 0–6–0 tanks had all been despatched and UK industrial business had evaporated. The Haifa Harbour and Southampton Docks contracts provided desultory activity for a while, as can be seen from the tabulated details of standard locomotives, and there was the solitary despatch of no. 1506 to the Pontop and Jarrow Railway. What solace there was came from a handful of export orders. Nos. 1653/4 were two more large 4–6–0 tender locomotives for Paiana Plantations. Nos. 1655/6 were two 2–6–2 tank locomotives for Kenya–Uganda Railways, and no. 1671 was an identical machine for the Tanganyika Railway (exhibited before shipment at Wavertree). Two A3 class 4–8–0 tender locomotives, nos. 1661/2, for Ceylon Government Railways and eight 0–8–0 tank engines, nos. 1663–70, for Gold Coast Railways, completed the year's production.

The Gold Coast Railways locomotives were significant in that they were the first of fifty-seven examples, all to 3 ft. 6 in. gauge, that were built over a period of 24 years, mainly for the railways of the Gold Coast (now Ghana) and Nigeria but with a small number for African industrial users.

Towards the end of 1930 Hunslet purchased the goodwill and stock-in-trade of the Stoke-on-Trent locomotive manufacturer Kerr, Stuart & Co. The apparent reason for Kerr Stuart's demise and the impact of this somewhat opportune purchase on the future diesel locomotive activity at Jack Lane are described in chapter 6. Steam locomotive design was not influenced by the acquisition, but there were a number of advantages nevertheless. Meantime, 1931 had to be survived.

It was an odd year. No locomotives were despatched until the pair of 0–6–0 tanks for the London Underground, nos. 1674/5, left the works on 9 February. Then came a gap of another two months until a stock 15 x 20 in. 0–6–0 saddle tank, HE 1678 *Joe*, with cab specially cut down to allow it into the retort house, ventured three miles up the Midland main line to begin its thirty-year life at the Robin Hood coke works of J. & J. Charlesworth on the East and West Yorkshire Union Railway. Morale-boosting royal visits to the works took place: Prince George of Denmark came in March and Prince George of England (later HM King George VI) in May. A flurry of activity then saw fifteen of the year's total production of nineteen locomotives despatched between 15 May and 24 August. Only one locomotive – a Howrah–Amta Railway 'Eva' class 0–4–2 tank, no. 1693 – went thereafter, on 1 December.

The fifteen peak-period despatches comprised the eight 14 x 20 in. outside-cylinder saddle tanks for Southampton Docks, the last two Haifa inside-cylinder 14 x 20 in. locomotives, a Kerr Stuart order for two 5 ft. 6 in. gauge standard 'Moss Bay' 0–4–0 saddle tanks (nos. 1680/1) for Bombay, a Kerr Stuart 0–6–4 tank (no. 1679) for the Bengal–Nagpur Railway, a Hunslet-designed 0–6–4 tank (no. 1676) for British Guiana, and finally the second British industrial locomotive of the year – a neat 12 x 18 in. 0–4–0 side tank, no. 1684, for Hall and Company's sand quarry at Coulsdon.

One could be forgiven the slightly cynical thought that a certain amount of creativity must have gone into ensuring a full shop for visiting VIPs (a practice also carried out occasionally for commercial reasons in the author's time in the 1960s); comments passed down from workers in the thirties confirm that stock locomotives were assembled and dismantled, lifted from place to place, wheeled and dewheeled frequently to maintain the appearance of bustling activity whenever 'strangers' were present.

That the purchase of Kerr Stuart took place at all suggests a relatively strong financial position – which was perhaps just as well, for 1932 could only be described as an investment for the future. Much managerial attention was

The two 0–6–0 tanks supplied to London Transport in 1931 were smart, workmanlike machines. No. L30, HE 1674, is seen here shunting at Kensington Olympia in 1959. ANDREW NEALE COLLECTION

Most locomotive-building men of a certain age, including the author, suffer from 'engineer's ear'. Prince George of Denmark tries to make a point while John Alcock stands bemused in the background. The erecting shop in March 1931.

YORKSHIRE POST

The first two 12 x 18 in. standard locomotives became Manchester Ship Canal Company Nos. 83 and 84. No. 84 (HE 1695/32) still looked brand-new despite its twenty-two years when photographed at Mode Wheel depot on 18 May 1954. Built at the same time as prototype diesel no. 1697 (chapter 6) this design lent much to the diesel, which was in effect an adaptation of the 12 x 18 in. frame, running gear and wheels with a diesel engine and gearbox taking the place of the boiler and cylinders.   C. A. APPLETON/J. A. PEDEN COLLECTION

HE 1718/33, originally built for the Mysore Railway, is seen at Birur in southern India, as Southern Railway of India No. 31045 on 4 March 1971.   D. TREVOR ROWE

focused on development of the diesel locomotive, and export orders had temporarily ceased to exist. One minuscule 'Wren' class 0–4–0 saddle tank, no. 1702, was despatched to Siamese Tin Mines on 21 March, but this was almost certainly brought over complete, or substantially so, from Kerr Stuart's works. The only other steam locomotives of the year were the final Welsh slate quarry unit, HE 1709 *Michael*, on 27 September, and five standard industrial locomotives: two 12 x 18 in. for the Manchester Ship Canal, one 15 x 20 in. for the Yorkshire Electric Power Co., and two 16 x 22 in. for Airedale and South Leicester Collieries (see tables).

In production terms 1933 was equally dismal. Standard 15 x 20 0–6–0 saddle tank HE 1699 went to Bickershaw Colliery on 23 January; 0–8–0 tank locomotive HE 1716, of which more later, departed on 23 March; and a 4–6–0 passenger tender engine went to the Mysore State Railway three months later.

Apart from the diesels covered in chapter 6, there was nothing until 3 November, when the first of ten large tender locomotives was despatched to China. Hunslet had been an active exporter to most areas of the world but never in China. Kerr Stuart, on the other hand, had done well there; they had built some very neat standard-gauge

4–4–2 passenger engines for Chinese National Railways and some 4–4–0s for the Shanghai–Nanking Railway. In 1910 the latter railway took delivery of some locomotives very similar to the 4–4–0s but this time with a 4–2–2 wheel arrangement. With 7 ft. 0 in. diameter driving wheels and 18 x 26 in. cylinders, and weighing 100 tons with tender, these four were the largest, the heaviest and the last 'single' express locomotives built anywhere. It is reasonable to assume that this Kerr Stuart Chinese connection played a large part in securing the orders for the remaining Hunslet 1933 production.

The previously mentioned 0–8–0 tank locomotive HE 1716 was a huge machine specially designed for loading and unloading the train ferry across the Yangtse Kiang river between Nanking and Pukow, which commenced operation on 22 October 1933. The locomotives (the second of which, HE 1754, was completed on 29 November 1934) travelled across on the ferry. The vessel had a traverser at the stern to release the locomotives, and each train was loaded in three sections. The locomotives and the ferry remained in operation until 1 October 1968 when a 1600 metre (5248 ft.) bridge was opened, rendering them redundant.

Even as late as 1954 the Hunslet Chief Draughtsman could be heard asking, 'What did we do on China Ferry?' when faced with a difficult design problem. 'China Ferry' was a reference to two 4 ft. 8¹/₂ in. gauge side tank locomotives, nos. 1716 and 1754, built in 1933/4 for hauling heavy trains on and off the Nanking ferry boats. Rolt (A Hunslet Hundred, 1964) claims that these were the largest and most powerful tank locomotives in the world, and perhaps they were in conventional terms, but this author would not be so bold. Certainly the 22¹/₂ x 26 in. cylinders, 38,710 lb. tractive effort and 83 tons weight in working order put them well above the norm. A 132-ton tender engine variant, HE 1755/35, was built to 5 ft. 6 in. gauge for the Tata Iron and Steel Company at Tatanagar in India.

Ten superheated 4–8–0 tender engines (HE 1727–32 and 1750–3) for the standard-gauge Hangchow–Kiangshan Railway followed; the first six despatched between 3 November 1933 and 8 January 1934 and the last four from 20 December 1934 to 25 January 1935. The tender version of the 'China Ferry' locomotives, one locomotive only (HE 1755), for the Tata Iron & Steel Company in India, appeared in April 1935, but there was little else of note.

Times were hard. Kitsons ceased locomotive production in 1938 and their drawings, plus those of Manning Wardle, passed temporarily to Robert Stephenson & Hawthorns at Darlington.

Another victim of the times was the Avonside Engine Company of Bristol. Like Fenton, Murray & Jackson they had supplied (as Slaughter, Gruning & Co.) broad-gauge locomotives to the Great Western Railway. Like Kerr Stuart they had ventured into diesels and geared steam locomotives, and like both Kerr Stuart and Hunslet they had built standard and special locomotives of all types. The depression forced Avonside into voluntary liquidation on 29 November 1934 and Hunslet quietly agreed to purchase the goodwill and the drawings on 10 July 1935. Like Kerr Stuarts' before them the Avonside goodwill and designs were to provide a useful top-up to the Hunslet order book in the years from then until the end of steam.

Hunslet had again managed to survive difficult times. By the time of the Avonside acquisition the diesel locomotive business was established and steam locomotive production had stabilized, although, with one exception, orders were for only one or two locomotives at a time. Between the despatch of the last Chinese 4–8–0 and declaration of war on 3 September 1939, just over five-and-a-half years, seventy-two steam locomotives were produced in among the diesels, and were of no less than thirty different designs.

Of these seventy-two locomotives a total of twenty-six were Hunslet standard industrial units as detailed in the accompanying tables, twenty-five were of Hunslet specialist customized designs, eleven were Kerr Stuart designs and the remaining ten were Avonsides.

The most notable of the Hunslet specials were six 2–6–0 superheated tender locomotives built in 1936 for the metre-gauge Assam–Bengal Railway, bearing works nos. 1776–81. They were fitted with poppet valves: three of them had Caprotti valve gear, and the remainder had the Lentz gear. Eight more of the African 0–8–0 tank design were produced: nos. 1783–5 and 1819–20 for the Gold

Coast Railway, nos. 1892 and 1900 for Pretoria Portland Cement Ltd, and no. 1893 for the Rand Water board. There was no. 1795, a massive 3 ft. 0 in. gauge 2–8–0 tender engine for the Trujillo Railway in Peru; no. 1749, a stylish standard-gauge 2–6–2 tank for the St Madeleine Sugar Co. in Trinidad; two more 'Eva' class 0–4–2s for Howrah Amta; a 3 ft. 0 in. gauge 0–6–0ST *Belvoir* for Stanton Ironworks; and sundry narrow-gauge and industrial units completed the tally.

The eleven locomotives to Kerr Stuart designs include three tiny 'Wrens', six 0–4–2 tanks and standard-gauge 'Moss Bays' for Pilkingtons at St Helens and Palmer Mann (of Sifta Salt fame) of Sandbach. Named *Sifta*, this latter was another 'lost' engine. The allocation of number 1677 suggests that it was complete, or nearly so, on the acquisition of Kerr Stuart's business, and it was probably the first of the batch that included the Bombay pair; however, it was not sold until 1937. Almost certainly these were partially built from stock items brought up from Stoke-on-Trent. The 'Moss Bay' design had originally been produced in 1899 for the Moss Bay Iron & Steel Co. (later the Workington Iron & Steel Co. Ltd) and more were to follow during and after the war. The author assembled the last one in 1952 and we still had a small quantity of pre-1930 parts available at that late date.

Good photographs of the Hunslet-built 'Moss Bay' saddle tanks are rare. Moss Bay No. 5, Kerr Stuart no. 683, was the progenitor of forty-five Stoke-built examples, and was supplied to the eponymous iron and steel company at Workington in 1900. Fifty-two years later, when the author was putting together HE 3773/52, the differences were marginal. The Ramsbottom safety valves were replaced by Ross 'pops', an upper backplate was added to the cab and parallel-shank buffers were substituted for the originals, but there was nothing more than this.

3 ft. 0 in. gauge 2–8–0 tender engine No. 108, HE 1795/36 pauses amid rugged Peruvian mountain scenery on 10 December 1982 at Huancavalica. Originally supplied to the Ferrocarriles de Trujillo it moved to the Huancayo–Huancavalica line when the former closed in 1966/7.   D. Trevor Rowe

Only six of these inside-cylinder 12 x 18 in. class appeared, and they were smart, tidy little engines when built. HE 1855/37 had changed by the time this picture was taken at Yorkshire Amalgamated Products, Warmsworth, Doncaster, on 3 December 1955. The firm specialized in carbonization by-products, so perhaps a water softener was appropriate; but why did it have to make Owd Bill so ugly?   ILS, Bernard Mettam collection

The last steam locomotive out of Hunslet before the outbreak of war was HE 2005, a 2 ft. 0 in. gauge two-cylinder geared steam locomotive of Avonside design despatched to Sezela Sugar Estates in Natal on 16 August 1939. It was not photographed before despatch but was identical to this earlier, Bristol-built example. The steam engine placed amidships can clearly be seen, and the design opportunity for the later Illovo diesel version is only too obvious.

The Avonside ten in 1937 included two standard 0–6–0 saddle tanks for Stanton Ironworks (HE 1848 and 1853) and two 0–4–0 saddle tanks, HE 1851 for Partridge Jones and John Paton's Caerleon tinplate works, and HE 1872 for the County of London Electric Supply Co. Ltd at Creeksmouth power station, Barking. There was one 0–6–0 side tank (HE 1899) for the African Manganese Corporation and an 0–4–2 for Zebedelia Estates in Natal; but the most notable examples were a massive 4–8–2 superheated side tank (1839) for Rand Leases in South Africa and three articulated geared steam locomotives for sugar estates in Natal. These last were of Heisler pattern carried on two 4-wheeled bogies. A two- or four-cylinder 90 degree vee engine mounted under the boiler drove through universal joints and cardan shafts to a worm gearbox on one axle of each bogie and thence by coupling rods to the second axle. Four-cylinder locomotives nos. 2003/4 went to Illovo Sugar Estates on 19 July and 2 August respectively. A two-cylinder version. no. 2005, went to Sezela Estates on 16 August and thus gained the doubtful distinction of being the last steam locomotive despatched from the Hunslet Engine Works before war was declared on 3 September 1939.

INSIDE-CYLINDER CLASSES

Table 5.2

	16 x 22 in. cylinders, 0–6–0 saddle tank, 3 ft. 9 in. diameter wheels		
Works No.	Date of Delivery	Customer's Name or No.	Original Customer
1438	28.2.23	Fitzwilliam	South Kirby Featherstone & Hemsworth Collieries Ltd
1451	15.5.24	Holly Bank No. 3	Hilton Main & Holly Bank Coal Co. Ltd
1452	17.12.24	South Kirby No. 7	South Kirby Featherstone & Hemsworth Collieries Ltd
1495	12.11.27	—	The Peruvian Corporation Ltd
1496	12.7.26	The Dean	Oxfordshire Ironstone Co. Ltd
1589	19.12.29	Newstead	South Kirby Featherstone & Hemsworth Collieries Ltd
1590	25.4.32	Stella	South Leicestershire Colliery Co. Ltd
1651/2	7.2.30– 19.2.30	—	The Peruvian Corporation Ltd (Southern Railway of Peru)
1707	27.8.34	Jean	Grassmoor Colliery Co. Ltd
1708	23.9.35	Enterprise	Shipley Collieries Ltd
1799	8.4.36	Jubilee	Grassmoor Colliery Co. Ltd
1800	15.6.36	Robert Nelson No. 4	Hilton Main & Holly Bank Collieries Ltd
1821	30.7.36	Carol Ann No. 5	Hilton Main & Holly Bank Collieries Ltd
1822	31.1.38	Markham	Staveley Coal & Iron Co. Ltd
1825	13.6.38	—	Robert Hudson Ltd (Port Kembla harbour improvements, Australia)
1826	6.2.39	Tom	Doncaster Amalgamated Collieries Ltd
1953	2.10.39	Jacks Green	Naylor Benzon & Co. Ltd
1954	18.12.39	Kinsley	South Kirby Featherstone & Hemsworth Collieries Ltd
1982	27.2.40	Ring Haw	Naylor Benzon & Co. Ltd
1983	18.5.40	Clement	Doncaster Amalgamated Collieries Ltd
2081	5.9.40	D.A.C.	Doncaster Amalgamated Collieries Ltd
2082	24.10.40	No. 49	Richard Thomas & Baldwins, Ebbw Vale
2374	30.7.41	Spencer	Oxfordshire Ironstone Company Ltd
2375	30.1.42	John Shaw	South Kirby Featherstone & Hemsworth Collieries Ltd
2688	24.2.43	Tom	Doncaster Amalgamated Collieries Ltd
2689	12.10.43	Llynfi	Edmundsons Elec. Corporation
2704	27.7.45	No. 3	Carlton Collieries Association
2705	13.8.45	Beatrice	Ackton Hall Collieries

Table 5.2 continued...

Works No.	Date of Delivery	Customer's Name or No.	Original Customer
3507	1.3.48	—	Peruvian Corporation
3593	22.9.50	No. 5	Frickley Colliery
3594	5.10.50	No. 1	Rossington Main Colliery
3690	26.9.50	—	Peruvian Corporation
3713	30.3.51	Hickleton No. 4	Hickleton Main Colliery
3714	12.4.51	Thorne No. 1	Thorne Colliery
3715	29.1.52	Primrose No. 2	Victoria Colliery
3716	7.2.52	Alex	Oxfordshire Ironstone Co. Ltd
3782	29.9.53	Arthur	Markham Main Colliery
3783	20.9.53	Darfield No. 1	Darfield Colliery
3804	9.10.53	Thorne No. 2	Thorne Colliery
3805	27.10.53	Darfield No. 2	Darfield Colliery
3855	30.11.54	Glasshoughton No. 4	Glasshoughton Colliery
3856	1.3.56	53	Cortonwood Colliery
3872	17.11.58	Frank	Oxfordshire Ironstone Co. Ltd

Table 5.3

	15 x 20 in. cylinders, 0–6–0 saddle tank, 3 ft. 7 in. diameter wheels		
Works No.	Date of Delivery	Customer's Name or No.	Original Customer
1439	19.3.23	Leigh	West Leigh Colliery Co., Leigh, Lancs.
1440	23.7.23	Airedale	Airedale Collieries Ltd
1441	23.7.23	Diana	J. & J. Charlesworth Ltd, Wakefield
1442	20.12.23	Dora	Waterloo Main Colliery Ltd, Leeds
1449	8.4.28	Wheler	Airedale Collieries Ltd
1450	15.4.35	Mary	R. H. Longbottom & Co., London for Shipley Collieries Ltd, Derby
1458	7.6.25	Y.E.P. Co. No. 2	Yorkshire Electric Power Co., Leeds (Ferrybridge power station)
1459	25.7.27	Colonel	Fletcher Burrows & Co. Ltd
1678	15.4.31	Joe	J. & J. Charlesworth Ltd
1698	8.7.32	Bawtry	Airedale Collieries Ltd (Wheldale Colliery)
1699	23.1.33	Bickershaw	Bickershaw Collieries Ltd; Ackers Whiteley & Co. Ltd
1725	30.8.35	Jubilee	Airedale Collieries Ltd
1726	16.9.35	Jubilee	Waterloo Main Colliery Co.
1809	8.4.36	Lyon	Bickershaw Collieries Ltd
1810	4.5.37	Coronation	Airedale Collieries Ltd
1827	29.11.37	M.D.H.B. No. 6	Mersey Docks & Harbour Board, Liverpool

Table 5.3 continued…

Works No.	Date of Delivery	Customer's Name or No.	Original Customer
1828	29.11.37	M.D.H.B. No. 8	Mersey Docks & Harbour Board, Liverpool
1901	20.7.38	Bill	Bickershaw Collieries Ltd
1902	27.6.38	Mexborough	Airedale Collieries Ltd
1955	28.12.39	Airedale No. 1	Airedale Collieries Ltd
1956	28.3.40	Airedale No. 2	Airedale Collieries Ltd
1984	6.2.40	M.D.H.B. No. 9	Mersey Docks & Harbour Board, Liverpool
1985	28.3.40	M.D.H.B. No. 11	Mersey Docks & Harbour Board, Liverpool
2079	25.3.41	M.D.H.B. No. 30	Mersey Docks & Harbour Board, Liverpool
2080	31.3.41	M.D.H.B. No. 31	Mersey Docks & Harbour Board, Liverpool
2408	24.4.42	Warsop	Staveley Coal & Iron Co.
2409	20.5.42	King George	Linby Colliery Co. Ltd (Gedling Colliery)
3509	15.7.47	Astley	National Coal Board, Fryston Colliery

Table 5.4

14 x 20 in. cylinders, 0–6–0 saddle tank, 3 ft. 4 in. diameter wheels			
Works No.	Date of Delivery	Customer's Name or No.	Original Customer
1482	3.2.25	Edith	J. & J. Charlesworth Ltd
1499	20.9.26	Cecil Levita	C. J. Wills, Chadwell Heath
1500	6.2.28	Lionheart	C. J. Wills, Chadwell Heath
1585	13.8.29	H.H.W.D. No. 1	Palestine Haifa Harbour Works*
1586	23.8.29	H.H.W.D. No. 2	Palestine Haifa Harbour Works*
1649	28.1.30	H.H.W.D. No. 7	Palestine Haifa Harbour Works*
1650	28.1.30	H.H.W.D. No. 8	Palestine Haifa Harbour Works*
1657	26.2.30	H.H.W.D. No. 9	Palestine Haifa Harbour Works*
1658	26.2.30	H.H.W.D. No. 10	Palestine Haifa Harbour Works*
1659	5.9.30	H.H.W.D. No. 11	Palestine Haifa Harbour Works*
1660	6.9.30	H.H.W.D. No. 12	Palestine Haifa Harbour Works*
1672	26.9.30	H.H.W.D. No. 13	Palestine Haifa Harbour Works*
1673	6.10.30	H.H.W.D. No. 14	Palestine Haifa Harbour Works*

Table 5.4 continued…

Works No.	Date of Delivery	Customer's Name or No.	Original Customer
1682	18.6.31	H.H.W.D. No. 15	Palestine Haifa Harbour Works*
1683	18.6.31	H.H.W.D. No. 16	Palestine Haifa Harbour Works*
1691	28.10.35	Haweswater	Shap Granite Co. Ltd
1692	16.3.36	Austin II	Austin Motor Co., Longbridge
1814	13.9.37	Austin III	Austin Motor Co., Longbridge
1815	8.6.40	M.D.H.B. No. 13	Mersey Docks & Harbour Board, Liverpool

Table 5.5

18 x 26 in. cylinders, 0–6–0 side tank, 4 ft. 0½ in. diameter wheels			
Works No.	Date of Delivery	Customer's Name or No.	Original Customer
1506	16.6.30	No. 15	John Bowes & Partners Ltd

Table 5.6

12 x 18 in. cylinders, 0–6–0 saddle tank, 3 ft. 0 in. diameter wheels			
Works No.	Date of Delivery	Customer's Name or No.	Original Customer
1694	12.1.32	No. 83	Manchester Ship Canal Co.
1695	10.6.32	No. 84	Manchester Ship Canal Co.
1704	3.3.37	S.L.P. No. 58 Lily	Sir Lindsay Parkinson & Co. Ltd (Royal Ordnance Factory, Chorley)
1705	3.3.37	S.L.P. No. 59 Aussie	Sir Lindsay Parkinson & Co. Ltd (Royal Ordnance Factory, Chorley)
1855	29.10.37	S.L.P. No. 195 Alma	Sir Lindsay Parkinson & Co. Ltd (Royal Ordnance Factory, Chorley)
1856	29.10.37	S.L.P. No. 196 Anzac	Sir Lindsay Parkinson & Co. Ltd (Royal Ordnance Factory, Chorley)

Table 5.7

	18 x 26 in. cylinders, 0–6–0 saddle tank, 4 ft. 0½ in. diameter wheels		
Works No.	Date of Delivery	Customer's Name or No.	Original Customer
1849	23.2.37	No. 17	Guest Keen Baldwins Iron & Steel Co. Ltd (East Moors)
1873	30.8.37	No. 18	Guest Keen Baldwins Iron & Steel Co. Ltd (East Moors)
1874	30.8.37	No. 19	Guest Keen Baldwins Iron & Steel Co. Ltd (East Moors)
2077	28.5.40	No. 7	Guest Keen Baldwins Iron & Steel Co. Ltd (East Moors)
2123	26.9.40	No. 20	Guest Keen Baldwins Iron & Steel Co. Ltd (East Moors)
2261	12.12.40	No. 23	Richard Thomas & Co. Ltd, Redbourn Works, Scunthorpe
2262	14.1.41	No. 24	Richard Thomas & Co. Ltd, Redbourn Works, Scunthorpe
2376	4.2.41		Richard Thomas & Co. Ltd, Ebbw Vale Works, Ebbw Vale, Mon.
2377	28.2.41		Richard Thomas & Co. Ltd, Ebbw Vale Works, Ebbw Vale, Mon.
2411*	27.11.42	No. 24	Stewarts & Lloyds Ltd
2412*	10.12.41	W.D. No. 65	Ministry of Supply (Long Marston)
2413*	30.12.41	Gunby	Stanton Ironworks Co. Ltd
2414*	27.1.42	W.D. No. 66	Ministry of Supply (Long Marston)
2415*	25.2.42	Hellidon	Parkgate Iron & Steel Co. Ltd
2416*	27.3.42	W.D. No. 67	Ministry of Supply (Long Marston)
2417*	17.4.41	Geddington	Stanton Ironworks Co. Ltd
2418*	11.5.42	Grantham	Stanton Ironworks Co. Ltd
2687	30.11.42	No. 6	Guest Keen Baldwins Iron & Steel Co. Ltd (East Moors)
2841	22.3.44	No. 25	Richard Thomas & Baldwins Ltd, Redbourn Works, Scunthorpe
3277	3.5.45	No. 26	Richard Thomas & Baldwins Ltd, Redbourn Works, Scunthorpe
3709	18.12.50	No. 27	Richard Thomas & Baldwins Ltd, Redbourn Works, Scunthorpe
3710	13.1.51	No. 28	Richard Thomas & Baldwins Ltd, Redbourn Works, Scunthorpe
3812	23.11.53	No. 29	Richard Thomas & Baldwins Ltd, Redbourn Works, Scunthorpe
3813	38.11.53	No. 30	Richard Thomas & Baldwins Ltd, Redbourn Works, Scunthorpe

* Note: 50550 class; all others were 48150 class.

OUTSIDE-CYLINDER CLASSES

Table 5.8

	18 x 24 in. cylinders, 0–6–0 saddle tank, 3 ft. 9 in. diameter wheels		
Works No.	Date of Delivery	Customer's Name or No.	Original Customer
1456	9.5.24	Joseph	Bridgewater Collieries Ltd, Manchester
1475	18.6.24	Bridgewater	Bridgewater Collieries Ltd, Manchester

Table 5.9

	16 x 24 in. cylinders, 0–4–0 side tank, 3 ft. 9 in. diameter wheels		
Works No.	Date of Delivery	Customer's Name or No.	Original Customer
1488	8.4.25	West Riding	Pope & Pearson Ltd
1557	7.11.27	Jessie	Tyldesley Coal Co. Ltd
3665	16.5.49	No. 9	Cadbury Bros. Ltd

Table 5.10

	14 x 20 in. cylinders, 0–4–0 side tank, 3 ft. 4 in. diameter wheels		
Works No.	Date of Delivery	Customer's Name or No.	Original Customer
1509	17.3.26	—	Bradford Corporation Gas Department

Table 5.11

	15 x 22 in. cylinders, 0–6–0 saddle tank, 3 ft. 7 in. diameter wheels		
Works No.	Date of Delivery	Customer's Name or No.	Original Customer
1446	16.7.29 (steamed 10.1.27)	Treasurer	Oxfordshire Ironstone Co.

Table 5.12

	16 x 24 in. cylinders, 0–6–0 side tank, 3 ft. 9 in. diameter wheels		
Works No.	Date of Delivery	Customer's Name or No.	Original Customer
1637	30.7.29	Birch Coppice	Morris & Shaw Ltd, Tamworth
1791	26.8.35	Tir John No. 1	County Borough of Swansea, Tir John Power Station

Table 5.13

	14 x 20 in. cylinders, 0–6–0 side tank, 3 ft. 4 in. diameter wheels		
Works No.	Date of Delivery	Customer's Name or No.	Original Customer
1643	26.10.29	H.H.W.D. No. 3	Crown Agents (Palestine Haifa Harbour works)
1644	4.11.29	H.H.W.D. No. 4	Crown Agents (Palestine Haifa Harbour works)
1645	14.11.29	H.H.W.D. No. 5	Crown Agents (Palestine Haifa Harbour works)
1646	25.11.29	H.H.W.D. No. 6	Crown Agents (Palestine Haifa Harbour works)
1647	15.6.31	Southampton	John Mowlem & Co. Ltd (Southampton Docks contract)
1648	29.5.31	Millbrook	John Mowlem & Co. Ltd (Southampton Docks contract)
1685	13.7.31	Nuttall	John Mowlem & Co. Ltd (Southampton Docks contract)
1686	13.7.31	Mowlem	John Mowlem & Co. Ltd (Southampton Docks contract)
1687	20.7.31	Grosvenor	John Mowlem & Co. Ltd (Southampton Docks contract)
1688	20.7.31	Southern	John Mowlem & Co. Ltd (Southampton Docks contract)
1689	27.7.31	Trafford Park	John Mowlem & Co. Ltd (Southampton Docks contract)
1690	27.7.31	Cunarder	John Mowlem & Co. Ltd (Southampton Docks contract)

Table 5.14

	16 x 24 in. cylinders, 0–6–0 saddle tank, 4 ft. 2 in. diameter wheels		
Works No.	Date of Delivery	Customer's Name or No.	Original Customer
1674	9.2.31	—	London Electric Rly
1675	9.2.31	—	London Electric Rly

Table 5.15

	12 x 18 in. cylinders, 0–4–0 saddle tank, 3 ft. 4 in. diameter wheels		
Works No.	Date of Delivery	Customer's Name or No.	Original Customer
1684	24.8.31	—	Hall & Co., Coulsdon

CHAPTER 6

Diesel dawn

From the previous chapter it will have been seen that the Hunslet Engine Co. came through the great depression of the twenties and early thirties in better shape than some of its competitors and neighbours. Indeed, Manning, Wardle & Co. Ltd, separated from Hunslet only by the thickness of a brick wall, had teetered to an end in 1927, and more locomotive firms, including Kitsons, were to tumble long before the end of the next decade.

But, as we have chronicled, the output from 125 Jack Lane had remained steadfastly steam, varied in content, advanced in design and good on quality, but steam nevertheless. Across Jack Lane, Hudswell Clarke produced a massive 330 h.p. 2 ft. 6 in. gauge diesel-mechanical locomotive for a nitrate railway in the Chilean Atacama Desert in 1930, and in the same year John Fowler and Company in nearby Leathley Road supplied a standard-gauge shunter for the Chesterfield Tube Company. Both firms had previously produced modest numbers of petrol locomotives, as indeed had Manning Wardle and others, but by 1930 Hunslet had not publicly shown any move towards the internal combustion engine.

Throughout this book there is an underlying hint of enterprise yet caution in the financial and business approach of the Hunslet management to new ventures, and this was combined with a truly patriarchal and hierarchical regime. Perhaps Hunslet was letting others put their collective toe in the water, on the basis that when the time was right to move forward into the newer form of traction they would build on the experience of others, get the formula right first time, and stay in there. The term *canniness* springs readily to mind, for events were to show that this is in fact what happened.

John Frederick Alcock graduated from Clare College, Cambridge with a master's degree in Science at the age of 22 in 1927. He joined the Hunslet Engine Company with a clear mandate from his father Edgar, the then patriarch and Managing Director, to concentrate on the development and production of diesel locomotives. His early years at Hunslet were spent in gaining practical shop floor experience combined with widespread travelling studying new manufacturing methods and machinery in preparation for the expected change in work pattern caused by the transition from steam to diesel locomotives. The first major projects were the installation of flame cutting, welding and heat treatment facilities. These were followed by the design of the first pre-selector four-speed gearbox with air-actuated sliding dog clutches, which, when coupled with an improved main clutch of dry multi-plate type, enabled a pattern of production that ultimately ensured over fifty years of continuous diesel locomotive manufacture.

While this preparation was going on, an apparently unconnected comment in *The Locomotive Railway, Carriage and Wagon Review* for 15 October 1930 advised that the Hunslet Engine Company had purchased the goodwill of the Stoke-on-Trent based locomotive firm of Kerr, Stuart & Co. Ltd, recently gone into liquidation. This in itself was an irony, for Kerr Stuart had also been performing well and was only brought down by the unofficial business dealings of its chairman.

The bizarre tale of the Lucanesque disappearance of the Kerr Stuart chairman, of the discovery of his hapless secretary shot dead through the brain alongside hastily burned papers, and of the titled lady Labour MP totally un-interested in the shattered lives of the Stoke-on-Trent work-force is graphically chronicled in L. T. C. (Tom) Rolt's autobiography *Landscape with Machines*. Rolt, serving an apprenticeship with Kerr Stuart at this time, also describes how his adoptive uncle, W. K. Willans, had eight years earlier created a geared steam locomotive by marrying the engine and boiler of a second-hand Sentinel steam wagon with the frame of a Manning Wardle 0–4–0 saddle tank locomotive. In 1924 Willans became one of the first three directors of Sentinel Industrial Locomotives (England) Ltd, where he stayed until 1927 when he joined Kerr Stuart as development engineer – in the event not the most enduring of career moves. He produced five narrow-gauge geared steam locomotives at Stoke in the final years.

From the Hunslet viewpoint the purchase of the Stoke business was timely, for Kerr Stuart had, from 1928, produced a markedly successful range of diesel-mechanical chain-driven locomotives of 30, 60 and 90 h.p. under the guidance of Willans. Twenty-six examples of the early integrated Kerr Stuart diesel locomotive range are recorded as having been built at Stoke on Trent prior to liquidation, comprising twelve 30 h.p., eight 60 h.p. and six 90 h.p. units. Five of the 60 h.p. 0–6–0 examples went to the Soudan: four for the 60 cm gauge Sudan Gezira Light Railway, of which more in a later chapter, and one for the 3 ft. 6 in. gauge Sudan Government Railway lines at Port Sudan, while two operated back-to-back in tandem on the 75 cm gauge Central Railway of Ecuador. The 30 h.p. machines were of 0–4–0 type, all 2 ft. 0 in. gauge or narrower; two went to Haifa, one each to Greece and

Mauritius, and the rest to quarry companies and contractors in the UK.

The 90 h.p. machines were a little more varied, comprising two of standard 4 ft. 8½ in. gauge and four of 2 ft. 6 in. gauge. The 2 ft. 6 in. examples were six-wheelers and the last to be built at Stoke-on-Trent, from where they went to the Associated Manganese Mines of South Africa. The first 90 h.p. was also of the six-wheeled type, delivered on 2 December 1929 to the Ravenglass and Eskdale Railway for use on the standard-gauge quarry lines at Murthwaite. It was sold on closure of the quarry in 1953 and later worked with the National Coal Board in Durham and at Rom River Reinforcements Ltd at Lichfield before finally reaching preservation at Foxfield, Staffordshire.

HE 1700/32 was indistinguishable in appearance from the Kerr Stuart 30 h.p. design. Given that a batch of six 30 h.p. locomotives, put in hand for stock by Kerr Stuart before their demise, had not been built at Stoke it is reasonable to assume that Hunslet 1700 and 1701 were partly, if not wholly, constructed from inherited components.
ANDREW NEALE COLLECTION

Those who have stayed with this apparent diversion thus far will have noticed that two of the twenty-six locomotives are as yet unaccounted for, and these provide the necessary Kerr Stuart–Hunslet link.

The prototype of the Kerr Stuart diesels, no. 4415, was ordered for stock on 5 May 1928 as a 60 cm gauge 60 h.p. machine with a McLaren Benz diesel engine. (All the Kerr Stuart diesels had McLaren engines of two, four or six cylinders. The 30 h.p. were hand-started but the 60 and 90 h.p. versions had auxiliary petrol engines for starting purposes.) By August 1928 no. 4415 had arrived at Dinas on the Welsh Highland Railway, which used it for some time on the Bryngwyn branch, *Beddgelert*'s old stamping-ground, before in March 1929 it went to the Festiniog Railway, which used it as a shunter until it returned to Kerr Stuart's Stoke works by August 1929. It went on its travels again to work on Sir Lindsay Parkinson's East Lancashire Road contract, but by 16 December it was on its way across the Irish Sea after having been back to the works for rebuilding to 3 ft. 0 in. gauge for the Castlederg and Victoria Bridge Tramway. Underpowered for the Irish railway, it was returned to Stoke to remain until Hunslet records of December 1930 show it as one of the four locomotives booked in as received at Jack Lane from Stoke for 'stripping, re-examination and re-erecting'. Rebuilt back to 60 cm gauge at Leeds, the final destination of Hunslet K4415, as it had become, was Mauritius, against an order received on 19 March 1932 from Robert Hudson Ltd on behalf of the Union Vale Sugar Estate. At the time of checking this manuscript, K4415 has just travelled yet again to return to its original North Wales haunts.

Also taken to Leeds were two steam locomotives (no. 4331 *Wren* and 'Tattoo' class no. 4391) plus ex-demonstration 90 h.p. standard gauge 0–4–0 diesel no. 4428. Built in 1929 and hired briefly to Gibbs and Canning Ltd in Tamworth between December 1929 and March 1930, no. 4428 was alleged to have serious transmission problems. Hunslet took the opportunity of fitting their newly developed air-operated constant-mesh gearbox, new control system and multiple-plate dry clutch, and at the same time turned the superstructure round to put the cab at the gearbox end instead of over the radiator and starting engine as had been standard Kerr Stuart practice. Early in 1932 it went on hire as Hunslet K4428 to the Air Ministry at Cranwell, and it was sold later in the year to Eastwoods Flettons Ltd at Kempston Hardwicke Brickworks. It is now at the Buckinghamshire Railway Centre at Quainton Road.

Coincidentally with this rebuilding of K4428, Hunslet began to allocate numbers for new-build diesels in the same sequential series as its steam locomotives. The first three numbers allocated were 1697 for a 150 h.p. standard-gauge machine and 1700 and 1701 for two narrow-gauge units, classified variously in different sales pamphlets as 35, 36 and 37 h.p. but in fact exact repeats of the Kerr Stuart 30 h.p. type.

The *Locomotive Railway, Carriage and Wagon Review* for 15 March 1932 carried a report on the recent British Industries Fair held at Castle Bromwich from 22 February to 4 March, which stated that

The Hunslet Engine Company, in conjunction with their agents, Robert Hudson Ltd, 38A Bond Street, Leeds, showed, under working conditions, a HUNSLET 150 *h.p. diesel–oil locomotive attached to a train of either side tipping wagons . . . along with a Ruston excavator . . . and at their stand, inside, the Hunslet Engine Co. showed a 35 h.p. diesel locomotive for the 24 inch gauge.*

Robert Hudson Ltd was a well-established builder and supplier of light railway equipment with a large factory in Gildersome some three miles distant from their head office in Leeds. They could, and would, supply anything from a track spike to a complete railway. The railways in Nepal, for example, were built in their entirety by Hudsons, which had large branch factories in India, South Africa, Rhodesia and elsewhere. However, apart from a handful of early petrol-engined units, Hudsons did not build locomotives. Prior to 1930 any locomotives required as part of a Hudsons light railway package had been purchased from an appropriate manufacturer, Hunslet included, but in 1911 an agreement was concluded with Hudswell Clarke for Hudswells to design and build a range of standard small steam locomotives capable of being supplied at short notice or in some cases from stock. Hudsons had, however, handled the sale of the Kerr Stuart narrow-gauge diesels, and this led to a long-standing arrangement with the Hunslet Engine Company when the Kerr Stuart business was acquired by Hunslet. Under this agreement Hudsons had exclusive sales rights for Hunslet diesel locomotives up to approximately 100 h.p. for light and narrow-gauge railways. Larger locomotives, and those for mainstream industrial and railway companies, were outside this agreement, although in some circumstances special arrangements were made.

Thus the two firms came to be sharing the star billing at Castle Bromwich, with Ruston Bucyrus playing a supporting role in loading the tipping wagons.

The 35 h.p. locomotive at the show would have been either no. 1700 or no. 1701. The latter is the more likely; it is shown in the Hunslet records as having been built, but not sold, and later dismantled, some parts being used for no. 1763, which was supplied in 1935 to the African Manganese Company together with a much larger 147 h.p. locomotive. No. 1700 was sold via Hudsons to the Greek Portland Cement Company in Athens, and so became the first purely Hunslet-built diesel to be despatched against a firm order. The order was placed on 14 January 1932 and the locomotive was sent out on 23 March of that year.

A series of photographs from the same spool of film, taken by John Alcock and undated but certainly from late 1931 or early 1932, show in turn K4428 finish-painted and ready for despatch, two sets of narrow-gauge frames set up for final erection and a standard-gauge diesel locomotive almost complete. The narrow-gauge frames are identical in appearance to those used by Kerr Stuart and may in fact have come from Stoke-on-Trent. They are unlike any

Mixed parentage – by Stoke-on-Trent out of Leeds. The standard-gauge Kerr Stuart 90 h.p. diesel no. 4428 stands at the exit to the Hunslet erecting shop after rebuilding as Hunslet no. K4428 and prior to going to Cranwell early in 1932. In the centre of the picture is the cab for Hunslet's pioneer main-line diesel shunter no. 1697, with the half-built locomotive to the right. With a solitary boiler symbolically positioned as a defiant gesture, the remainder of the steam locomotives are manoeuvred into the background. Original picture by John Alcock from his personal album.

AUTHOR'S COLLECTION

Hunslet no. 1697, successively LMS No. 7401 and later No. 7051, W.D. no. 70225, John Alcock on the Middleton Railway and finally, after a spell in the National Railway Museum at York, returning to Middleton justifiably fêted in full LMS glossy black livery with shaded letters and numbers. This is the original works photograph taken prior to the British Industries Fair in 1932. No handrails on the engine casing ('bonnet' to the layman), original cab side openings, air horn, exhaust silencer at front of frame and total lack of lamp brackets. Original, if unimaginative, grey livery. It worked, though!

purely Hunslet frames and can only have been for nos. 1700 and 1701. The standard-gauge diesel is unmistakably no. 1697, which had been put in hand as a stock demonstrator, and much midnight oil had been burned to finish it in time for the opening of the British Industries Fair.

Let us dwell for a time on no. 1697 and its immediate successors, for they were a marker for the subsequent history of the Hunslet diesel shunter, and for that of other companies as well.

At the close of the fair no. 1697 was also returned to the makers' works in Leeds, and operational trials were conducted with heavy coal trains on the two-mile-long branch line of the nearby Waterloo Main Colliery. The London, Midland and Scottish Railway then agreed to a week's trial in ordinary service on their metals, and so successful was this that a further 10 weeks' operation began, on 15 August 1932, at the former Midland Railway yard at Hunslet Lane. The last six weeks of this period entailed working continuously from 5 a.m. each Monday until 6 a.m. the following Sunday. On 30 October no. 1697 was returned to the makers for examination and then resumed its duties at Hunslet Lane until January 1933, when it was repainted so that it could again be exhibited and demonstrated at that year's British Industries Fair.

It was at this time that the London, Midland and Scottish Railway decided to experiment with other designs of diesel shunter. Orders were accordingly placed for nine locomotives from five manufacturers: one each from Armstrong Whitworth, Harland & Wolff and the Drewry Car Co. Ltd, two from Hudswell Clarke, and four from the Hunslet Engine Co. The Armstrong Whitworth locomotive was to be a diesel-electric; the rest were diesel-mechanicals.

These nine locomotives were allocated running numbers 7400–7408, very quickly amended to 7050–7058. The four Hunslets were to be nos. 7401–4, later nos. 7051–4. The first three actually received the original numbers and were later renumbered; the fourth entered service as no. 7054.

The four Hunslet locomotives differed in the type of diesel engine and transmission adopted – this being the nature of the experiment – in addition to possessing differences in external appearance. The livery was in all cases black overall with red buffer beams and the standard LMS gold lettering of the period.

No. 7401 (7051) was the original BIF demonstrator no. 1697 slightly modified by altering the cab side cut-outs to accommodate a wooden arm-rest forward of the door opening and by filling in the semi-circular cut-out to the rear of the door opening. The air horn was replaced by an air-operated bell whistle, and a ballast weight was added under the leading drag plate. The German-built MAN engine developed 150 b.h.p. at 900 r.p.m. (165 b.h.p. at 1000 r.p.m. on overload). A Hunslet patent friction clutch and a Hunslet-designed four-speed gearbox built by David Brown transmitted the power through a jackshaft and coupling rods to six 3 ft. 0 in. diameter wheels. The weight in working order was 21 tons.

Nos. 7402 and 7403 (7052 and 7053), Hunslet nos. 1721 and 1723, were substantially identical to each other in size and appearance, but were physically larger than no. 7401, both weighing 26 tons with six wheels 3 ft. 4 in. in diameter on a 9 ft. 0 in. wheelbase. No. 7402 however had a 150/165 h.p. McLaren–Benz eight-cylinder oil engine running at 1000 r.p.m. driving through a multi-plate friction clutch and a two-speed constant-mesh gearbox, while no. 7403 received a Brotherhood–Ricardo sleeve-valve engine of similar power output running at 1200 r.p.m., a Vulcan Sinclair fluid coupling and a David Brown two-speed epicyclic gearbox.

The last of the four, no. 7054 (Hunslet no. 1724) was much larger; it weighed 29 tons and had a Paxman–Ricardo engine of 180/200 h.p. at 900 r.p.m. coupled to a Vulcan Sinclair coupling, a Humfrey Sandberg rocking-type freewheel clutch and a three-speed constant-mesh gearbox. The locomotive was higher and wider than its predecessors, the cab having a high arched roof and three large plate-glass windows fore and aft. As in the second and third of the set, sliding wooden shutters were provided inside the cab to close off the side openings, and hinged cab side doors to waist height were fitted.

The official LMS purchase orders for the four Hunslet locomotives were received by the maker on 1 April 1933, and no. 7401 (7051) took up its official railway company duties at Hunslet Lane goods yard six weeks later on 15 May; the intervening period was taken up in carrying out the above-mentioned minor modifications requested by the customer and in repainting in the railway company's colours. The locomotive had already proved its worth at Hunslet Lane between the two British Industries Fairs, and subsequent events were to suggest that this original design, despite its small size, was the best of the bunch, fully vindicating its Hunslet-designed clutch and gearbox. The Hunslet Engine Company subsequently developed its own in-house gear-cutting and gearbox manufacturing

Three of the four Hunslet LMS diesels together with the two Hudswell Clarke units being inspected by the LMS top brass in Hunslet Lane goods yard when newly commissioned. From the front, Nos. 7051, 7052, 7054, 7055 and 7056. Note the bell whistle on No. 7051. The rest had organ-pipe whistles or, perhaps more correctly, hooters – more in keeping with the new man Stanier. No. 7051 is still a bit short on lamp brackets: Nos. 7052 and 7054 fare a little better with brackets over the buffers. Note the cast-iron ballast weight between the frames on No. 7051 in the place previously occupied by the exhaust silencer. The exhaust now passes out via a pipe to the top of the engine casing.

facility, and this basic jackshaft-mounted gearbox and its derivatives were to be the foundation of the Hunslet diesel shunter range until the axle-mounted final drive (again produced in-house) became the norm for diesel-hydraulics in the early 1960s. Indeed, the jackshaft drive gearbox was still occasionally being used at the customer's request (by the Sudan Gezira Board, for example) as late as 1986. In the words of Dick Hardy, who was District Motive Power Superintendent at Stratford in 1955, referring to the then new Ipswich-based DJ12 204 h.p. Hunslets in an article forty years later, 'they were indestructible.' They needed to be – changing gear on a Hunslet diesel-mechanical locomotive was an acquired art. Nevertheless, from 1935 onwards no Leeds-built Hunslet shunter was fitted with a 'foreign' gearbox.

This gearbox manufacturing capability was unique among British locomotive manufacturers, and, combined with the other machine shop improvements, gave a flexibility that allowed the firm to ride out future economic and political storms and take full advantage of any possibilities for diversification.

The 'Diesel Railway Traction' supplement of the *Railway Gazette* for 3 November 1933 describes no. 7401 (7051) as 'an outstanding success'. The locomotive replaced 17 in. and 18 in. 0–6–0 steam tank locomotives. In a speed trial from Leeds to Bradford via Shipley (locomotive and brake van only) the 29 miles were covered at the planned average speed of 30 m.p.h. start to stop. General shunting duties usually involved sixteen or twenty wagons, say 180–200 tons, but one of the regular duties at Hunslet lane was hauling the nightly Leeds–Carlisle freight – around 650 tons including a main-line goods engine – for about a mile before the train changed direction to head north toward the Settle and Carlisle line. Despite less than sensitive handling by crews who were accustomed to lever-reverse steam locomotives, clutch wear is said to have been only $^1/_{32}$ inch in 6000 hours' operation.

The second locomotive, no. 7402 (7052) was delivered on 15 January 1934 and put to the same duties. Its greater weight (26 tons) and higher tractive effort (12,000 lb. as against 9560 lb.) were better suited to general shunting work, but the two-speed gearbox, giving a maximum running speed of 8.23 m.p.h., was a disadvantage when moving around the yard and on the Carlisle turn. This had a considerable influence on the subsequent tendency to fix the optimum speed at around 15 or 16 m.p.h. for new designs of shunter, four-speed gearboxes being used on diesel-mechanicals and, later, single-speed gearboxes with Twin Disc hydraulic transmission.

The next of the series, no. 7403 (later 7053), was reported by the *Railway Gazette* for 20 April 1934 as having been 'recently delivered to Hunslet Lane. No. 7401 (7051) was to go to Chester when the newcomer was 'run in', but it appears that no. 7403 had to go back into the works for a few months before the LMS would accept it. For no. 7403 was indeed a 'bake': the sleeve-valve engine gave trouble, and the freewheel and the David Brown epicyclic gearbox were replaced in 1936 by a French-designed, British-built Cotal electromagnetic four-speed epicyclic unit – surprisingly compact at less than two feet in diameter and a foot long. The locomotive does not appear in any of the group shots taken at Hunslet Lane after delivery of no. 7054 and the two Hudswell Clarkes, nos. 7055/6. It does appear in the official works view of no. 7053, clearly carrying the old number 7403. The same photograph when used on the Hunslet catalogue sheet has been retouched to show the new number, suggesting that a return for modifications is a distinct possibility.

Little else is known about the pre-war activities of this varied collection of locomotives. No. 7051 was transferred to Chester; no. 7052 is believed to have remained at Hunslet Lane; no. 7053 is reported to have been seen at Plaistow on the L. T.& S.; while no. 7054 was variously at Derby, Ditton Junction and Chester. After a spell of military service, becoming WD Nos. 27, 28, 224 and 225 (later 70027, 70028, 70224 and 70025), all four loco-motives were repurchased by the maker at intervals between 1942 and 1945.

In 1943 the former no. 7052 was rebuilt by Hunslet to protected flashproof standard, but retaining its original engine and gearbox, and sold to the Royal Naval Armaments Depot at Broughton Moor, where it became Yard No. 87. In 1966 it was purchased by Birds Commercial Motors Ltd of Long Marston, who eventually scrapped it in 1969.

The remaining three locomotives were still on the Works siding when the author joined Hunslet early in 1949, taking a low priority behind a new-build order book reaching well into three figures and a queue of war-weary locomotives awaiting urgent repair. In due course no. 1697, the original BIF demonstrator, was re-engined with a second-hand McLaren MR6 unit, repainted in light Brunswick green with black and yellow lining and despatched to London & Thames Haven Oil Wharves Ltd, Stanford-le-Hope, Essex, where it remained on hire until a new Hunslet locomotive was delivered in 1951. On its return to Hunslet from Thames Haven no. 1697 became works shunter, with occasional outings on hire. One such job was in connection with the relining of Bramhope tunnel, where its specially fitted water-bath exhaust proved invaluable in minimizing toxic fumes. In September 1960 it was handed over to the Middleton Railway, which later painted it blue and named it *John Alcock* after its designer, by then the Railway's benefactor and Chairman and Managing Director of the Hunslet group. For a time it was on loan to the National Railway Museum (repainted black as LMS no. 7051), but is now back at Middleton.

The unfortunate no. 1723 was considered prohibitively expensive to repair and was scrapped in March 1954. No. 1724 however was given a light repair and sent on loan to Brodsworth colliery near Doncaster, where it acquired the unofficial but affectionate nickname of 'Dog'. It was eventually purchased by the National Coal Board and was rebuilt by the maker in 1960 as a diesel-hydraulic with a Rolls-Royce C6N engine and Twin Disc transmission. It returned to Brodsworth painted chocolate brown with black-and-yellow diagonal stripes at front and rear, and ended its days during the seventies at Hickleton Main colliery.

Hunslet was breaking new ground with the LMS diesels, and very well they were doing it too, but caution was still the underlying theme. Hunslet Lane goods yard, where by the end of 1934 at least three of the four were working round the clock, was only half a mile from the works; and a fitter could walk there in ten minutes.

In real terms the rate of delivery of diesel locomotives in the years 1932–34 was slow. Including no. 1700 and the LMS shunters, only seven, eight and seventeen diesels, respectively, were produced in those three years. These included one 35 h.p. special inside-frame 3 ft. 0 in. gauge machine for Burmah Oil in India; twenty-three eminently forgettable 20 h.p. chain-driven units with transversely-mounted two-cylinder Lister engines, six of them for a land reclamation scheme at Mucking in Essex; and, of more interest, two 75 h.p. locomotives and one of 112 h.p. which were to set down a further marker for another niche market – that of very specialized, robust, high-performance, 'full-size' narrow-gauge machines for difficult conditions.

Despite wide differences in appearance and application these three narrow-gauge units followed the basic Hunslet standardization that has already been seen in the later

steam locomotives. In common with the second LMS locomotive they had locally-built McLaren–Benz engines started by Scott petrol engines through a clutch connection to a Bendix pinion engaging a gear ring on the main engine flywheel. The range of Benz engines of the period produced by J. & H. McLaren had a cylinder bore of 135 mm and a piston stroke of 200 mm, and ran at 1000 r.p.m. Hunslet used two-, three-, four-, six- and eight-cylinder versions of this engine for most of its pre-war 37, 56, 75, 112 and 150 h.p. locomotives.

The first of the trio to appear, at the end of February 1934, was no. 1722 *Albert*, a decidedly neat locomotive of 75 h.p. carried on two four-wheeled bogies, all four wheels on each bogie being driven by coupling-rods from a worm-driven jackshaft. Despite running on a track gauge on only 18 in., *Albert* had a full-height driving cab with duplicated controls, air and hand brakes, and weighed 13 tons 5 cwt. The second, no. 1747, delivered in August 1934, was another 75 h.p. machine with virtually the same traction

Albert, HE 1722/34, at 75 h.p. and 13.25 tons, was a large engine for 1 ft. 6 in. gauge. Supplied to the Royal Arsenal Railway, Woolwich, it was also quite advanced for its time, with cardan-shaft drive to each bogie from a central frame-mounted gearbox. It could haul 236 tons at 8 m.p.h. on level track. Another locomotive of similar type named Carnegie *was supplied in 1954.*

arrangements, but arranged as a 9 ton 13 cwt. six-wheeled chain-driven unit for the Lobitos oil fields in Peru. The chain drive arrangement, with its torque reaction and tensioning rods, was almost identical to that of the Willans-designed Kerr Stuart 60/90 h.p. pattern; this had been the Sentinel arrangement, and most British chain-driven locomotives inherited this feature by one route or another.

Two months later, on 30 October, came no. 1740, a stubby 75 cm gauge 0–4–0 driven by coupling-rods, ordered for passenger work on the Egyptian Delta Light Railways. This had a top speed of 20 m.p.h., and with 112 h.p. at 1000 r.p.m. (continuous) installed in a locomotive weighing only 11 tons 5$\frac{1}{2}$ cwt it conferred a higher power-to-weight ratio than any previous Hunslet diesel. Despite the relatively high power, strict axle-loading limitations made it necessary for special designs to be prepared in order to embody Hunslet features, including preselective gear change, without exceeding the weight limit. The axle limit, wheelbase and weight were specified to make the locomotive conform to existing steam locomotive practice on the line in question.

The plate-type locomotive frame carried McLaren's six-cylinder Benz engine, which was equipped with CAV–Bosch injection apparatus and started by a Scott petrol engine in the manner that had become standard on Hunslet diesel locomotives. Since high working temperatures would be encountered, a radiator considerably larger than standard was employed; this was of the Serck sectional type with removable sections, some of which were used for cooling the engine oil.

Welding was used to a large extent in the construction of the gearbox and certain other parts of the locomotive. Transmission from the engine included a Vulcan-Sinclair traction-type hydraulic coupling, and the effort was passed on through a special Hunslet device affording a quick release to speed up gear changes; this considerably reduced the time lost in waiting for the engine to accelerate or decelerate before the actual gear change could be effected.

Thus by the end of 1934 a pattern had been established for the future direction of the company in its diesel locomotive development. The four LMS locomotives were a base for the larger surface shunters; the 75/110 h.p. designs covered the middle, mainly narrow-gauge, range;

Two generations of the same basic locomotive. Central Railway of Peru No. 505 in the foreground is HE 1988/39, while No. 507 is HE 4176/50. The former is identical in size and power to the prototype Hunslet diesel no. 1697 and has a McLaren 150/165 h.p. engine. The driving wheels are 3 ft. 0 in. in diameter on an 8 ft. 0 in. wheelbase. No. 507, on the other hand, is a standard post-war 204 h.p. unit with Gardner 8L3 engine and 3 ft. 4 in. diameter wheels on a 9 ft. 0 in. wheelbase. Photographed at Callao depot on 22 January 1970.   D. TREVOR ROWE and the 20/35 h.p. contractors' types established Hunslet firmly in the higher-quality end of the cheap-and-cheerful small narrow-gauge market. The only way now was up, and the designs developed accordingly. Only the flameproof underground mining locomotive five years hence was to add a new dimension to the locomotive scene, and of this we shall see much later.

Without doubt the four LMS diesels had, between them, evolved a successful design for surface shunting, even if the LMS was ultimately going to decide on the larger diesel-electric, and variations on the same basic theme were to continue for the next thirty years. Locomotives virtually identical in dimensions and appearance to LMS no. 7051 were produced in 1935 for the Egyptian Phosphate Company, for the Sinai Mining Co., also in Egypt, and for the Piata Piura Railway of the Peruvian Corporation in 1939. These retained the four-speed gearbox but with a top speed of 20 m.p.h. instead of no. 7051's 30 m.p.h., and the by now reliable McLaren 8MDB (as used on LMS no. 7052) replaced the German MAN unit.

The 3 ft. 0 in. diameter wheels and 8 ft. 0 in. wheelbase were retained. While the Peruvian unit was standard-gauge and required little other than a change of drawgear, the Egyptian machines were for track gauges of 2 ft. 6 in. (Sinai) and one metre (Egyptian Phosphate) and adopted outside frames and fly cranks for the coupling rods. The 3 ft. 0 in. diameter wheels and 8 ft. 0 in. wheelbase standard-gauge frame assembly was used again in 1937 on locomotive no. 1846, WD no. 1, for the Army ammunition tunnels at Corsham in the area adjacent to Box tunnel on the GWR main line. This was a significant development in that it was the first application of the Gardner 8L3 diesel engine in a locomotive and it also pioneered the use of flameproof electric starting and lighting for use in hazardous areas such as ammunition depots, oil refineries, etc. A two-speed gearbox was provided, 15 m.p.h. being adequate for the duties involved, and a low overall height of 9 ft. 0 in. was needed for use in the underground sidings, as was a water-bath exhaust conditioner with exhaust flame traps. The flameproofing equipment on this

The first-ever application in a shunting locomotive of the Gardner 8L3 diesel engine was in W.D. No. 1, HE 1846/37, a 155/170 h.p. 0–6–0 for the ammunition tunnels at Corsham in Wiltshire. Cut down in height and fitted with exhaust conditioner for tunnel work, it was otherwise dimensionally similar to the original LMS diesel No. 7051. Two more were supplied to Corsham; the Gardner engine was uprated to 204 h.p. and, when married to a larger chassis based on LMS No. 7054, provided the basis for what was to become the classic Hunslet shunter.

The controls of the Corsham locomotive. The Hunslet patent hydraulically-operated preselective gearbox had three speeds, selected by the handwheels on each side and engaged by operating the lever under them while at the same time clutching and declutching using the large vertical hand lever.

The Gardner 8L3 engine and Broom and Wade compressor installed in HE 1846/37. An early form of flame trap is fitted to the engine air intake manifold as a precaution against explosion should the engine backturn. The Hunslet patent friction clutch can be seen at extreme left.

The best-looking of the early 37 h.p. diesels must have been Percy, HE 1774/35, and Sir William, HE 1981/38. Used for serving the open-hearth furnaces of the Park Gate Iron and Steel Company, Rotherham, and painted in full main-line style, Percy lasted to the end of the internal works 3 ft. 0 in. gauge system in 1970.

For £839.2s.6d. you could have this 'de luxe' 37/41 h.p. narrow-gauge locomotive with coupling-rod drive instead of chains. HE 1838/36 supplied to Balfour Williamson for a contract in South America.

Already well over twenty years old, A.M.Co. No. 11, HE 1762/35, hauls a train of manganese ore on the 2 ft. 6 in. gauge line of the African Manganese Company in the Gold Coast (now Ghana) in the mid-fifties.

By 1936 the Hunslet gearbox had matured, and future development was to concentrate on gear-selection techniques and how to transmit the higher powers that were inevitably to be demanded. This is a two-speed gearbox for a 56 h.p. metre gauge 0–6–0 with coupling-rod drive, HE 1807/36, supplied to the Director of Army Contracts for the Singapore Naval Base.

locomotive was of a rudimentary type by later standards; nevertheless it was the first comprehensive attempt to prevent the emission of sparks from either the diesel engine inlet and exhaust or the electrical equipment. It provided the base from which the underground mines locomotives mentioned in later chapters were developed. Two further locomotives of the same type were built for Corsham in 1940.

The 35 h.p. chain-drive locomotive had been refined into a very robust and chunky 37/41 h.p. machine with the two-cylinder McLaren 2MDB engine, and eleven of these locomotives were built at intervals between 1935 and 1938. Most went overseas: four to Balfour Beatty's Kut Barrage contract in Iraq, two (with coupling-rod drive) to a similar Balfour Williamson contract in South America, two to Elders & Fyffes in the Cameroons and one to the Sudan Light and Power Co. Two of them, fitted with full cabs and fully lined-out, with brass name-plates, *Percy* and *Sir William*, worked the 3 ft. 0 in. gauge internal system serving the open-hearth furnaces of the Parkgate Iron and

Steel Company's works in Rotherham until the system closed in the late 1960s. One of the Kut Barrage quartet came back to England via a Balfour Beatty contract in the Orkneys to become *Taff* of the Penmaenmawr & Welsh Granite Company Ltd, in 1947, and soldiered on, albeit with a replacement Gardner 4LW engine, for a further twenty years.

The Kut Barrage also gave rise to a very neat 2 ft. 6 in. gauge coupling-rod drive 0–6–0 locomotive with a 65 h.p. three-cylinder McLaren 3MDB engine derated to 57 h.p. and with detachable frame side-screens (rather like the Wisbech & Upwell tram engines) for desert conditions. Six of these locomotives were ordered in 1935 (a seventh was ordered a year later) on behalf of Mr D. C. M. Brooks of Cairo, who had the subcontract responsibility for the transport of materials for 18 miles over arid country from the quarries to the river barge transhipment station, whence the materials continued by barge to the actual site of operations.

The early 20 h.p. locomotive with transversely mounted Lister 18/2 engine was built at intervals from 1932 to 1939 in either basic form or with a variety of optional extra equipment. This is HE 1741/34, a 60 cm. gauge machine with electric lighting but retaining hand starting, supplied through Robert Hudson Ltd. to Hiram Craven for plantation duties in Kenya. Some other examples had a canopy over the driving position.

The Kut Barrage 0–6–0s had an open cab, but two very similar locomotives with fully-enclosed cabs were built: a one-off on metre gauge for the Director of Army Contracts, Singapore Naval Base, in 1936; and another one-off on 2 ft. 6 in. gauge for Swartkop Chrome Mine, South Africa, in 1939.

Following the Kut Barrage 0–6–0s in August 1935 came an 'intermediate' 0–6–0 of 146 h.p. on 2 ft. 6 in. gauge with a Ruston diesel engine for the African Manganese Company Ltd in the Gold Coast (later Ghana). By this time the coupling-rod drive 0–6–0 had become rationalized on four frame configurations each having a particular combination of wheel diameter and wheelbase, and a similar pattern was soon to emerge for four-wheeled coupling-rod drive machines also. These were:

Wheel arrgt.	Wheel diameter	Wheelbase	Nominal weight	Horsepower
0–6–0	2 ft. 0 in.	5 ft. 3 in.	8 tons	56/62
0–6–0	2 ft. 9 in.	7 ft. 0 in.	20 tons	122/146
0–6–0	3 ft. 0 in.	8 ft. 0 in.	22 tons	150/165
0–6–0	3 ft. 4 in.	9 ft. 0 in.	28 tons	180/204
0–4–0	2 ft. 4 in.	5 ft. 0 in.	11–12 tons	70/77, 75/82
0–4–0	2 ft. 9 in.	5 ft. 6 in.	10–13 tons	40/44, 50/55, 70/77, 80/88
0–4–0	3 ft. 4 in.	5 ft. 6 in.	20–23 tons	102, 153

The small 20 h.p. locomotive with the Lister 18/2 engine, first seen in 1932 for the Essex scheme, was to be built at intervals, usually as stock batches, until by July 1939 a total of forty-five had been delivered. They were for a variety of track gauges from 45 cm up to one metre, but there were two freaks. Both of these, which had longitudinally-mounted engines, were delivered within three days of each other, although they were from two distinct batches of stock parts, and very few parts common to other members of the batches could have been used. The smallest standard-gauge diesel shunters ever produced by Hunslet, they weighed 6 tons 12 cwt. No. 1737 went to Eldon Brickworks on 27 July 1935 and no. 1786 to the brewers Courage at Alton, Hampshire, on 30 July. The latter can still be seen at the Middleton Railway. A further oddity was the yard locomotive no. 1850 produced a few years later using, to quote the internal works order, a 'standard broad gauge 20 h.p. frame as the Eldon and Courage locos and fitted with a 10 h.p. loco reverse gear'. It also took the petrol engine from Edgar Alcock's

Armstrong Siddeley limousine, which produced 25.5 h.p. at 1200 r.p.m. How long this hotch-potch of engine and transmission lasted is not recorded, but the locomotive itself survived until 1987, by which time it had been weighted up to over 8 tons and had acquired a Perkins P4 50 h.p. engine and a gearbox to suit.

Despite the protestations of political apologists the drums of war were sounding ever more loudly, notwithstanding tea with 'that nice Mr Hitler' and, later, Neville Chamberlain's infamous 'Peace in Our Time' declaration. Industry was unmistakably setting out its stall, and the ensuing pages of the Hunslet Engine Company's order books (and no doubt those of many other companies) are testimony to that. With very few exceptions the emphasis was either on direct contracts for government and military establishments or for civil contracts associated with strategic installations. Of the thousand diesel locomotives built by Hunslet in the ten years to the end of World War II less than a hundred came outside these two categories, and even some of the exceptions fell into a doubtful 'grey area'.

March 1937 saw the first two of a batch of six 20 h.p. 3¼ ton locomotives of a totally new design. Although the vast majority of Hunslet's small and medium locomotives had been sold through Robert Hudson as agents, these six were the first to carry the name 'HUDSON HUNSLET' on the radiator, thereby setting the trend for small narrow-gauge units that was to continue for almost fifty years. The locomotive frame was a one-piece high-grade iron casting, and the gearbox casing was built up from castings of the same material. The chain drive was to four chilled iron wheels, and the Ailsa Craig CF2 diesel engine had a massive cast-iron flywheel on which was mounted the Hunslet friction clutch. All the castings were supplied by the local Hill Brothers foundry (not to be confused with Thomas Hill of Rotherham), which also supplied castings for steam locomotive cylinders, wheels, etc. to other locomotive builders (including Hudswell Clarke) and did general foundry work for other customers including local authorities. Hill Brothers later became a wholly-owned subsidiary of the Hunslet Engine Company, and so Hunslet found itself, indirectly, supplying castings to other locomotive builders.

The first two 20 h.p. Hudson Hunslet locomotives (nos. 1830/1) went to Sir Lindsay Parkinson & Company, who were building the Royal Ordnance Factory at Chorley, Lancashire, and the remainder of the first batch (nos.

Twenty of the rather basic 40/44 h.p. design were built between 1936 and 1943 for the War Office and essential industries. This posed shot shows the prototype, HE 1840/36, with the second example of the class, HE 1847/36, loaded on to a well wagon. The Fowler-Sanders four-cylinder diesel engine was used. No. 1840, for the Conisboro' Cliff Co. Ltd near Rotherham, had hand starting for the engine and hand brake only, while no. 1847, for the Royal Ordnance Factory, Kings Meadow, fared better, with electric lighting and starting and an electric horn. The driver of no. 1840 must have clapped his hands if anyone got in the way. The location is the Hunslet Engine Works yard with the erecting and boiler shops blocked out of the background. Both locomotives were despatched on 15 December 1936, which dates the photograph.

The engine compartment of HE 1847/36 emphasises the simplicity of the 40/44 h.p. design.

HE 1832/37, the third locomotive in the first batch of the ubiquitous Hudson-Hunslet 20 h.p. class, of which vast numbers were built and several are still extant. The basic design changed little over the years, although the superstructure was restyled in the fifties to allow a seated driving position.

1832–5) were supplied against an army contract for use in the labyrinthine underground network of ammunition storage tunnels at Corsham in Wiltshire – a job for which they were fitted with exhaust gas conditioners. Many others were to follow.

The design lent itself to rapid assembly by one man, and vast numbers were built, including almost six hundred by the middle of 1944 for the Ministry of Supply and twenty-five for UNRRA. Other batches were authorized but not built, while some war surplus units were returned to the makers and later resold under new numbers, making it difficult to establish with certainty the exact number that were built.

The Ailsa Craig CF2 engine was superseded in 1939 by the RF2 – basically the same engine but with a Ricardo-pattern indirect-injection cylinder head. Both 20 h.p. and 25 h.p. versions were built – the extra power being

obtained by increasing the engine speed from 1200 to 1500 r.p.m.; the road speed increased in proportion. The weight and haulage capability were the same in both cases. A sales pamphlet issued in November 1937 describing the 3¼ ton 20 h.p. Hudson Hunslet states: 'Today's price £365, delivery from stock.' As time went on, 25 h.p. 4¼ ton, 30 h.p. 5 ton, 30 h.p. 6¼ ton, 40 h.p. 6½ ton and 50 h.p. 7½ ton versions, all looking very similar but getting progressively larger in physical dimensions, were added to the range, but they did not sell in quite the same numbers as the 3¼ ton machines. Lister engines were used in the 30 h.p. machines and Perkins engines in the 40 and 50 h.p. examples. McLarens were used in the 4¼ ton 25 h.p.

The biggest diesel event in the months leading up to the war was undoubtedly the despatch, on 1 February 1939, of a large metre gauge 0–6–0 shunter for the Guaqui–La Paz Railway owned and operated by the Peruvian Corporation.

For some obscure reason this 50 h.p. Hudson-Hunslet, HE 1896/37 just had the name 'HUNSLET' cast in the radiator header and was fitted with standard makers' plates without Hudson's name incorporated. There were other examples from time to time. Nevertheless the order came from Hudsons, and the locomotive went to the Société des Sucreries d'Égypte at Hawamdieh for sugar plantation duties.

The Woolwich Arsenal design was developed to produce this 2 ft. 0 in. gauge 88 h.p. version (HE 1877/38) for the Doornkop sugar estates in South Africa. The McLaren diesel engine was derated to 82 h.p. to suit the ambient conditions at site.

This neat little 70/77 h.p. 2 ft. 6 in. gauge locomotive, HE 2020/39, for the Zwartkop chrome mine is typical of Hunslet styling, and was the inspiration for the six- and eight-wheeled locomotives supplied in large numbers to the Sudan Gezira Board after the war. The Gardner 6L2 diesel engine was derated by 20 per cent to allow for the temperature and altitude in the Transvaal.

The Guaqui La Paz locomotive HE 1904/39 arrives at Chijini station during its delivery run from Antofagasta.

With this locomotive the dawn of the diesel had truly broken. The order had been placed on 14 September 1937 and the destination was La Paz.

La Paz, the capital of Bolivia, although at an altitude of 11,500 feet, lies in a deep valley almost 1500 feet below the level of the high plateau across which the three main railway routes approach the city. The Guaqui–La Paz Railway line climbs out of the city by a bank $4^{1}/_{2}$ miles long, the 7% (1 in 14.3) gradient having sharp uncompensated curves cut into the hillside in a series of figure 8 turns. This severe climb had made steam traction impracticable, and electric traction had been employed since the line was built in 1905. The cost, which had been high even at the outset, was steadily rising. In 1937, after a very careful study of the whole situation, the Peruvian Corporation, in collaboration with their consulting engineers Messrs Livesey & Henderson, placed an order

with the Hunslet Engine Company for a diesel locomotive to work the route, knowing of Hunslet's already extensive experience with diesel traction.

In developing the design, Hunslet and the Peruvian Corporation kept in view the conditions on the Corporation's other railways, particularly the Central Railway of Peru, which is the highest standard-gauge railway in the world, rising from sea level to a height of 15,806 feet in a distance of 110 miles. Thus the locomotive for the Guaqui–La Paz Railway was designed not merely to give the very best results on that line but also to yield information for diesel locomotives working at altitudes of up to 16,000 feet above sea level.

The engine selected was the Mirrlees Ricardo 5 U.D. type. This engine was normally capable of developing 330 b.h.p. at 900 r.p.m.; but owing to the extremely high altitudes in question, it was supercharged to increase the

HE 1904/39 with a freight train on the Guaqui La Paz Railway main line. Note the brakeman on the last van standing perilously close to the overhead electrification wires.

power output at sea level by 50 per cent, so that output should not drop below 242 b.h.p. even at the highest altitude contemplated – the derating amounting to 51 per cent. Owing to the low air density at high altitude it was also necessary to increase the size of the radiators, brake compressors and various other auxiliaries.

The severe gradients also had to be allowed for. In addition to the hand brake, Westinghouse automatic and straight air brakes of the self-lapping type were provided. The trains also were fully braked.

The locomotive was shipped out to South America in the charge of the Hunslet Service Manager, Albert Hancock, whose job it was to superintend its progress to La Paz and there supervise its testing and commissioning. The ship experienced severe storms on the way to Antofagasta, and the locomotive, which had been shipped as deck cargo, sustained considerable damage in the heavy seas.

After a week spent in Antofagasta repairing the damage, presumably using local labour, the locomotive with its train of sleeping car, dining car and tool van made the long run of 700 miles from Antofagasta to La Paz over the Antofagasta and Bolivia Railway, climbing from sea level to a maximum altitude of 12,976 feet at Ascotan and on over the plateau to La Paz without a hitch, handling its load on the steepest grades and at the highest altitudes with ease. One notable feature of this run was that Mr Hancock, with no more in front of him than a gradient profile and the distances between stations, was able to arrange a timetable with the railway's traffic department to which he was able to adhere strictly, making all crossings at passing loops on the single line, without a single delay either to his own train or to any other traffic on the line. The Chief Mechanical Engineers of the Peruvian Corporation and the Antofagasta and Bolivia Railway, and several other officials, travelled with the train and were delighted with the general performance. On arrival at La Paz, general inspection of the locomotive prior to running tests on the Guaqui–La Paz Railway revealed no defects whatsoever.

When it was put on test, the locomotive not only hauled its full guaranteed tonnage up the extremely steep gradients of the 'Bajada' but was able to restart the same load on the 90 metres radius curves that occur without compensation on the 7% grade referred to above. The locomotive also made several test runs between La Paz Alto station and Guaqui, the port on Lake Titicaca, hauling heavy freight trains with remarkable success. Keith Alcock reported on one of his visits in 1973 that no. 1904 was still working at La Paz.

CHAPTER 7

The Second World War and its immediate aftermath

The last chapter opened by remarking on the cautiousness with which Hunslet had approached the development of the diesel locomotive, but went on to describe the pace at which development took place once the decision had been made.

By the outbreak of war it was clear that diesel and steam were to be at least equal partners at Jack Lane, and in course of time the former was to eclipse the latter entirely.

In the closing years of peace a further development, allied to the diesel, was to take place that would, in post-war years, give Hunslet a third mainstream product to ensure its ongoing prosperity – the flameproof underground diesel mining locomotive.

The first Flameproof Approval Certificates for complete locomotives were awarded early in 1939 by what was at that time the Government's Mines Department, after exhaustive testing by the Safety in Mines Research Establishment's Testing Station at Buxton.

In order to record the early stages of mining locomotive development it is necessary to go back to 1937. Diesels were already well established in metalliferous mines where there was no hazard from methane gas, or fire damp as it was sometimes called, but the Mines Department was far from satisfied that they could be safe in British gassy mines. German manufacturers had developed flameproof diesel locomotives and had obtained approval for their use from their own authorities. By 1939 France and Germany each boasted of upwards of 250 of these locomotives, and there were 100 in the Saar and about fifty each in Holland and Belgium. By 1948 the NCB also had forty-four electric battery locomotives, but no trolley or compressed-air locomotives as used in continental Europe. The British authorities, however, did not consider the German flameproof devices robust enough to remain safe for any length of time; and the problem, therefore, was to produce, among other things, a flame trap that was not only flameproof but was so robust and foolproof that it could be stripped for cleaning daily and reassembled with precision, without any fear of damage during the process or incorrect assembly due to carelessness. Furthermore, the material

had to be such that the service conditions and the frequent handling would not impair its effectiveness over a lengthy period of service. Hunslet and Ruston & Hornsby competed keenly; both firms obtained their initial Approval Certificates in 1939 and both delivered their first locomotives that same year.

The principal objectives of flameproofing locomotives were to provide measures whereby sparks or high temperatures could not be introduced into a hazardous environment. In this context the term 'flameproof' is used to describe a complete system that has been designed, built, tested and certified against the stringent regulations imposed by the Government's Mines Department, later the Health and Safety Executive. The terms 'sparkproof', 'protected' and 'flashproof' refer to systems protected to standards considered safe for applications other than coal-mining (e.g. petrochemical plants, ammunition depots, etc.), where explosive materials are present but more readily controlled, and not necessarily certifiable.

Sparks and flames could be emitted from the diesel engine exhaust, or from the air inlet manifold if the engine backfired; therefore flame traps were required on both inlet and exhaust. The exhaust was cooled by passing it through a water bath conditioner, or scrubber, before reaching the exhaust flame trap. The exhaust gas conditioner also removed the irritants (aldehydes) from the exhaust, and after dilution by air from the cooling fan, an acceptable underground environment was possible.

Batteries for electric engine starting motors were not considered safe for use underground; therefore air starting was used, fed from large storage reservoirs on the locomotive, and the pressure maintained by an engine-driven compressor after the initial shore supply charge.

Electrical equipment in general was kept to a minimum, restricted to only approved head and tail lights directly fed from a flameproof dynamo and connected by very heavily armoured cable.

The engines themselves were specially chosen, and adapted internally in many cases, to prevent an internal engine explosion from reaching the atmosphere. This

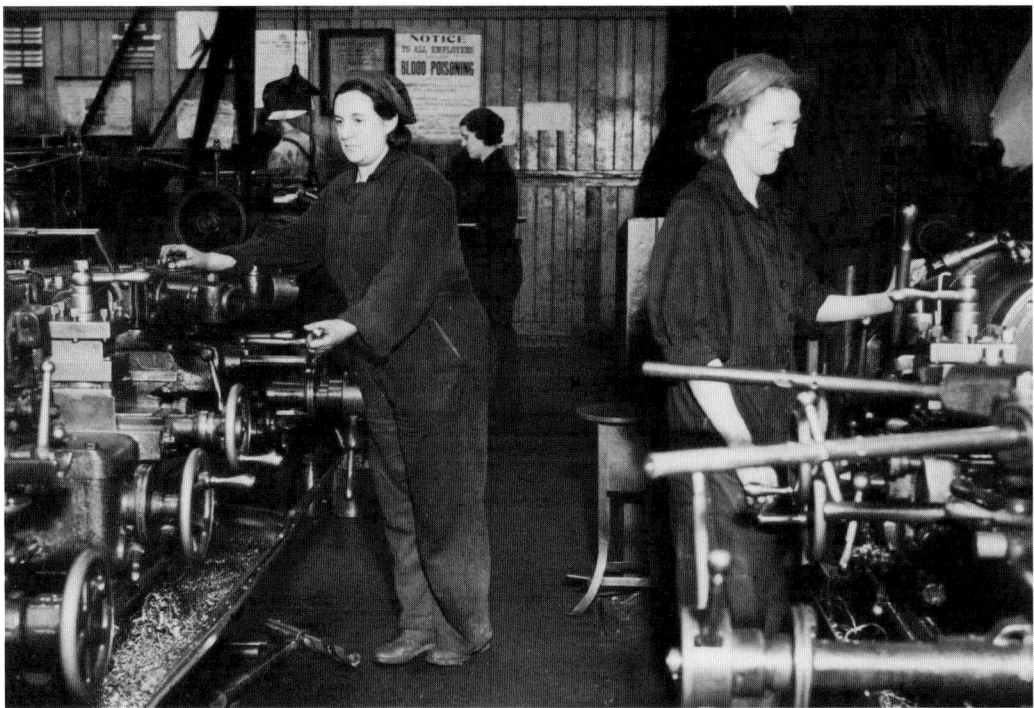

In both wars Hunslet was in the forefront when it came to employing women as machinists and fitters. Here is a typical World War II example of women operating turret combination lathes. Note the explanatory notice in the background regarding the dangers of blood poisoning. No gloves, however; no guards, no masks; just care and an instinct for survival.

YORKSHIRE POST

Women were employed for the most precise of machining tasks, as demonstrated by this example of a gearbox shaft being finished on a cylindrical grinder.

YORKSHIRE POST

Probably posed as a propaganda item but interesting nevertheless. A woman striker operates one of the medium-sized Massey pneumatic hammers in the blacksmiths' shop some time during World War II.

The original 23 h.p. flameproof locomotive, HE 2008, is seen here with its man-riding train at Rossington colliery shortly after delivery. The reduction in length as compared with the 50 h.p. unit will be noted. In the event a greater number of these first-generation Hunslet mining locomotives went to naval armaments depots than to coal mines, the Navy taking nine out of the sixteen 23/25 h.p. and twenty-seven out of a total of forty-five 50 h.p. machines. This example is now in the Leeds Industrial Museum.

Although differing considerably in size, the first two designs of Hunslet flameproof mining locomotives had similar arrangement layouts for the equipment, and both were widely used by collieries and naval armaments depots. The smaller 23/25 h.p. unit used the two-cylinder Gardner 2L2 engine, while the four-cylinder 4L2 powered the 50 h.p. version. One of the latter (HE 2397/41, for Comrie colliery in Fife) is seen here with the engine casing doors removed. In turn, from the cab forward, can be seen the diesel engine, clutch, compressor compartment, exhaust gas conditioner, fan compartment and radiator. These were the only Hunslet designs to have the gearbox at the leading end, a position long favoured by competitors Hudswell Clarke across the road.

meant that the whole of the cylinder head design, including the cylinder head joint, injector joints, valve stems, valve covers, manifold joints and indeed all joints throughout the inlet and exhaust systems had to be of a type and robustness that would stop the most penetrating type of flame.

Stainless steel was used throughout for the exhaust gas conditioner, exhaust pipes, flanges, etc., with phosphor bronze for flame trap (flame arrester) bodies. After experimenting with metal cuttings (swarf) from machine tools in the early flame arresters, the standard became stacks of two-inch-wide grille plates accurately positioned 20 thousandths of an inch apart in the phosphor-bronze bodies. These not only constituted a heat sink, in the manner of a Davy lamp, but also enforced laminar flow and so minimized any tendency to ignite. Sparks that might have formed were extinguished by impingement on the sides of the grille plates.

By 1948 the basic principles had been established, but an ongoing programme of research and development went on over the next forty years to cater for ever-increasing safety requirements and demands for greater efficiency and higher power in smaller spaces. As the years went by the engineering complexity was to become extreme.

Although Rustons obtained the first flameproof locomotive certificate issued by the Mines Department, Hunslet just managed to be the first to have a locomotive in service. This was a nominally 4½ ton coupling-rod drive four-wheeler with a Gardner 2L2 engine developing 23 h.p. at 1200 r.p.m., maker's number 2008, which commenced man-riding duties at Rossington colliery near Doncaster towards the end of July 1939.

This 23 h.p. locomotive had been designed as one of a pair of designs with a larger 50 h.p. 8½ ton 50 h.p. machine with the four-cylinder Gardner engine, which obtained its certificate in June 1941. The 50 h.p. engine ran at 1300 r.p.m., and later 23 h.p. engines also were speeded up accordingly and redesignated 25 h.p. Four-speed gearboxes were fitted, giving top speeds between 12.25 and 14.8 m.p.h. according to the particular customer's requirements.

The Director of Navy Contracts did in fact beat the colliery owners by taking a number of both types for various Royal Naval Armaments Depots throughout the country, and in Singapore – the merits of flameproof locomotives in ammunition tunnels were clear – from April 1939 and throughout the war, and most of these survived

for well over forty years and long after their contemporary mining brethren had been superseded.

With the demand on engineering resources becoming more and more military-biased, development of the mining locomotive was naturally seriously handicapped during the early war years; but new interest was shown during 1943 and 1944, and with the publication of the Reid Report in March 1945, activity increased very considerably. In 1946, Hudswell Clarke built their first flameproof mining locomotives, this being the first 100 h.p. and the first six-wheeler to receive Buxton certification. With the nationalization of the coal mines in January 1947 there was, for a time, a slight lull as far as orders were concerned; deliveries however continued to increase, covering the backlog of those orders placed by the private colliery companies prior to nationalization. One of the first decisions the National Coal Board made was that diesels were to be used to an even greater extent, and bulk orders were placed on the industry by NCB headquarters in London covering something like 250 locomotives. In addition the Ministry of Supply took over progressing of the work, production was stepped up, and it was generally understood by manufacturers that they were not expected to export these particular types of locomotive until the NCB's needs had been fully met.

Under these conditions output increased until, by mid-1945, production had reached eight locomotives a month, half of them coming from Hunslet alone. By the end of 1948 the NCB had 191 flameproof underground diesels, of which Hunslet had provided 121. The Hunslet locomotives were of seven different types, these being the 23/25 h.p. and 50 h.p. four-wheelers already mentioned, a 50 h.p. two-speed coal hauler and then the integrated second-generation range of 65 h.p., 70 h.p. and 100 h.p. general-purpose units. A 24 h.p. 'pit pony' recently introduced completed the range at that time. The position in 1948 is summarized in tables 7.1 and 7.2.

While the development of the mining locomotive during wartime was a significant activity, it was only a fraction of the Hunslet Engine Works' war effort. Locomotives were in fact a very low-priority item; the whole of the former Manning Wardle erecting shop was given over to the mass production of 8-inch howitzers – a short gun for firing shells at high angles and low velocities – hence the sobriquet 'the Gun Shop' that was used universally from then on to describe Manning Wardle's former shop.

Nevertheless, locomotives *were* produced, and in very

The first 50 h.p. mining locomotives were initially for hauling man-riding trains, but generally found themselves involved in heavy coal haulage. The weight of the locomotives varied between 6 ³/₄ and 8¹/₂ tons (usually dictated by the capacity of the mine cage winding gear), and the four-speed gearbox gave maximum speeds of either 12.25 or 14.8 m.p.h. In 1945 Comrie colliery took delivery of Hunslet no. 3200, the first 50 h.p. 'coal hauler', weighing 8 tons 14 cwt. This had a new design of frame comprising a one-piece high-grade iron casting with the gearbox bottom section incorporated within the wheelbase to provide a centre jackshaft and coupling-rod drive. The Gardner 4L2 diesel engine was retained, and the gearbox gave two speeds of 4 and 8 m.p.h. Twenty-eight were built, the last being no. 3520 in 1947. By this time the 4L2 engine was out of production. The design was developed into the 65 h.p. general-purpose locomotive by increasing the frame weight and substituting the newer Gardner 4LW engine. A total of 136 of the 10 ton 65 h.p. model were produced between 1947 and 1968.

100 h.p. HE 3476/47 makes an interesting comparison in size with the 14 x 20 in. 0–6–0 saddle tank, Peckett no. 1310 of 1914, during a demonstration on the surface at Blackhall colliery, Co. Durham, in April 1947.

This close-up of the Gardner 4LW engine in a 93/102 h.p. chain-drive locomotive, HE 2071/40, shows the simple, no-nonsense design with everything get-at-able. It was built to 2 ft. 6 in. track gauge for Umfolozi Consolidated Sugar Planters Ltd in Natal (telephone number Matubatuba Two as late as 1981, for those with that kind of humour).

African Manganese Co. Nos. 18 and 19 (HE 2257/8 of 1941) looked larger than their older sisters, mainly on account of a larger fuel tank which raised the bonnet top. They were of the same power and weight, though, as the original 1935 unit.

large numbers, under a carefully-controlled government regime. Between the declaration of war in September 1939 and the end of 1948, over 1100 diesels and almost two hundred steam locomotives left the Jack Lane works. Granted, the diesels included some 870 narrow-gauge machines in the $3^1/_2$ to 5 ton range, most of them with the Ailsa Craig two-cylinder engine, some with 25 h.p. and 30 h.p. McLaren power units, built in large batches against block government contracts; but there were a number of other designs and one or two surprises. Effectively, civilian contracts were forbidden in wartime, yet a few escaped this restriction.

The biggest of the surprises – in retrospect, although there must have been good reason – was no. 2016, the first British-built rigid-frame eight-coupled diesel, a 252/277 h.p. 34 ton 18 cwt 0–8–0 with a 1000 r.p.m. Paxman 6R2 engine ordered by Crown Agents on 8 November 1938 and despatched to Trinidad Railways on 12 April 1940. Other overseas examples included the first 93/102 h.p. chain-

drive estates locomotive for Umfolozi Co-operative sugar planters in Mtubatuba, Natal, on 17 July 1940 and a number of Perkins Leopard-engined Hudson Hunslet 50 h.p. chain-drive machines for plantation use, mainly sugar, in Antigua, Burma, Egypt, India, Tanganyika and even China. A home market order that was posted overseas was a 3 ft. 0 in. gauge 70/77 h.p. 0–6–0 ordered on 6 January 1941 by the Penmaenmawr Welsh Granite company and diverted on completion on what would have been Bonfire Night in 1941, if bonfires had not been banned, to join other Hunslet locomotives at the Anglo-Iranian Oil Company's Abadan refinery. Two 0–4–0 standard-gauge locomotives of the same power, nos. 2069 and 2070, did however manage to reach Penmaenmawr in October 1940, while a 3 ft. 0 in. gauge new four-wheeler went to Penmaenmawr in October 1944 as a substitute for the requisitioned unit.

The emphasis was on locomotives for the armed forces and their support industries. Mining-type locomotives for

The small 70/77 h.p. standard-gauge flameproof locomotive design boasted seven examples, all for Royal Naval Armaments Depots. Nos. 2389–92 went to Trecwn in 1941, and nos. 2641–2 to Crombie and Broughton Moor in 1942. They were based on Fox and Heather, nos. 2069 and 2070, supplied to Penmaenmawr in 1940.

the ammunition tunnels at the naval armaments depots have already been mentioned, and these were augmented on the surface by a number of 0–4–0 diesels at the depots and dockyards. The Navy purchased two 22 ton 80/88 h.p.locomotives with Paxman 6RQT engines for the Portsmouth Dockyard. These were nos. 2062 and 2076, which were delivered in June and July 1940. Later examples of the 22 ton design had the Gardner 4L3 102 h.p. or 6L3 153 h.p. engine. Six 70/77 h.p. 13½ ton 0–4–0s with Gardner 6L2 engines, nos. 2389–92 and 2641–2, were built for various naval armament depots: four for the depot at Trecwn, one for Broughton Moor and one for Crombie. These locomotives were mechanically the same as the standard-gauge Penmaenmawr pair but with sparkproof electrical equipment for use in hazardous areas. The War Office took several of the 40/44 h.p. Fowler Sanders-engined nominal 10 tonners for the ordnance factories throughout the country. Seven Gardner 6L3 engined 153 h.p. 22 tonners (nos. 2065–8, 2078 and 2121–2) went to the larger depots at Branston, Donnington and Wrexham. The Navy ordered two batches of similar locomotives but with 102 h.p. 4L3 engines and full flameproof equipment for Portsmouth. Four of these, nos. 3130–3, were delivered in August and September 1944, but the second batch, five locomotives this time, missed the war and arrived on duty between October and November 1946.

Some of the standard 40/44 h.p. diesels found their way to civilian owners, owing no doubt to the strategic importance of concerns such as the South Durham Steel and Iron Co., Cargo Fleet Ironworks, Baker Bessemer, Firth Brown, etc., while much correspondence passed between Hunslet, the customers and the Ministry of Supply before Sir Robert McAlpine and Sons Ltd and the Mersey Docks and Harbour Board each received a 26-ton 204 h.p. shunter in 1944, on which we shall comment more fully later.

Three export steam locomotive orders comprising thirteen locomotives were in hand at the outbreak of war. All had been placed at very much the same time. A further standard-gauge 2–6–2 tank, no. 2055, was despatched on 12 October to the St Madeleine sugar estate in Trinidad, while six more of the large and elegant 'M' class 4–6–0 tender locomotives (nos. 2049–54) began their hazardous sea journey to India between 17 November 1939 and 2 March 1940. These had been ordered by Robert White and Partners, consulting engineers to the Jaipur State

Railways. Six 14 x 20 in. 0–6–0 side tanks for the Sudan Government Railways (nos. 2056–61) were somewhat delayed, but they too took their chance with the German navy between 11 October 1940 and 8 December 1941. All these, it will be noted, were repeat orders.

Materials for items other than weaponry were becoming difficult if not impossible to obtain, and Hunslet's practice of authorizing stock building of standard designs of steam locomotives was to pay dividends in wartime. Not only did it mean that a number of locomotives were either already well advanced in production, or at least that the material was already available, but it relieved pressure on the design and drawing offices, whence a high proportion of the staff had already been drafted into the armed forces. Anything that could be put together quickly without affecting either the armaments production, or the diesel locomotives on order by the various military establishments, was built up and sold. Even the customers were carefully controlled by the Ministry of Supply, and, predictably, were those who had an important ancillary role in war production.

As can be seen from the tables of standard locomotives in chapter 5, the Mersey Docks and Harbour Board took one 14 in. and four 15 in. standard 0–6–0s. The rest, a mixture of 15, 16 and 18 in. locomotives, were distributed around steelworks, ironstone quarries, collieries and power stations, while one 14 in. machine went to Brookes quarry at Halifax.

Four 0–4–0 saddle tanks of Kerr Stuart's 'Moss Bay' design were supplied for use at various Royal Ordnance Factories in September and October 1941. This was a strange choice considering that most other ordnance factories were being supplied with diesels and that one recipient factory, Thorpe Arch, north-east of Leeds, was involved with hazardous materials. All four were initially destined for Thorpe Arch, but presumably wiser counsels prevailed, and all but the first went to less dangerous locations.

On 6 January 1941 the company received an order from Stewarts and Lloyds Ltd for eight locomotives intended for the Islip Orefield Development, a government scheme. At £4,895 0s. 0d. each these locomotives were 18 x 26 in. 0–6–0 saddle tanks based on the earlier 48150 class designs for Guest Keen & Baldwins and Richard Thomas. They became known as the 50550 class, and, like the Richard Thomas examples, had a full-depth buffer beam and straight frames also of full depth for the full length. Designed to a very high specification, they had cleaning

The Hunslet shell lathe was produced in large numbers in wartime. Some survived into the fifties.

The former Manning Wardle erecting shop justifies its later name of the 'Gun Shop' at the height of eight-inch howitzer production in 1941. Gun parts abound, as do turrets for shell lathes and a mixture of Ailsa Craig and Hunslet Ricardo diesel engines in the front right-hand corner.

manholes of large diameter under the boiler barrel and adjustable gib-and-cotter big-end bearings on the coupling-rods. The 4 ft. 0½ in. diameter driving wheels were of cast steel and the full-length saddle tank extended from the cab to the front of the smokebox.

The Islip scheme was cancelled at the last moment. Of the eight locomotives, nos. 2411–8, only no. 2411 went to Stewarts and Lloyds, being directed to their Corby quarries. The remaining seven were reallocated by the Ministry of Supply. Three of them (nos. 2413, 2417 and 2418) were purchased by the Stanton Ironworks Company Ltd for their Buckminster, Glendon and Harlaxton quarries; Parkgate Iron and Steel Co. Ltd took no. 2415 for its Charwelton Quarry; and nos. 2412, 2414 and 2416 became War Department Nos. 65–7. All eight locomotives

were completed between 27 November 1941 and 11 May 1942.

On 21 April 1942, eight weeks after no. 2415, *Hellidon*, had been despatched to Charwelton, the Hunslet Service Manager, Albert Hancock called at the quarry to investigate a reported minor water gauge leak. In the event, the driver had solved the problem, but Albert reported: 'All appear to be well satisfied with the loco. but it is very evident that it is much too big for the job. The track is most unsuitable, being very irregular and indeed poor.' No. 2415 soon moved to the nearby Oxfordshire Ironstone Company, where is was more suited to that company's long main-line duties. The three War Department locomotives went to the Royal Engineers depot at Long Marston on the Great Western Railway's Stratford-on-Avon to Cheltenham line.

The eight '50550' class 18 x 26 in. engines HE2411–18 were ordered for Stewarts and Lloyds' Islip orefield development but diverted to various locations owing to the abandonment of the scheme just after the first locomotive had been delivered to Corby. A de luxe version of the Guest, Keen & Baldwin 18 in. 0–6–0, they had brass tubes, copper firebox, boiler manhole, cast-steel wheels, gib-and-cotter coupling-rod ends and an extended saddle tank to give greater water capacity. All these features except the larger tanks were discarded for economy on the Austerity locomotives that followed.

As with the diesel locomotives there were one or two 'quirks' for which there must have been good strategic reasons. There was no. 2384, a solitary 6 x 9 in. 0–4–0 saddle tank locomotive, supplied to the National Smelting Corporation at Avonmouth, on 30 June 1941, two batches of 0–4–2 pannier tanks to the Tata Iron and Steel Co. in India (nos. 3140–3 in late 1944, nos. 3278–80 in June 1945), and another 0–4–2, no. 2647, to Chakas Kraal in March 1943.

These nine locomotives were all 2 ft. 0 in. gauge machines and were of the Kerr Stuart 'Wren' (0–4–0ST) and 'Tamar' (0–4–2PT) classes. These locomotives, like the 'Moss Bay' class, were extremely simple and cheap to put together, the drawings were easy to issue, and there may have been parts left over from Stoke-on-Trent.

A batch of six large 5 ft. 6 in. gauge 0–6–2 side and back tank locomotives (nos. 2378–85), ordered at the outbreak

of war by consultants Rendell, Palmer & Tritton for the Calcutta Ports Commissioners, were delayed somewhat and appeared between March 1945 and January 1946. A 12 x 20 in. 0–4–0 saddle tank (no. 3289), of Avonside Engine Co. design, was built for Pilsley colliery in June 1945.

But the locomotive destined to engrave the name 'Hunslet' on the minds of railway operators and enthusiasts more than any other design was the Austerity 0–6–0 saddle tank.

With massive air strikes on strategic railway locations on both sides of the Channel, and with the planned Allied invasion of Europe, there was clearly a need for large numbers of main-line freight locomotives and heavy-duty shunters from 1942 onwards.

The LMS class 3F 0–6–0 side tank was the initial choice for fulfilling the Ministry of Supply shunter requirement, but after much discussion, enthusiastically prosecuted by

Edgar Alcock, it was decided that an industrial locomotive, with its shorter wheelbase and simple, straightforward design, would be a more easily maintained and more versatile machine. The Company was now solely in the hands of the Alcock family, Edgar having become Chairman and (with John Alcock) joint Managing Director after the death of Alexander Campbell in March 1941. The design of the Austerity saddle tank can be summarized as an amalgam of the standard 48150 class Hunslet 18 x 26 in. cylinder steelworks shunter, first built in 1937 for Guest Keen and Baldwins' East Moors works at Cardiff, and the later 50550 class locomotives of similar power built to the order of Stewarts and Lloyds in 1941. Both designs were a natural progression from the 0–6–0 side tank, no. 1506, that had been sold in 1930 to the Pontop and Jarrow colliery railway in Durham. The boiler of the Pontop and Jarrow locomotive was of similar proportions

to those of the LMS class 3F mentioned above, of which Hunslet had built ninety between 1924 and 1930.

In essence the Austerity took the frame of the Pontop and Jarrow locomotive, the boiler of the Guest Keen and Baldwins machine and the extended saddle tank of the Stewarts and Lloyds variant. A larger coal bunker was fitted and the cab roof had rounded eaves to provide more universal loading gauge clearance. This 'trimming' of the cab was to some extent necessitated by the increase in wheel diameter from 4 ft. $0\frac{1}{2}$ in. on the previous designs to 4 ft. 3 in. on the Austerity to provide greater clearance above rail level and to allow the use of easily replaceable under-hung springs.

Other modifications were dictated by wartime material difficulties and the need to cut down man-hours and total production time. Steel castings were virtually eliminated and replaced either by iron castings or by welded steel fabrications. Wheel centres were of cast iron, and steel

The original Austerity, Hunslet no. 2849 of 1942, as it emerged from the Works on 1 January 1943. Apart from the later removal of the Furness lubricators, all the locomotives built before 1949 were identical except for the handrail pillars, as explained in the text. From 1950, new locomotives had double buffer-beam gussets, and older locomotives returning to Hunslet for overhaul were usually retrofitted with the second gusset. Otherwise, the only observable variations throughout the 485 locomotives as built were thicker buffer beams, gravity sanding and repositioned injectors on the handful of steelworks examples, and twelve-spoke cast-steel wheel centres on some of those supplied new to steelworks and NCB Durham area.

boiler tubes replaced brass. Cast-iron coupling rod bushes were tried at first, but they were an economy too far and were replaced by gunmetal, phosphor bronze or brass as available.

The first 'Austerity' left the works on 1 January 1943, less than six months after the initial order for fifty locomotives had been placed in July 1942. Despite this achievement it became apparent that the quantities required by the Ministry of Supply were greater than the rate of four or five a month that Hunslet could supply. Robert Stephenson and Hawthorns, Hudswell Clarke, Bagnall and, later, Vulcan Foundry and Andrew Barclay were supplied with sets of drawings; and of the 377 Ministry locomotives supplied between 1943 and 1946 Hunslet built 120, RSH ninety, Bagnall fifty-two, Hudswell Clarke fifty, the Vulcan Foundry fifty and Andrew Barclay fifteen. Five more ordered from Hunslet were cancelled owing to the cessation of hostilities, and there was some duplication of allocated makers' numbers as orders were transferred from one maker to another to facilitate production. Hunslet acted as parent concern over the whole contract and ordered a good deal of the material in bulk.

On completion the locomotives were drafted to ordnance depots, factories, docks, etc., and after the Normandy invasion many were sent to the Continent and to North Africa. Several were taken into the stocks of the Netherlands and Tunisian railways, and further examples found permanent work in Dutch steelworks and coal mines and with French industry and light railways. A number were sent new to large UK outcrop (open-cast) coal sites, and in 1947 the newly-formed National Coal Board purchased over forty that had been lying idle in France and Belgium since the cessation of hostilities. During the war years the allocation of materials for engineering production was very closely controlled, and the likelihood of a private concern's obtaining a new locomotive was very remote indeed. Two additional Austerities did however slip through the net of officialdom, and these went direct to Rothervale and Manchester Collieries.

The correspondence of the time affords a remarkable insight into the triangular relationship between supplier, customer and officialdom, and the trust sometimes placed on gentlemen's agreements, this last commodity so lacking in the enlightened later years of the twentieth century.

A typical example of this was when, on 11 July 1945, Guest Keen Baldwins asked Hunslet whether they might buy two Austerities. At that time there were eighty-eight brand-new Austerities stored on Ministry of Defence property, and Hunslet had parts for at least five more that had been cancelled subsequent to the end of the war in Europe two months earlier.

Patriotic to a fault, Edgar Alcock replied for Hunslet that 'perhaps it would be as well if you sent us an order for the two [locomotives] and we think in the country's interests [!] we should endeavour to get two of these [88 stored] engines transferred to you rather than build them.'

This solution was agreed. However, although the Ministry had not paid the manufacturer, it required payment in full before letting them leave storage on the Longmoor Military Railway. A bizarre exchange of correspondence ensued, including the following gems.

21 August 1945. Hunslet to Guest Keen Baldwins:

Further to ours of the 13th we are now pleased to confirm that we are a stage further with the Ministry and that there is a prospect of getting hold of these locomotives in the very near future.

We enclose a copy of a letter which we have received from Contracts department [Ministry of Supply] this morning and we have already applied for permission to get the locomotives off site and despatched to you. It may be that we cannot manage this until cash has been paid in accordance with their instructions. We think you will agree with us in thinking that the Ministry are somewhat premature in asking for payment, particularly as they have not paid us for the locomotives which were delivered to them some weeks ago.

We will do our best to get them to fall into line with more normal business practice. If on the other hand you are prepared to let us have a cheque for the amount in question, we will arrange to pass it on to them, but in any case you may rely upon us doing the best we can to get on with the job and we will write you again as soon as we have anything more to report.

23 August 1945. Guest Keen Baldwins to Hunslet:

We are obliged for your letter of the 21st instant, reference JFA/VW, ext. 22, and are pleased to note that you have reached a further stage with the Ministry, and that there is prospect of getting hold of these locos in the near future. . . .

It is rather amazing that they are asking you to pay for these locos when you have not received payment for them.

However, we are desperately anxious to have these locos and now enclose our cheque for £10,600. which we shall be glad if you will pass it on to the Ministry of Supply, and we hope then that we can really get the matter moving.

In true gentlemanly fashion the deal was done, and on 1 September 1945 Hunslet advised Guest Keen Baldwins in the following terms:

We are pleased to inform you that we have received from the Ministry this morning a permit for us to take over the two locomotives at Longmoor.

We are therefore arranging to send one of our leading fitters down to Longmoor on Monday to make the necessary arrangements with the Railway Company and the Commanding Officer at Longmoor. We think he will probably be able to make arrangements to leave there with the two locomotives on Tuesday or thereabouts and that you may expect the engines turning up at your Cardiff works sometime during the week.

We have asked our man particularly to make sure that the engines are complete with tools exactly as we supplied with the locomotives.

'Our man' was Sam Childs, and on 14 September Hunslet 'wired' him ten pounds (telegraphic transfer in

---

HUNSLET ENGINE Company Ltd.

Great Austerity Pantomime
"Cinders in the Cellar"
by FRED VERITY

JANUARY 14th, 1944

Director	- - J. F. ALCOCK	Ballet	- Tracing Office Trio
Producer	- - E. ALCOCK		DORA HORSFALL
Script	- - D. PULLAN		MAVIS WILKINSON
Scenario	- - H. E. DEAN		MARY ROSE
Costumes	- C. BROOK	Jokes	- JAMES SMITH
(Cheap and serviceable)		Specialities	A. R. SCHOLES & CO.
			(Bonus, Income Tax, etc.)

Musical Director - - J. M. KNOWLES (without baton)
Production Manager - - G. S. GUEST
New Talent discovered by A. W. COOPER

PROGRAMME 3d. each (or more)
Sold in aid of Soldiers' and Sports Fund

There was no lack of humour even in the darkest days of wartime, as these Works Pantomime programmes show.

THE HUNSLET ENGINE Co. Ltd.
JACK LANE, HUNSLET - - - LEEDS 10

By Authority of
THE SHOP STEWARDS
(Conductor : W. MIDDLETON)

A SUPERB SHOW
"Babel in the Works"

FEBRUARY 12th, 1943

NEW EDITION - 55 HOURS WEEKLY
SUNDAYS by arrangement

The Show produced in no time at all, with no help from anyone, by the Millwrights

Delays specially arranged by the Management in consultation with the Production Committee

This Programme is sold in aid of a good cause
THE SOLDIERS' FUND   Price 3d. or more

---

today's parlance) to pay his expenses whilst in Cardiff and to pay for his railway ticket home having successfully commissioned locomotives nos. 3219 and 3220.

After the war, 106 further Austerity locomotives were built, seventy-seven for the National Coal Board, fifteen for the steel industry and fourteen more for the Army. With the exception of two from Robert Stephenson and Hawthorns Ltd and eight from the Yorkshire Engine Company these all came from Hunslet. They will be discussed more fully in chapter 9.

It is worth returning to discuss in detail the two 204 h.p. diesel shunters briefly mentioned earlier in this chapter as being supplied to Sir Robert McAlpine & Sons Ltd and the Mersey Docks and Harbour Board in 1944. These were, after all, to set the standard for many such locomotives produced more or less continuously over the next 18 years.

Design-wise they were an amalgam of the frame and running gear of the three larger LMS shunters of 1934, with 3 ft. 4 in. driving wheels on a 9 ft. 0 in. wheelbase, with an improved four-speed gearbox and powered by the Gardner engine as first used in the three Corsham locomotives of 1937 and 1940.

Both customers had to wait patiently for their locomotives. The Mersey Docks requirement was triggered by one of the Luftwaffe bombing raids on Liverpool in the spring of 1941, which destroyed a 14¹/₂ x 18 in. Barclay fireless steam locomotive that had been working at the

R. E. (Dick) Ketley standing alongside the first 16 h.p. mining locomotive. Fitted with a two-cylinder Lister diesel engine, the design was later produced in flameproof form in small numbers for British coal mines. These flameproof units used, first, the 24 h.p. Hunslet Ricardo engine and, later, the 22 h.p. Ailsa Craig. The driver of locomotive no. 2420/43 is Hughie Wilkinson. Dick Ketley went on to form Hunslet Africa Pty (later Hunslet Taylor Consolidated) in Johannesburg and thus create a new industry in South Africa. Over two thousand Hunslet locomotives were subsequently built in South Africa. It was Hughie who took the 500 h.p. 0–8–0 on its trials in north-west Yorkshire and eventually to Peru. He later became Service Manager and finally Training Officer until his retirement in December 1988.

Dingle Oil Installation at the south end of the dock complex. In response to the Board's enquiry of 12 August Hunslet submitted a quotation on 20 August, apologizing for the delay in replying (!) and offering the alternatives of the 70/77 h.p. and 153 h.p. 0–4–0 diesel locomotives as already built for the various Government departments. The customer's Engineer in Chief was a cautious man, and, while accepting the theory of increased efficiency and higher tractive effort obtainable from a diesel locomotive, he submitted that 'it is strongly recommended that we must be libéral in our assumptions as to the power required, much better to be well above than slightly under power'. His Superintendent of Outside Machinery added his two penn'orth by retailing his experience with a Fowler 150 h.p. locomotive that 'was in the repair shops more time than in service owing to transmission troubles, particularly clutch'. Albert Hancock, back from La Paz and not unworldly with regard to the practical aspects of customer relations, got the message and persuaded John Alcock to offer the six-wheeled 204 h.p. machine even though the drawing office, depleted by the demands of military service, would be hard pressed to meet the demand. Much time was spent, to no effect, between November 1941 and April 1942 on a misguided attempt to hire one of the four LMS diesels as an interim measure.

Meanwhile McAlpines had issued an enquiry on 9 December 1941 for a diesel locomotive of 80/100 h.p. 'for short distance shunting purposes' for an unspecified destination. By February 1942 this had developed into a requirement for a six-wheeled locomotive, preferably with Gardner engine, for the Royal Naval Propellant Factory at Caerwent in Monmouthshire for which McAlpines were the construction contractors. Again John Alcock hedged his bets, on the one hand quoting for the 204 h.p. machine but adding on the other that:

Generally speaking, we are in a very awkward position for carrying out new design work at the present moment. Any such new design would have to proceed very slowly, with the result that deliveries would probably be from 18 months up to 2 years and naturally we are therefore anxious for you to accept locomotives which have previously been built and for which all drawings are in existence.

It would be churlish to suggest that his anxiety was due in part to discussions that may have been going on with the Ministry of Supply regarding the Army's steam locomotive requirement that resulted in the order for Austerity saddle tank locomotives in July 1942 and which would show a better return on design office costs.

The first 'real' Hunslet 204 h.p. Gardner-engined shunter, no. 2699 (McAlpine no. 2697 was the second) at Princes Half Tide Dock as Mersey Docks & Harbour Board No. 32 shortly after delivery in April 1944.

3 ft. 0 in. gauge 80/86 h.p. HE 3129/44 Vixen at Penmaenmawr on 14 June 1968.

INDUSTRIAL LOCOMOTIVE SOCIETY/BRIAN WEBB COLLECTION

Fitters put the finishing touches to a pair of 93/102 h.p. Gardner 4LW-engined flameproof surface shunting locomotives in October 1946. Ten of this design were built, nos. 3130–33 in 1944, no. 3282 and nos. 3393–7 in 1946. No. 3282 went originally to the Royal Naval Armament Depot in Milford Haven, the others to RNAD Bedenham, near Gosport. The steam locomotive in the background is HE 3386/46, No. 34 Sir Gordon, a standard-gauge 16 x 22 in. 4–6–4 tank for the British Guiana Railway. The boiler for no. 3387/46, No. 35 Sir John, is in front. Just in the picture, bottom left, is one of the first Hunslet 100 h.p. flameproof underground mining locomotives.

In the event, McAlpines booked their order first on 3 April 1942 at a price of £5350 carriage paid home. The MD&HB order followed on 7 May 1942. MD&HB had no objection to electric starting for the diesel engine, but McAlpines were wary and opted for compressed-air starting fed by a hand-started Ruston & Hornsby petrol-engined compressor set. Otherwise the locomotives were identical.

Then came the problems of official authority to build. For some reason the McAlpine locomotive, ostensibly for a military contract, was considered non-essential; the Board of trade refused authorization on 4 May 1942, and the matter was still outstanding at the end of the year. Clearance to buy the steel was obtained from the Board of

Trade on 18 February 1943 and the diesel engine was ordered on 13 April 1943.

MD&HB fared better. They obtained Ministry of War Transport authorization for steel on 28 May 1942. Nevertheless the Austerity steam locomotives took priority and the assembly of the diesels was slow. On 31 march 1944, nearly three years after the air raid, 21-year-old Hughie Wilkinson took HE 2699/44 to Liverpool. The lack of formality in transportation by rail in those days is epitomized by the request dated 30 March 1944 from the Hunslet Engine Company to the London, Midland and Scottish Railway Goods Agent at Hunslet Lane Goods Station, which read as follows:

Tomorrow we shall consign a six wheels coupled Diesel Locomotive to Messrs. Mersey Dock & Harbour Board, Princes Loco Shed, Princes Dock, Liverpool, dead on its own wheels, and we would thank you to arrange a gauger to visit us to gauge it. Our man H. Wilkinson will be travelling with the locomotive to destination and we shall be obliged if you will advise us the service you are able to fix up.

If the Locomotive is required to travel on any other Railway Co's system we leave it to you to arrange the necessary gauging.

Indemnity form will be sent down to you by hand first thing in the morning. The weight of the locomotive is about 25 Tons.

This confirms telephone conversation with Mr. Crump today.

How different it would be in the hyper-safety-conscious 'new railway' of the late 1990s.

The McAlpine locomotive, HE 2697/44, had fallen behind no. 2699, and when it was offered to them for inspection early in June McAlpines replied to the effect that no inspection was possible because 'Mr Kenneth McAlpine is now in the Air Force'. Furthermore they requested that the locomotive be sent to 'McAlpine Sidings, Newton Tony, Amesbury Branch' for 'our Boscombe Down contract', to which it duly went on 17 June 1944 by LMS Railway from Hunslet Lane and, presumably, Great Western and Southern, suitably gauged,

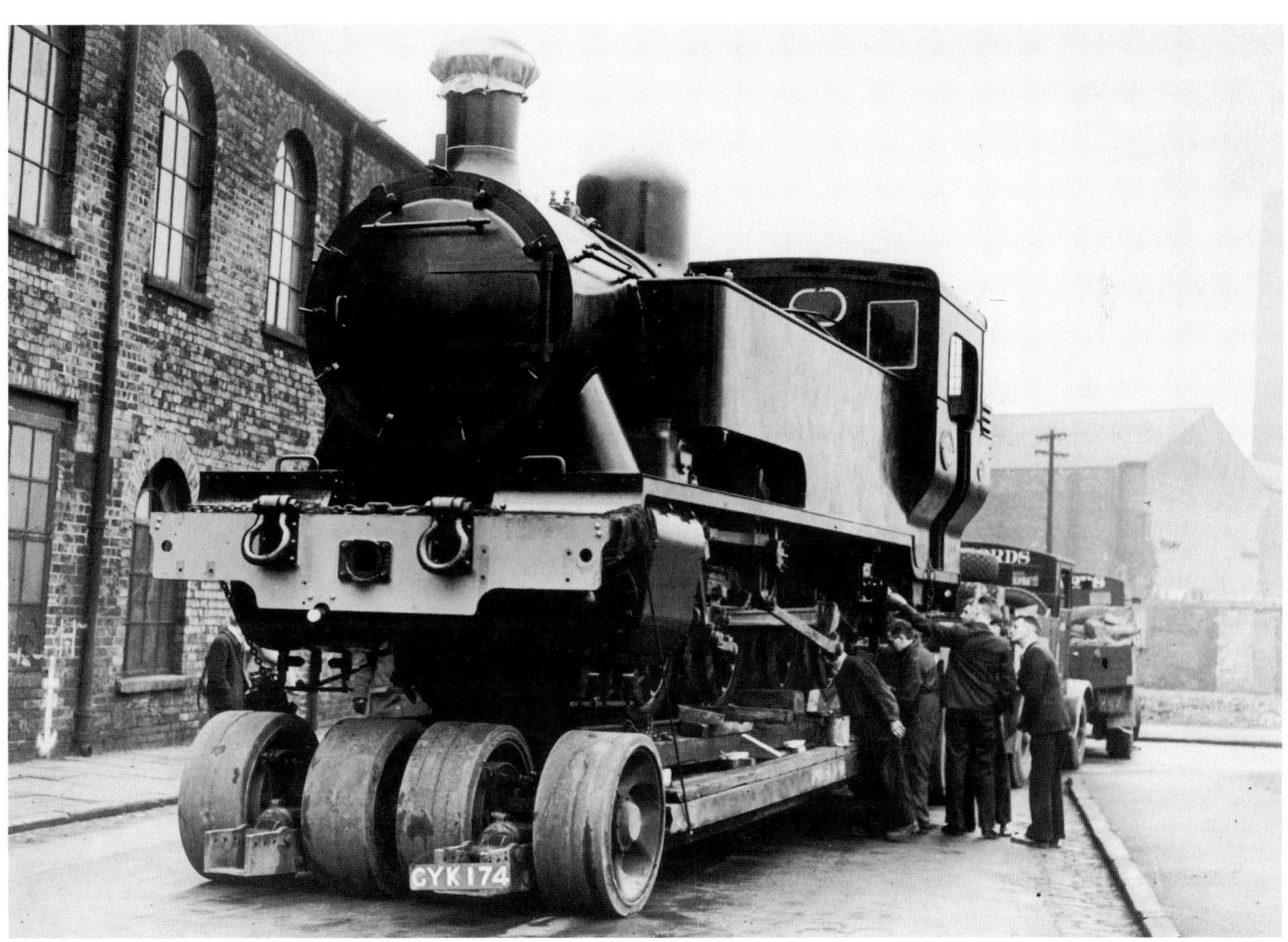

Nigerian Railways No. 74, Hunslet No. 3345/46, loaded ready for transfer by road to Birkenhead West Float and thence to Lagos on 8 January 1946. Two of Pickford's typical Scammell tractors of the period provide the motive power. Fifty-seven of this standard 'African' design were produced at intervals between 1930 and 1954: twenty-five for Nigerian Railways, twenty-seven for Gold Coast Railways, two each for the Rand Water Board and Pretoria Portland Cement, and one for Northern Rhodesia Copper Mines. The railway company examples were shipped complete as shown through Lagos, Port Harcourt and Takoradi; but at least one of the industrial machines, no. 3383 for the Rand Water Board, was fully dismantled for shipment via Durban on 25 February 1947 and re-erected on arrival. They were large machines: 18 x 23 in. cylinders, 48 tons 8 cwt in working order, and although of only 3 ft. 6 in. track gauge they were 12 ft. 4 in. high and 9 ft. 5 3/4 in. wide. Six further examples for Nigerian Railways appeared in 1955 but with six-wheeled tenders, giving a total weight of 84 tons 12 cwt. YORKSHIRE POST

Edgar Alcock, second from left, and John Alcock, right, proudly demonstrate a 65 h.p. flameproof mining locomotive to three well-heeled Cuban government ministers during a works visit on 3 February 1948.

to reach the vast military hinterland on the Hampshire–Wiltshire border to the east of Salisbury Plain.

Victory in Europe Day, 8 May 1945, heralded some semblance of return to normality, but there was still the conflict with Japan, which was to continue for another three months, and it was some time before workers returned from the armed forces. The Company had been at the forefront of the movement to employ women on the shop floor, in all categories of skilled, semi-skilled and unskilled jobs, and this was to continue, at least for a while. Large numbers of the 20 and 30 h.p. Hudson Hunslet locomotives on order for the War Office were cancelled and the parts put in stock, some to emerge later as complete locomotives and some to be sold as spares. To add to the difficulty in keeping an accurate check on exactly how many of these small locomotives were built, many of them were returned to the works unused in their packing cases. These were either checked over and resold as and when required, in which case they retained their original numbers, or they were classed as remanufactured

and given new numbers. This process went on for at least four years.

The last new 20 h.p. Hudson Hunslet locomotive was no. 4304, one of two supplied to the Belgian Congo on 31 July 1951. These two units were from a batch of five of which the first two, nos. 4300/1, were the prototypes of the 'next generation' of the marque referred to as the '21 h.p.' The power was in fact the same, the change in designation being made purely for identification in the way that was common when the same diesel engine was used in two or more classes of Hunslet locomotive. The real difference was that the 21 h.p. had a seated position for the driver, was of lower overall height, and had sloping bonnet sides with removable engine compartment doors. In time, the 21 h.p. became 24 and then 29 h.p., finally living on to the end as 43 h.p. but with a single-speed gearbox (all the rest had had two-speed gearboxes) and with the choice of Ford or Perkins diesel engine when the Ailsa Craig went out of production.

The height of mining locomotive production. Delegates from the Harrogate Mining Convention pose with a selection of Hunslet 100 h.p., 70 h.p., 65 h.p. and 24 h.p. flameproof mining locomotives in June 1948.

The four 2 ft. 6 in. gauge 132/146 h.p. locomotives nos. 4156–9 supplied to the Kelani Valley Railway in Ceylon during 1950 were virtually the same as the 1941 African Manganese examples. Shorn of its Hunslet name-plates, one of the quartet arrives at Avissawella with the 13.00 hours Padukka–Colombo 'express' on 22 February 1984.

D. W. WINKWORTH

Table 7.1

Complete list of official approval numbers for first generation mining locomotives issued by the Ministry of Fuel and Power after type tests for flameproofing at the Buxton Testing Station.

Approval number	Date	Manufacturer	Type and h.p.
1	25.3.39	Ruston & Hornsby	48 and 40
2	21.7.39	Hunslet Engine Company	25
By letter (added to Approval No. 1)	29.3.41	Ruston & Hornsby	30 and 20
3	13.6.41	Hunslet Engine Company	50
4	7.7.45	Hunslet Engine Company	50 mark II
5	10.5.46	Ruston & Hornsby	48 and 40 Modification
6	20.8.46	Hudswell Clarke	100
7	28.10.46	Hunslet Engine Company	100
8	28.10.46	Hunslet Engine Company	102 (4 ft 8^1/$_2$ in gauge)
9	15.1.47	Hudswell Clarke	68
10	20.6.47	Hunslet Engine Company	65
11	23.1.48	Hunslet Engine Company	24
12	5.7.47	Hunslet Engine Company	70
13	7.7.48	North British	100
14	22.10.48	Ruston & Hornsby	100

Table 7.2

Flameproof diesel locomotives in British coal mines at 31 December 1948.
They are shown in the order in which they were first supplied, with other Approved Locomotives being built at that time but not in service.

Maker	*B.h.p.*	*Number delivered to NCB*	*First supplied to*			*Gauge options*		*Engine*		*Wheel arrgt.*	*Wheel (tons)*	*Max speed (m.p.h.)*	*Tractive effort in bottom gear (lb.)*	*Final drive*	*Wheelbase*	*Length over buffer beams*	*Overall height*	*Overall width for narrowest gauge*
			Pit	*Date*	*Gauge*	*Min.*	*Max.*	*Maker*	*Type*									
Ruston	48	31	Comrie	6.7.39	2 ft. 8 in.	1 ft. 11 in.	3 ft. 6 in.	Ruston	4VRHL	0–4–0	7	9.5	3430	Chains	4 ft. 3 in.	14 ft. 6 in.	5 ft. 4¼ in.	4 ft. 2 ½ in.
Hunslet	25	7	Rossington	19.7.39	2 ft. 0 in.	1 ft. 11½ in.	3 ft. 0 in.	Gardner	2L2	0–4–0	4.5	12.5 or 14.3	2450	Rods	3 ft. 0 in.	11 ft. 6 in.	5 ft. 0 in.	3 ft. 4 in.
Hunslet*	50	21	Comrie	16.6.41	2 ft. 8 in.	1 ft. 11½ in.	3 ft. 6 in.	Gardner	4L2	0–4–0	7	12.2 or 14.8	5300	Rods	4 ft. 0 in.	13 ft. 10 in.	5 ft. 0 in.	3 ft. 11 in.
Ruston	20	5	Pennyvenie No. 4	17.3.45	2 ft. 0 in.	1 ft. 6 in.	3 ft. 6 in.	Ruston	2VSHL	0–4–0	3.5	8.75	1750	Chains	2 ft. 7½ in.	9 ft. 10½ in.	5 ft. 0¾ in.	3 ft. 5¼ in.
Hunslet*	50	25	Comrie	25.6.45	2 ft. 8 in.	1 ft. 11½ in.	3 ft. 6 in.	Gardner	4L2	0–4–0	9	8	3980	Rods	4 ft. 3 in.	12 ft. 8 in.	5 ft. 3 in.	3 ft. 10 in.
Hunslet	100	26	Lofthouse	19.10.46	3 ft. 0 in.	2 ft. 6 in.	3 ft. 6 in.	Gardner	6LW	0–6–0	15	15	8000	Rods	5 ft. 3 in.	14 ft. 3 in.	5 ft. 6 in.	4 ft. 6 in.
Hudswell	100	10	Moor Green	21.10.46	3 ft. 6 in.	1 ft. 11½ in.	3 ft. 6 in.	Gardner	6LW	0–6–0	15	14	8300	Rods	5 ft. 6 in.	15 ft. 11 in.	5 ft. 6 in.	4 ft. 0 in.
Ruston	30	7	Lochhead	6.3.47	2 ft. 1 in.	1 ft. 6 in.	3 ft. 6 in.	Ruston	3VSHL	0–4–0	4.5	8.75	2250	Chains	2 ft. 7½ in.	10 ft. 0½ in.	5 ft. 0¾ in.	3 ft. 5¼ in.
Hudswell	68	17	Fryston	21.3.47	1 ft. 11½ in.	1 ft. 11 in.	3 ft. 6 in.	Gardner	4LW	0–4–0	10	14	5250	Rods	3 ft. 9 in.	14 ft. 5 in.	5 ft. 3 in.	3 ft. 4 in.
Hunslet	65	37	Beoch	18.9.47	2 ft. 10 in.	1 ft. 11½ in.	3 ft. 6 in.	Gardner	4LW	0–4–0	10	5–9	4500	Rods	4 ft. 3 in.	12 ft. 10 in.	5 ft. 3 in.	3 ft. 10 in.
Hunslet	24	1	Valleyfield	2.7.48	2 ft. 0 in.	1 ft. 6 in.	2 ft. 6 in.	Hunslet	2HRW	0–4–0	2.8	6.25	1225	Chains	2 ft. 0½ in.	8 ft. 0 in.	4 ft. 0 in.	3 ft. 3 in.
Hunslet	70	4	Frickley	23.7.48	2 ft. 1½ in.	1 ft. 11½ in.	3 ft. 6 in.	Meadows	4DT420	0–4–0	10	5–15	5250	Rods	4 ft. 3 in.	12 ft. 9 in.	5 ft. 3 in.	3 ft. 10 in.
Ruston	40	—	—	—		1 ft. 6 in.	3 ft. 6 in.	Ruston	3VRHL	0–4–0	6	10.25	2580	Chains	3 ft. 4¾ in.	11 ft. 11½ in.	5 ft. 4¼ in.	4 ft. 0¼ in.
North British	100	—	—	—		2 ft. 6 in., 3 ft. 0 in., 3 ft. 6 in.		Paxman Crossley	6RQE 5BWL	0–4–0	15	15	8400	Rods	4 ft. 7 in.	15 ft. 1 in.	4 ft. 5 in.	4 ft. 0 in.
Ruston	100	—	—	—		2 ft. 6 in., 3 ft. 0 in., 3 ft. 6 in.		Ruston	6VRHL	0–6–0	15	12.5	7850	Rods	5 ft. 10 in.	18 ft. 7⅛ in.	5 ft. 3 in.	4 ft. 0 in.

* Superseded by 1948 owing to the L2 type of Gardner engine going out of production.
Comrie, Valleyfield, Beoch, Pennyvenie and Lochhead pits were in NCB Scottish Division.
Fryston, Frickley, Lofthouse and Rossington pits were in NCB Yorkshire Division.
Moor Green pit was in NCB Nottinghamshire Division.

Table 7.3

Summary of Austerity Locomotives Built

Purchaser	Original number	Maker	Maker's number	Year built	Remarks
Ministry of Supply	75000–49 75050–79 75080–99 75100–49 75150–79 75180–99 71437–56 71462–66 71467–76 71477–86 71487–506 71507–26 71527–69 75250–71 75222–81 75282–231	Hunslet RSH HC Hunslet WB RSH Hunslet Andrew Barclay HC RSH HC RSH Andrew Barclay WB RSH Vulcan Foundry	2849–98 7086–115 1737–41, 1745–62 3150–99 2738–67 7130–49 3201–20 2211–15 1785–94 7286–95 1763–72, 1774–83 7161–80 2181–90 2773–94 7202–11 5272–321	1943/4 1943 1943/4 1944 1944/5 1944 1944/5 1946/7 1945/6 1945 1944/5 1944 1944–6 1945/6 1945 1945	377 locomotives, all virtually identical and to a basic wartime design. Single buffer-beam gussets.
United Steel Companies	—	Hunslet	3134	1944	Two wartime civilian locomotives, virtually identical to the 377 Ministry of Supply engines.
Manchester Collieries	—	Hunslet	3302	1945	
National Coal Board	*Various*	Hunslet	3685–9	1948/9	As nos. 3134, 3302
GKB	3	Hunslet	3691	1949	As earlier locomotives except for 3 in. thick buffer beams and with injectors moved to firebox backplate.
National Coal Board	*Various*	Hunslet	3692–701	1950	As nos. 3685–9 except double buffer-beam gussets fitted to nos. 3700–1.
GKB	14 and 24	Hunslet	3717–8	1950	As no. 3691 but with mechanical instead of steam sanding.
National Coal Board	*Various*	Hunslet	3767–72	1951/2	As nos. 3700–1.
National Coal Board	*Various*	Hunslet	3776–81	1952	As nos. 3700–1.
National Coal Board	*Various*	Hunslet	3784–89	1953	As nos. 3700–1 except that nos. 3784–5 had 12-spoke cast-steel wheels instead of 14-spoke cast iron.
Ministry of Supply	190–203	Hunslet	3790–803	1953	Fourteen post-war military engines virtually identical to the wartime order but with double gussets and vacuum pipes.
National Coal Board	*Various*	RSH	7751–2	1953	As standard wartime units.
National Coal Board	*Various*	Hunslet	3806–11	1953/4	As nos. 3700–1.
United Steel Companies	*Various*	Yorkshire Engine Co.	2566–73	1954	Virtually identical to nos. 3700–1 but with 12-spoke cast-steel wheels as nos. 3784–5.
Stock batch for NCB except nos. 3848 (GKB) and 3850 (Stewarts & Lloyds)	*Various*	Hunslet	3816–51	1954–62	Stock batch identical to nos. 3700–1 except that no. 3848 was modified as nos. 3717–8 and no. 3851 was built with underfeed stoker (modified chimney cowl and protector plate below rear buffer beam).
National Coal Board	65–6	Hunslet	3889–90	1964	Final engines, identical to no. 3851.

Except for the smallest details all Austerities looked much the same when new. The most obvious differences (e.g. the cut-down 'Lambton' cab) were customer additions after delivery. Only after 1961 did the modifications, including stokers, extra steps, ladders, extended bunkers, etc., begin to appear (LNER Class J94 modifications excepted). The Army did, however, ring the changes with air and vacuum brakes, and one locomotive (*Brussels*) was provided with and used to test a number of special fitments.

The wartime locomotives were, in a manner of speaking, kit-built, with pooling of resources between manufacturers; therefore the components on any one locomotive could have come from one works or several. An illustration of this is evident in the handrail pillars on the saddle tanks. Some builders, notably RSH and Barclay, used simple round bar, while others employed shaped 'knobs'. Generally, the style of pillar would match the final build's style, but cases did arise in which firms helped another out, giving rise to anomalies.

HUNSLET

CHAPTER 8

Introduction to part two (In living memory)

The author must have been all of six years old when a glimpse of a polished silver fox atop a name-plate with the odd legend *The Percy*, and embellishing a smart green engine numbered 288, first acquainted him with the activity of train spotting. It was in the closing months before the outbreak of World War II and the locomotive was, of course, one of Nigel Gresley's 'Hunt' class 4–4–0s engaged at that time on semi-fast passenger workings on the Leeds to Wetherby line.

So began an inescapable fascination with railways, both full-size and miniature, that was to dominate my life, and that of my subsequent family, in a way that I could not possibly have foreseen. Nor could it have been expected that I should subsequently be taking part in the ending of an era that had seemed as firmly established and indestructible as the Pyramids themselves. Living as I was close by the former North Eastern line out of Leeds, my schooldays were dominated by the constant stream of LNER and NER classes, with the occasional visit to the west end of Leeds City station to see the LMS 'foreigners' working in from the former LNWR, LYR and Midland lines and, later still, ex-Great Northern and Great Central types at the inappropriately-named Leeds 'Central', now defunct.

Although I did not realize it at the time, what really set the seal on my ultimate destiny was a chance stroll one summer Sunday evening at the end of World War II through Templenewsam Park, four miles east of Leeds, and past the stately Tudor mansion where Henry, Lord Darnley, second husband of Mary Queen of Scots was born in 1545. Over the rise to the south, and just out of sight of the house, was a cluster of huts marking the start of operations of one of the first open-cast coal sites. The contractor was Sir Lindsay Parkinson, and alongside the huts was a siding containing a row of around thirty of the most fascinating little contractors' locomotives you could imagine. This was a new breed of locomotive to me. All were saddle-tanks with either four or six wheels; and, with one exception – a Black, Hawthorn named *Hardy*, all proclaimed 'Leeds' on their makers' plates. There were Manning Wardles, Hudswells and Hunslets in profusion,

some of them ancient, with only the most exiguous of weather-boards to offer the driver protection, but others were more modern, with cleaner lines and fully enclosed cabs. Most of them had names, *Risley*, *Jennie*, *Anzac* and *Aussie* to mention but four, while others carried the more prosaic legend 'SLP No…'. To my regret, I was not able to record them photographically, and they were dispersed shortly afterwards. For me, this chance find sparked off an increasing number of forays to local collieries and other establishments with private sidings, and by the end of my schooldays I had a pretty wide layman's knowledge of industrial railway systems.

When I began my full-time connection with the Jack Lane works in 1949, the approach was usually made from one of the two tram stops roughly a quarter of a mile away. Private cars were a rarity; there would be two or three directors' cars, and perhaps a service vehicle or two, but very little else. A large proportion of the work force lived within earshot of the works hooter and quite a few of us living further afield would walk from the city centre, having already made one tram or train journey into the city and not wishing to wait, or pay, for a connecting service, preferred to walk briskly up Leathley Road past Hunslet Lane goods yard.

The 1882 office block (actually proclaiming '1864' above its doorway) fronted, as it still does, on to Jack Lane – hence the address, '125 Jack Lane' – and one entered the works through a lean-to passageway at the right-hand end wherein was the time office window with the timekeeper maintaining an ever-watchful presence. To the right of this passageway was the main double gate with the rail connection leading across Jack Lane to pass Hudswell Clarkes and drop down the incline to connect with the ex-Midland Railway main line. The tiny works diesel shunter, with its cab open to the elements, would ply incessantly across the road, down the incline, back up the diverging left-hand track to cross Jack Lane again and into what was then known as the back yard, and would return sometimes with a locomotive for repair that had been put out of the

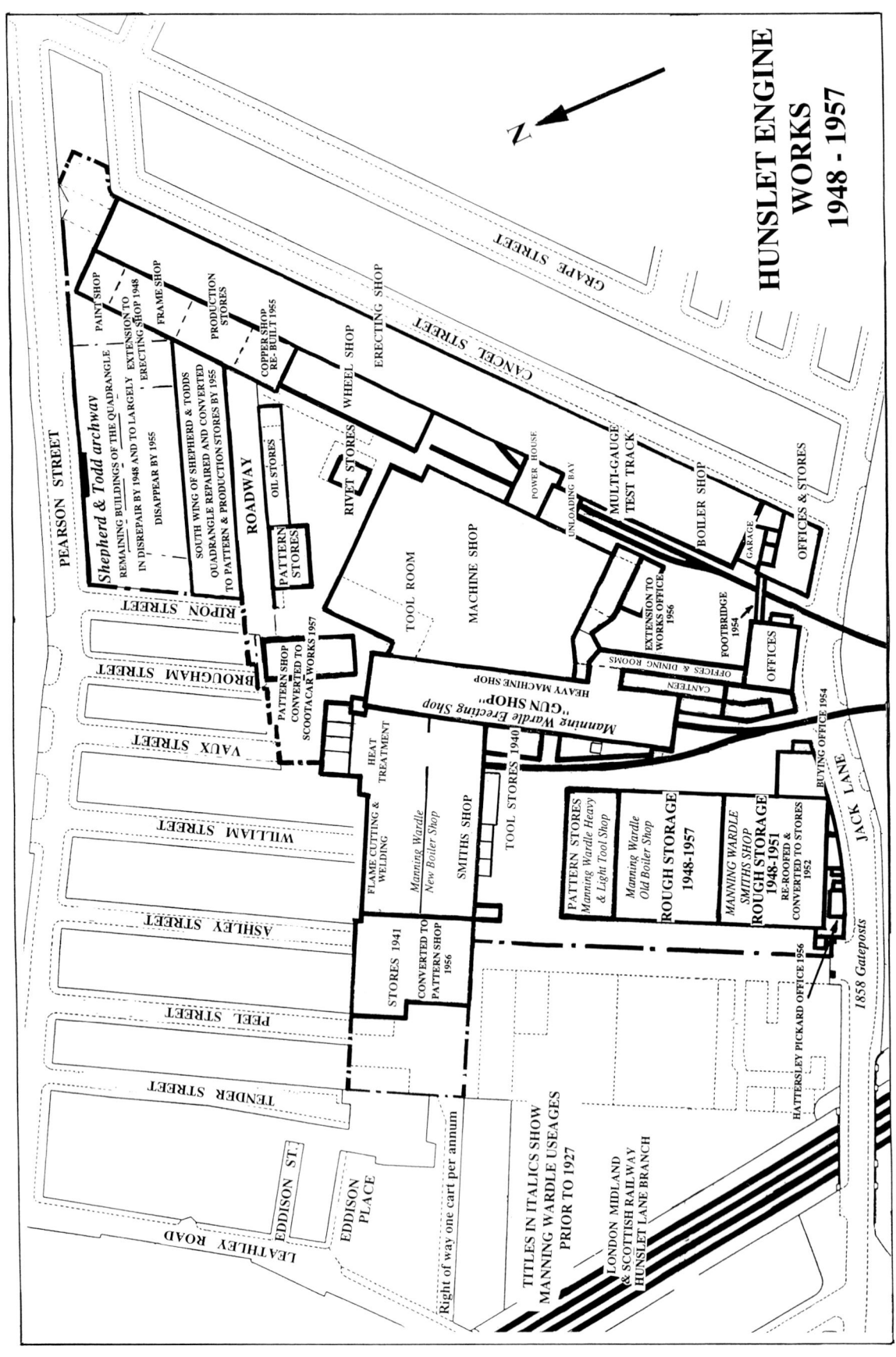

HUNSLET ENGINE WORKS 1948 - 1957

N

PEARSON STREET

Shepherd & Todd archway

REMAINING BUILDINGS OF THE QUADRANGLE
IN DISREPAIR BY 1948 AND TO LARGELY
DISAPPEAR BY 1955

SOUTH WING OF SHEPHERD & TODDS
QUADRANGLE REPAIRED AND CONVERTED
TO PATTERN & PRODUCTION STORES BY 1955

PAINT SHOP

EXTENSION TO
ERECTING SHOP 1948

FRAME SHOP

PRODUCTION
STORES

COPPER SHOP
RE- BUILT 1955

WHEEL SHOP

ERECTING SHOP

RIVET STORES

GRAPE STREET

CANCEL STREET

ROADWAY

OIL STORES

PATTERN
STORES

RIPON STREET

BROUGHAM STREET

PATTERN SHOP
CONVERTED TO
SCOOTACAR WORKS 1957

VAUX STREET

TOOL ROOM

MACHINE SHOP

POWER HOUSE

UNLOADING BAY

MULTI-GAUGE
TEST TRACK

BOILER SHOP

GARAGE

OFFICES & STORES

EXTENSION TO
WORKS OFFICE
1956

FOOTBRIDGE
1954

OFFICES

HEAVY MACHINE SHOP

Manning Wardle Erecting Shop
"GUN SHOP"

OFFICES & DINING ROOMS

CANTEEN

OFFICES

HEAT
TREATMENT

WILLIAM STREET

FLAME CUTTING &
WELDING

*Manning Wardle
New Boiler Shop*

SMITHS SHOP

TOOL STORES 1940

ASHLEY STREET

STORES 1941

CONVERTED TO
PATTERN SHOP
1956

PEEL STREET

PATTERN STORES
*Manning Wardle Heavy
& Light Tool Shop*

*Manning Wardle
Old Boiler Shop*

ROUGH STORAGE
1948-1957

*MANNING WARDLE
SMITHS SHOP*
ROUGH STORAGE
1948-1951
RE-ROOFED &
CONVERTED TO STORES
1952

TOOL STORES

TENDER STREET

Right of way one cart per annum

TITLES IN ITALICS SHOW
MANNING WARDLE USEAGES
PRIOR TO 1927

LEATHLEY ROAD

EDDISON ST.

EDDISON
PLACE

LONDON MIDLAND
& SCOTTISH RAILWAY
HUNSLET LANE BRANCH

1858 Gateposts

HATTERSLEY PICKARD OFFICE 1956

BUYING OFFICE 1954

JACK LANE

way for some reason, or with a main-line wagon loaded with steel plates or similar. Activity was everywhere.

Standing at the entrance, looking inwards, there was to the immediate right of the gate a large low store converted from the pre-1882 offices, and then the boiler shop and

erecting shop extending away into the distance for the full length of the site.

Immediately ahead and to the left of the railway track was the front yard, a portion of it still displaying the soft black sand floor of the now-demolished original smiths'

The Hunslet Engine Works as the author first made its acquaintance in 1949. The stages in development of the boiler and erecting shops alongside Cancel Street can be clearly seen, with the 1947 extension at extreme right contrasting with the blackened saw-tooth roof of the 1898 erecting shop next to it. Then comes the pitched roof of the 1920 connection with the boiler shop, the change of height in the brickwork plinth showing where the 1902 boiler shop proper began.

The 1882 office block is clearly seen, fronting on to Jack Lane at middle left. At right-angles to this can be seen the wartime observation tower on the roof of the gearbox drawing office and the rows of skylights above the otherwise windowless first-floor diesel drawing office. Behind these is the former Manning Wardle erecting shop (the 'Gun Shop'). Connecting the Gun Shop and the erecting shop are the various machine shop bays, the highest of which was the original Hunslet erecting shop. At far left is the recently re-roofed Manning Wardle boiler shop. The original Shepherd & Todd Railway Foundry is windowless but still largely intact in the centre of the picture, while beyond lie McLarens' (the former Kitson works) and Fowlers' Steam Plough Works, with Hunslet Lane goods yard just in the picture top left. Hudswell Clarke's is in the lower left-hand corner.

With the exception of the office block and the British Queen public house (just entering the picture at centre right), all this has now gone.

In 1949 the only way from any other part of the works to collect supplies was through the blacksmiths' shop. This is a scene typical of what would greet the young apprentice on his errand for the obligatory 'long stand' – and on his way back ten minutes later when the storekeeper decided he had stood long enough. Teamwork, dexterity and a good eye were prerequisites in producing this final drive gearwheel blank from a billet of nickel-chrome-molybdenum steel.

Cutting one's teeth in the locomotive business – literally. The combination of skilled machinist Fred Town and this Pfauter gear-hobbing machine provided the author with his first day's experience of gainful employment almost fifty years ago.

After the teeth were cut on the Pfauter or other appropriate machine and all burrs removed, the gears were degreased in a bath of trichloroethylene before being cyanide-hardened and ground. Fred Town is shown alongside the Orcutt gear-grinding machine. Fred and colleague George Walker, plus one apprentice, operated a bank of eight or ten specialist gear-cutting, grinding, splining and broaching machines in accordance with production demands, setting up each one in turn as the others went through their operating cycles.

A dapper George Cowell checks the weight on each wheel of a 100 h.p. mines locomotive. Moving on a few years later from being Hunslet's erecting shop foremen to the Consett Iron Company's locomotive engineer, George was still taking an active interest in locomotives at Tanfield fifty years later.

Albert Hancock (Service Manager), John Dickson (Inspection Superintendent) and Albert Burns (Works Manager) examine a 65 h.p. 10 ton mines locomotive destined for the newly nationalised National Coal Board. A St Rollox-trained Scot, John Dickson was to retire from the position of Works Director in December 1979 having earlier completed a stint as chairman of the Railway Division of the Institution of Mechanical Engineers.

shop. This portion was used for the receipt of castings from the foundry, and they were kept for a certain amount of weathering and destressing before going inside for machining. In more leisurely times these could sit around for years. To the left in the corner of the yard were bicycle sheds and the whole of the left-hand side was bordered by a two-storey building with the directors' dining room and the buying department on the ground floor, the gearbox drawing office, print room and photographic dark room on the first floor. Straight ahead, and forming the fourth side of the roughly square yard, was a relatively new brick building with ambulance room and garage on the ground level, works planning office and Works Manager's office above. A large sliding door at the top left-hand corner of the yard led direct into the large five-bay general machine shop, or, if one turned sharp left on entering, into the heavy machine shop (the 'Gun Shop'), formerly Manning Wardle's erecting shop. At the bottom left of the Gun Shop further doors led off to the smiths' shop and the heat treatment, shot-blasting, flame-cutting and welding shops. The main stores were at the far end of the smiths' shop. The pattern shop was reached either from a door in the bottom right-hand corner of the Gun Shop or by the bottom exit from the machine shop.

Returning to the main yard, a passageway roughly thirty feet wide separated the machine shop from the combined boiler and erecting shops.

Down the passageway continued the railway track, diverging into two tracks as it passed the corner of the garage immediately by the Works Manager's overlooking office window. The left-hand track turned off the main track by means of a sharp left-hand turnout followed by an equally sharp right-hand curve. The main track, once clear of the turnout, became multi-gauge for test purposes, and, after passing the full length of the boiler shop, turned to enter the erecting shop at an angle of roughly thirty degrees, terminating over an inspection pit with a movable rail section. The multi-gauge test track had been extended as part of (and funded by) the Ministry of Defence contracts in 1942, and could accommodate almost all track gauges from 18 inches to 5 ft. 6 in. Oddly enough, the Irish 5 ft. 3 in. gauge was not included, but in the event the only subsequent Hunslet locomotives for either Eire or Northern Ireland were the three 1350 h.p. diesel-electrics fitted out at Doncaster in 1971.

If one were to walk over to the right-hand side of the yard and enter the boiler shop from the top, and then walk the full length to the bottom of the erecting shop (say 1000 feet in all), the first port of call would be the plater's, where cabs, tanks and diesel engine casings (bonnets) were flame-cut, rolled, flanged and assembled on surface plates with jig-drilled locations in them to ensure reasonable interchangeability. Next would come the boiler plate rolls, tube-plate and boiler backhead flanging presses and furnace, boiler assembly, riveting, boiler mounting and testing. The latter would take place on three stands just before the test track entered from the left.

The erecting shop proper commenced at the far side of the test track and inspection pit. There was room for three parallel rows of locomotives under assembly, the left-hand row (looking almost due north down the shop) being exclusively for steam locomotives and the right-hand row for diesels. The centre row was a bit of a no-man's-land, and could be used for either as the work varied or for the odd locomotive for repair. The mainstream repair work was usually carried out at the far end of the shop or using the last bay in the steam row.

At the very far end (the Pearson Street end), the 1945 extension was some six feet lower at floor level than the main shop, and was reached by a gently-sloping ramp at the left-hand side. There was space here for, at a pinch, a further eight locomotives, and to the left a further extension was used for frame sub-assemblies and by the service engineers. The extension roof was at the same height as the main shop, and since the overhead crane girders ran the full length of the whole building there was much greater headroom here for lifting locomotives over the top of others, and for loading on to road vehicles for transit to the docks. These exited into Pearson Street, thereby avoiding the sharp turn into the narrow and, by 1949 standards, busy Jack Lane. At this time any new or repaired standard-gauge locomotive for a British customer, and any rail-worthy incoming repair job, would travel by rail. This custom continued with few exceptions into the 1980s, when British Rail's rates and restrictions militated against rail transportation. Increasing highway maintenance costs, and the complaints of bus passengers, who were being severely jolted by the two 'level' crossings, combined with BR's lack of interest in special traffic, resulted in the rail connection's being severed in 1984.

Leaving this new extension by a door at the centre left, one was immediately confronted by the remains of the original Shepherd and Todd Railway Foundry. At this time (1949), and clearly shown on a contemporary aerial view,

These extensions to the Hunslet erecting shop were added in 1947 and are seen shortly afterwards with the left-hand bay not yet occupied by the fitters who formed the frame assembly gang. The larger locomotives are three of a batch of ten 204 h.p. diesel shunters (Hunslet nos. 3559–68) for Algerian State Railways, while on the left is a 100 h.p. 0–6–0 flameproof mining locomotive. In the foreground are two 70 h.p. and one 65 h.p. mining locomotives, and the frames of two further Algerian shunters can just be discerned at the far left corner of the bay. Six Gardner diesel engines (five 100 h.p. six-cylinder and one 65 h.p. four-cylinder) are in the centre of the picture. To the left of the cab windows of the nearest shunter, two War Department 20 h.p. Hudson Hunslet locomotives can just be seen.

A photograph taken on the same day as the previous one but looking in the opposite direction up the erecting shop towards the boiler shop. The difference in levels between the old 1898 erecting shop and the 1947 extension can be clearly seen. Beyond the Algerian Railways 204 h.p. diesel shunter, (centre left), are further mines locomotives and a Hudson Hunslet 50 h.p. unit. The steam locomotive ready to go out on the test track is HE 3384/48, an oil burning 2-8-0 tender engine for the metre gauge Guaqui – La Paz Railway, which left the works on 15 October 1948. Just inside the doorway top right can be seen a 100 h.p. mines locomotive on test whilst the 'Austerity' saddle tank in the centre is most probably the first of the post-war N.C.B. examples, HE 3685/48, ordered for Silverwood colliery but delivered to Manvers Main.

ANDREW NEALE COLLECTION

The patriarch in his kingdom. John Alcock at his desk in 1948. Joint Managing Director with his father, Edgar, at that time, he was to assume full control on the death of Edgar Alcock in 1951.

Authority that brooked no argument. H. E. Dean (Chief Draughtsman) and Arthur Betteridge (Chief Diesel Locomotive Draughtsman) examine the arrangement drawing for the 65/70 h.p. mining locomotive clutch assembly, some time in 1948.

Hunslet no. 1643/29, originally Haifa Harbour Works Dept. No. 3, returned to the works in 1938 and was rebuilt for military service at W.D. Bramley, Hants as Bramley No. 4. Returned again to Hunslet by 1947 it was hired first to North Gawber colliery and then to the Ministry of Fuel and Power at Darton, both locations near Barnsley. Sold to the Peruvian Corporation and seen here awaiting repair and modification prior to despatch to the Southern Railway of Peru in 1952.
AUTHOR

The diesel drawing office at Christmas 1958, still very much unchanged since wartime. Situated on the first floor just to the side of the Gun Shop and later, from the mid-sixties, to be used as a machine shop for the apprentice training school. The author is seated second from right. Norman Deighton, fourth from left, had been Chief Engineer at McLarens' in the pioneering diesel engine days. The stacks of drawers on the left housed some of the sixty thousand-odd drawings at roughly a hundred to the drawer. Behind these was the dark room with Kerr Stuart and Avonside record books, photographs and Hunslet glass plate negatives dating back to locomotive no. 1 in 1865. These last were destroyed in the modernization of the offices in the 1960s.

the buildings were largely intact, but lacking most of the windows and the roof on the Pearson Street and Yarmouth Street sides. Only recently acquired, this old factory was being put to partial use as a production store and for the temporary accommodation of several 20 h.p. Hudson Hunslet diesel locomotives, still in their packing cases, returned as war surplus and awaiting checking over for civilian and overseas buyers.

The directors' and administrative offices of the company were on the ground floor of the main Jack Lane building, with the accounts department (called, in typically down-to-earth style, 'the cost office') to the left in what had been Manning Wardle's 'paint shop'. In those days the term 'paint shop' usually meant little more than a paint storage or paint mixing room: locomotives were painted by hand wherever was convenient. The first floor of the main building housed the Chief Draughtsman's office and the steam locomotive drawing office, from the far left-hand corner of which led a passageway to the tracing office and gearbox drawing office already mentioned. Between the gearbox office and the print room was a door into the diesel drawing office, and thence on to the spares, planning and production offices on the first floor of the right-angle between the Gun Shop and the main machine shop. At that time this warren of first-floor offices was an Aladdin's cave of old and new drawings, record books and glass-plate and film negatives. 'Tis pity that much has been lost the while, and that recollection only returns in the fitfulness of fickle memory.

The level of the equipment in use throughout the factory was a quaint combination of old and new, and was of a far more varied nature than would normally have been expected in a locomotive works. The armaments contracts of World War II had left a legacy of precision machines which, when combined with the steam locomotive facilities (smith's shop, boiler shop, etc.) and gear-cutting machinery meant that not only could every part of a locomotive be made in house if necessary but so could other products in the event of a shortfall on locomotive orders. This ability to diversify and adapt was to be a tremendous advantage over the next almost forty years.

Visual impact was one of constant bustle and productivity, but the general ambience impacted on all the senses. The machine shops emitted a continuous hum punctuated by the squeal of metal being cut and accompanied by a smell of hot oil and cutting fluid. The heavy hammers in the smiths' shop produced a steady *thump, thump* that could be felt as well as heard, and the boiler shop riveters joined in with machine-gun precision at frequent intervals. The erecting shop was generally quieter, relying on the background noise from elsewhere, but with the occasional shout of 'Up a bit,' 'Down a bit' or 'Whoa!' to the overhead crane drivers some twenty feet above. In winter, the atmosphere was eerie, with the tungsten lamps battling against the smoke from several coke braziers dotted around, and an acrid smell lay about, particularly in the early hours.

CHAPTER 9

Steam locomotive production, 1949–71

These were the circumstances that were to greet me when, shortly before 7.30 a.m. on an August morning in 1949, I was standing outside the Hunslet Engine Company's works in Jack Lane waiting for 'Scotty' (otherwise Fred Scott, the timekeeper) to introduce me to the delights, or otherwise, of locomotive-building. In common with several other young hopefuls, I was taken by Scotty to meet the Machine Shop Foreman, Jack Dobson, and we were each placed in the care of a charge-hand on a particular section. For the next twelve months or so we gradually progressed from machine to machine on a variety of production jobs – milling, turning, planing, gear-cutting, grinding, etc. Learning engineering from the sharp end is a method of training that has never been equalled, to my mind, despite the growth of wider educational facilities and much political posturing on initiatives for training, retraining and 'maintaining the skill base'.

In those days, each component progressed around the works accompanied by a job card setting out the route from start to finish. For example, a brake hanger bracket would be routed: Stores (issue material) – Smiths' Shop (forge blank) – Centre Lathes (turn shank) – Drillers (drill mounting holes) – Slot Driller (machine cotter slot) – Fitting Bench (deburr) – Stores (to await assembly). In addition to the time allocated for each operation the card would carry the drawing number, the part number and the order number. If the part was intended for new production, the order number would correspond to the locomotive maker's number and be prefixed by the letters *LS* if for a steam locomotive and *LD* if for a diesel locomotive. Spare parts orders were numbered in the 60,000 series if steam, 90,000 if diesel; while parts for locomotives in for repair carried a 70,000 series number. At this time the maker's serial numbers for new locomotives had just passed the 3600 mark, covering both steam and diesel production, and the decision was taken to initiate a new series for diesels, starting at LD4000. The remaining numbers in the 3600-plus series were left for further steam locomotive production; they continued up to LS3902, which was the final Hunslet steam locomotive, built in 1971.

On a previous visit to the works when still a schoolboy trying to establish a career policy, I had been impressed by the remarkable variety of locomotives on the shop floor at any one time. The erecting shop had space for, say, three or four steam locomotives, a similar number of 'large' diesel shunters (200 h.p. or more), and perhaps five or six small mining locomotives. On this first visit there were a number of 70 h.p. diesels for Borneo and Rhodesia; 204 h.p. diesels for South Africa, Algeria and the Consett Iron Company; two of a batch of twelve 3 ft. 6 in. gauge 18 x 23 in. 0–8–0Ts for Nigerian Railways; a 2 ft. gauge balloon-stacked 2–8–2 for Eastern Province Cement in South Africa (the largest rigid-frame 2 ft. gauge locomotive ever built); and an exquisite 0–4–0 tank in crimson lake for Cadbury's chocolate factory at Bournville.

This array was something close to heaven for a sixteen-year-old already smitten with the locomotive bug, and provided an *entrée* into a branch of locomotive engineering that was not covered at that time in the small number of railway books that were available. By studying the workshop drawings in conjunction with the order numbers on the job cards one could, if sufficiently interested, piece together a locomotive list and gain some idea of the locomotives and customers that featured in the Company order book at any one time – this information not officially being available outside the main admin. office. 'Piece work' was the normal procedure – the more you produced, the more you earned. Substandard work was rejected by a combination of watchful foremen and systematic inspection and deterred by the thought of having to replace the item at a less advantageous price.

The works canteen provided a free 'full English breakfast' for all foremen and above who arrived in time to eat it before 7.30 a.m., and immediately afterwards Edgar Alcock would appear at the top of the machine shop in white linen jacket and trilby hat to see that all was right with his world.

Thus the steam locomotive order book in 1949 came out as shown in table 9.1 (page 181).

Cadbury Bournville No. 9, makers' number 3665 of 1949, a standard 16 x 22 in. side tank beautifully hand-finished in crimson lake with black-and-yellow lining and with lettering in pure gold leaf outlined in vermilion. Varnished to perfection after lettering, the livery slowly mellowed in service to a rich chocolate brown, probably due in no small way to the coke used as fuel. The true visual effect of the paintwork is difficult to describe. It had a subtlety and softness – depth almost – most unlike modern glass-like precision finishes which in consequence never recreate the original appearance.

This represented a total of 102 steam locomotives (eighty-six still to deliver when I started at Hunslet in August of that year), of fifteen different designs and five track gauges, not to mention an equally heavy demand for diesel locomotives. Quite a tall order for a company employing around 500 people: it is not surprising that the major part of the GWR-designed '94XX' pannier tank order for twenty locomotives, including final erection, was subcontracted to the Yorkshire Engine Co. at Sheffield, who were already building locomotives of this type. During the years 1949–56, production from Hunslet varied from 76 to 128 locomotives annually, the split numerically being roughly one-third steam and two-thirds diesel. On a tonnage basis, the division between steam and diesel was more even, since the steam locomotives tended to be uniformly large types while the diesel list contained several small underground units weighing between $3^1/2$ and 10 tons.

My early days in 1949 and 1950 were spent mainly in the two machine shops. The second of these contained the heavier boring and planing machines and was still universally known as the 'Gun Shop' on account of its wartime activities. It will be remembered that until 1928 it had been Manning Wardle's erecting shop.

Parts were delivered to each machine in batches, theoretically according to urgency, and could be variously for steam or diesel locomotives. Take the slot drilling machine, for example: this was a double-spindle, horizontal milling machine for cutting oval cotter slots of either parallel or taper pattern. A normal day could involve cutting slots $1/8$ in. wide by $1/4$ in. long in diesel gearbox selector pins, cotter slots in crossheads, and finally the huge $1^1/4$ in. wide by 3 in. long taper slot in the gib end of a Nigerian 0–8–0T connecting rod.

Lunch breaks gave the opportunity to walk into the erecting shop and witness progress on the various locomotives under construction. There were two distinct gangs of men in this holy of holies. The steam gang, supervised by Jimmy Howcroft and Ken Cross, used the left-hand half of the shop for steam construction, while the diesel gang, led by Bob Riley and Lou Cotterill, operated down the right-hand side. The centre of the shop was used for the occasional rush job and for lagging boilers and weighing complete locomotives. The test track entered at an angle and terminated in an inspection pit. All track gauges from 18 in. to 5 ft. 6 in. were represented, with the exception of 5 ft. 3 in.

When I began my time with Hunslet, the last three Nigerian locomotives of a batch of twelve were virtually complete. Even by British main-line standards these were large shunters. Two outside cylinders, 18 in. diameter by 23 in. stroke, with piston valves, drove eight 3 ft. 6³/₄ in. diameter driving wheels on a 13 ft. 3³/₄ in. rigid wheelbase to give a maximum tractive effort of 23,532 lbf. The weight in working order was 48 tons cwt., the overall height 12 ft. 4 in. and the width 9 ft. 5³/₄ in., and the water capacity was 1000 gallons. Fifty-seven of these locomotives were built

in all, most of them for Nigerian Railways, but with a fair number for Gold Coast Railways (now Ghana), including a batch of seven in 1950 and three in 1954.

Next came the Barsi Light Railway 2 ft. 6 in. gauge 2–8–2 tender locomotives. Falling into the category of Indian Railways Class 'F' and weighing 61 tons 6 cwt. in working order, the design incorporated a larger superheated boiler with a wide Wootton-type firebox, two 15¹/₂ x 18 in. cylinders with piston valves, and Walschaerts valve gear. Unlike the Nigerian locomotives, which were shipped complete, those for Barsi were stripped down to frame, boiler, cab and other sub-assemblies, with all pipes and fittings carefully labelled for reassembly.

The three Barsi engines were completed on 25 October, 14 November and 16 December 1949, and of course as one locomotive left its building stand for the test track, the frame plates of another locomotive took its place. Already, attention had turned to a batch of six shunters for the Calcutta Ports Commissioners, and so slick was the change

Hunslet nos. 3667–9 of Indian Railways 2 ft. 6 in. gauge 2–8–2 'F' class, built in 1949, were the last steam locomotives supplied to the Barsi Light Railway west of Bombay. The Barsi system was engineered by E. R. Calthrop, of Leek and Manifold Valley Light Railway fame. Overnight trains included sleeping coaches, which was unusual on a 2 ft. 6 in. gauge line.

from one batch to the next that the first Calcutta engine left the works on 17 December, with all six being away by 20 March 1950. If the Nigerian locomotives were large, the only word for these dock shunters is 'massive'. At 10 ft. 6 in. wide and 12 ft. 6 in. high, they took full advantage of the Indian Railways' generous loading gauge for 5 ft. 6 in. gauge lines. A large-diameter boiler with a Belpaire firebox, pitched high on a fabricated steel smokebox saddle, was flanked by short, stubby side tanks and surmounted by a copper-capped chimney that would not have looked out of place on a Swindon-built express locomotive. Additional water was carried in a back tank under the coal bunker, and the all-up weight was just 65 tons.

Hard on the heels of the last Calcutta locomotive came a neat little 2 ft. 6 in. gauge 0–6–2T named *Shreebishnu* and carrying the running number 7 of the Nepal Government Railways. This was a very slightly modified update of an Avonside Engine Company design dating back to 1928. Avonside (absorbed by Hunslet in 1935) had built four locomotives of this type before its Bristol works were closed; the first one of these had also gone to Nepal Government Railways, while *Shreebishnu* was the seventh Hunslet example. A further six were to be built before the end of steam.

After running on the works test track, *Shreebishnu* was immaculately painted in glossy black, with red-and-white lining, before being carefully dismantled and packed in over a dozen crates and shipped to the foothills of the Himalayas.

The Nepal locomotive left the works on 28 March 1950, and we got on with seven 0–8–0Ts, works nos. 3677–83, for Gold Coast Railways. The first two were despatched on 24 May 1950; two more followed on 2 August, and the last one on 28 August. As far as I can remember, they were only distinguishable from the Nigerian Railways batch by virtue of the American-style standard type 'E' buckeye coupler that was fitted instead of the bell-mouth, link and pin arrangement that was used in Nigeria.

The home market finally got a look in on 26 July 1950 when the first of a batch of ten standard Austerity 0–6–0Ts was sent out to Old Fold Sidings on the old No. 3 St Helens Area of the National Coal Board. Another two went out to St Helens on 31 July and 12 October, and after a spell of feverish activity the 'famous five' (*Rodney, Respite, Renown, Revenge* and *Repulse*) left for Manchester Collieries' Walkden yard in the five weeks from 19 October to 21 November. Of the final two in the batch, one went to Llay

Main Colliery in North Wales and the other to Barnborough Main at Wath, two weeks before Christmas. To add a little variety, a solitary 16 x 24 in. standard-gauge 0–6–0 with no cab – only a roof supported on four pillars – a Weir feed water heater and Westinghouse pump went away on 28 September to Peru against an order from the Peruvian Corporation for use on their Southern Railway.

By August 1950 I had joined Jimmy Howcroft's steam locomotive erecting gang, and so took a much more active part in the assembly of all the Hunslet steam locomotives over the next two years. In times of extreme urgency, or if there was an imbalance between the steam and diesel order books, the steam gang would occasionally be called on to build a diesel locomotive; and from time to time it would also carry out a major repair on an old locomotive. Curiously enough, this practice never worked in reverse: the diesel and repair gangs never to my recollection assembled a new steam locomotive.

The Austerity 0–6–0T, or 'J94' as some enthusiasts rather loosely describe it, was a dream to assemble. Frame plates and other components were jig-drilled, and a gang of three men and two apprentices could complete a locomotive, from laying the frame plates to steam testing, in just over a fortnight. Being of diminutive stature in those days I always got the job of going in to the saddle tank to fix handrail pillars, caulk studs, etc. If the tank had been painted just previously with Apexia compound, the atmosphere was claustrophobic – the more so if some wag placed a coke brazier under the tank while one was inside. Apexia was a toxic primer containing red lead, red oxide in colour; it was used for the outside of boilers and the inside of tanks.

These coke braziers were the only form of heating in the vast erecting shop at that time, and the 10 a.m. tea break would see several thick chunks of bread impaled on toasting-forks improvised from folding two-foot steel rules. Each section had its own labourer who would go off to the canteen around 9.30 each morning with a dozen or so pint mugs and return with hot, sweet tea and piles of thick slices of bread plastered with beef dripping. My section's own labourer, Jimmy Mattison, was an old Hunslet stalwart who, although officially classed as unskilled, undertook a variety of jobs and probably had a wider practical knowledge of steam locomotives than many of the skilled fitters. He was always around to offer help and advice, and, on the days when a locomotive was on test, he would clock in around 4 a.m. to raise steam and

Sudan Railways No. 44, HE 3744/1951, was one of twenty-five 14 x 20 in. 0–6–0 side tanks supplied to this customer between 1927 and 1952. Of typically standard Hunslet design, they had many mechanical features in common with the outside-cylinder locomotives supplied to the Southampton Docks and Haifa Harbour contracts and to other overseas customers.

generally prepare the locomotive. He would also act as driver and/or fireman when the occasion demanded. A breed of man long extinct.

In 1950 there were also two modified Austerities, with cast-steel wheel centres (these had slimmer spokes than their cast-iron equivalents) and injectors on the firebox backhead instead of under the footplate, for Guest Keen and Baldwin's Cardiff East Moors steel works, and two standard 18 in. steel works locomotives for Richard Thomas and Baldwins at Scunthorpe. These last were similar in appearance to the Austerities, except that they had 4 ft. 0¹/₂ in. instead of 4 ft. 3 in. wheels, a smaller bunker with a sloping back and a saddle tank that stopped short at the smokebox.

By Christmas 1950, roughly fifty per cent of the order backlog had been cleared (most orders dated from 1947/8), but the '94XX' 0–6–0PTs were an embarrassment

and were hived off to the Yorkshire Engine Company, where their low priority, and the pressure of other work, ensured that they were not completed until 1954–6.

New Year 1951 started with the first of a batch of ten 14 x 20 in. outside-cylinder 0–6–0 side tanks for Sudan Government Railways, which left on 10 January. These machines had balanced slide valves and a 'saturated' boiler, and very neat lines with sloping side tanks and a large cab roof that overhung the cab sides by almost two feet to provide protection from the sun. In those days, overseas locomotives were fitted with an insulated double roof incorporating highly polished, top-quality mahogany tongued-and-grooved boards for the inner lining, with an air space beneath the outer steel plate.

The joiners' shop was impressive. Not only did it produce all the patterns for the various castings, but it also

Proof of the soundness of the original 1906 design, nineteen of these 2 ft. 0 in. gauge 8¹/₂ x 14 in. 0–4–2 tanks were supplied to Calcutta's Howrah–Amta Light Railway over a period of almost fifty years. This is no. 3866, the penultimate one, supplied with sister engine no. 3867 in February 1955.

turned out furniture of superb quality, including magnificent office desks and occasionally kitchen and bedroom furniture. It even produced the wooden frame for the ten-seater shooting brake body built on to a converted 1932 Bentley chassis, which at the same time acquired a Leyland engine with a fluid flywheel.

The Sudans were clear by the end of June 1951, but half-way through, in April and May, two more Austerities were inserted, for Nantgarw and Bargoed Collieries, respectively. Also in May came the first of three quaint 'Eva' class tanks.

The 'Eva' class design dated back almost to the turn of the century, the prototype having been built in 1906 for Martin & Co., engineers and contractors, Calcutta. Martins, later known as Martin, Burn & Co., operated several narrow-gauge railways in various parts of India, and the 'Eva' class was a 2 ft. gauge 0–4–2T with 8¹/₂ in. diameter cylinders and 2 ft. 6 in. diameter driving wheels, and they weighed 15 tons. The type was specially designed for the Howrah–Amta light railway in Calcutta. Two more

were built as late as 1955. Over the 49 years between the prototype and the last one built the only visible differences in design were the substitution of electric lighting, Ross 'pop' safety valves and welded side tanks in place of the oil lamps, Ramsbottom valves and riveted tanks of the original.

T. A. Martin, as the company was then known, was a regular Hunslet customer, and quickly followed the order for the three 'Evas' (placed on 2 June 1948) with orders for two 0–6–2Ts for the Arrah–Sasaram Railway and two 2–6–4Ts for the Shahdara–Saharanpur Railway; the latter was a line that ran for a distance of 92 miles due north from Delhi. Both sections of railway were 2 ft. 6 in. gauge, and the 0–6–2Ts were identical to Nepal No. 7. Indeed, of the seventeen locomotives of this class that were built between 1928 and 1962, the Martins' light railways took thirteen, the remainder going to Nepal. The 2–6–4T design dated back to two prototype locomotives that had been supplied in 1921, these in turn being an update of eight 2–6–2Ts supplied in 1907. Two locomotives of the class

The photographic grey paint applied to HE 3864/55 prior to its despatch to the Shahdara–Saranpur Railway complements the clean lines of the design. The family resemblance to the other Hunslet passenger tank locomotives such as the smaller Russell is emphasized. Nos. 3864 and 3865 were the last of the class to be built; steam locomotive production for the narrow-gauge railways of India, and elsewhere for that matter, was coming to an end.

Two of these graceful 2–8–4 tank locomotives, Hunslet nos. 3707/8, were built in 1951 for the 2 ft. 6 in. gauge Dholpur State Railway where they became Nos. 7 and 8 respectively. They were identical in all respects – even down to the oil headlamp – to the DSR's ZA/3 class built thirty years earlier by Kerr Stuart.

had been built in India in 1945/6 using spare frames and cylinders supplied from Leeds before World War II, while three more had been built at Hunslet in 1948. At 8 ft. 2^1/$_2$ in. they were exactly the same width as the standard-gauge Austerity, yet they were a slender, sleek design with side tanks running the full length of the boiler.

Although the first of the 'Eva' class was despatched on 31 May 1951, no more of the seven Martins' locomotives went away until early December. The second 0–4–2T went on 3 December, the third with an 0–6–2T on 18 December, and the first 2–6–4T on 21 December. In those days it was taboo for there to be any alcoholic drink on the works premises, but some Good Samaritan usually managed to pass a sack full of pint bottles of Tetley's (from the British Queen public house near by) through a hole in the fence by the erecting shop on the last working day before Christmas. Disposing of the empties was usually the problem, but the difficulty did not arise in 1951: I should like to have seen the faces of the Indian fitters at Shahdara when they found the firebox of 2–6–4T no. 3711 practically full of empty beer bottles! The remaining 2–6–4T and 0–6–2T left the works in the last week of January 1952.

In the period between the first and second 0–4–2 – from May to December 1951 – the steam gang had not been idle. Quite the reverse, for in the mean time there had appeared two standard 16 x 22 in. saddle tanks for the National Coal Board and two large 2–8–4Ts for the Dholpur State Railways in India, as well as a further six of the huge Calcutta Ports Commissioners' 5 ft. 6 in. gauge 0–6–2Ts.

The 'Dholpurs' were built strictly in accordance with the original 1920 vintage Kerr Stuart drawings, except that welding superseded riveting for the side tanks, cab and bunker. (At no stage was welding ever accepted for Hunslet boilers.) With 12 x 18 in. cylinders and running on 2 ft. 6 in. gauge track they tipped the scales at 36 tons. The total wheelbase was 26 feet exactly, 10 inches longer than an LNER 'K3' 2–6–0.

But it was perhaps the 'Calcuttas' that represented the highest level of achievement. All six of these 65-tonners were outshopped between 20 October and 29 November 1951, by dint of considerable overtime working including weekend working and, to my own personal knowledge, at least three continuous 24-hour shifts. Despite this, the steam gang managed to slip in a 204 h.p. diesel locomotive for Freddies Gold Mines in South Africa before recommencing the batch of Martins' locomotives. While the machine operators at the works were on

straightforward piecework pay, the erectors were paid a 'tonnage' bonus calculated against the total weight of locomotives despatched each month. Christmas 1951 was a prosperous one!

The year 1951 had produced twenty-seven steam locomotives and forty-nine diesels, but 1952 was a record year, with twenty-eight steam and 100 diesel. Granted, more than half the diesels were small 20 h.p. units, and the steam locomotives lacked the variety of the previous year, but they were nevertheless of eight different types. In addition to the two Martins' locomotives referred to earlier, the total comprised twelve Austerities, two more standard 16 x 22 in. engines (one for the National Coal Board, the other for Oxfordshire Ironstone), a repeat batch of five Sudan Railways 0–6–0Ts, two 5 ft. 6 in. gauge 0–6–0STs for Madras Port Trust, four little 0–4–2PTs and a solitary 0–4–0ST.

Two of the Austerities, modified with cast-steel wheels and repositioned injectors, went to Guest, Keen and Baldwin's East Moors Works, but all the remainder were for NCB locations in South Wales, the East and West Midlands and Lancashire. The two Madras locomotives had outside cylinders with Walschaerts valve gear and an open-backed cab. Standard side buffers and centre draw-hooks were fitted to suit the Indian Railways broad-gauge rolling stock, but, in addition, two sets of Norwegian-style 'chopper' couplers were mounted low down on the buffer beam at roughly 13 inches either side of the centre-line for use with metre-gauge rolling stock.

The smallest steam locomotives produced during my time on the shop floor were the four 60 cm gauge 0–4–2PTs that were supplied in April 1952 to Elders and Fyffes Ltd for their plantations at Tiko in the British Cameroons. Known as the 'Tamar' class, they were in effect a pannier tank version of the Kerr Stuart 'Brazil' class saddle tank that may still be seen at Sittingbourne and Whipsnade. They were very simple little machines to build, with Hackworth's valve gear to the 9 x 15 in. cylinders, a domeless boiler and Ramsbottom safety valves.

The remaining locomotive in 1952 was another Kerr Stuart design – a 15 x 20 in. 0–4–0ST of the 'Moss Bay' class, a mite prematurely named *Coronation*, which was despatched to United Glass Bottle Manufacturers, St Helens, on 1 August.

By late 1952 I had progressed to design work, mainly on diesels for British Railways, Woolwich Arsenal, NCB, Iraq and South Africa, and consequently lost my more

Calcutta Port Commissioners No. 19, HE 3750/51. Hunslet built eighteen of these massive locomotives in three batches between 1945 and 1951. Others, to the same design, were supplied by Henschel and Mitsubishi.

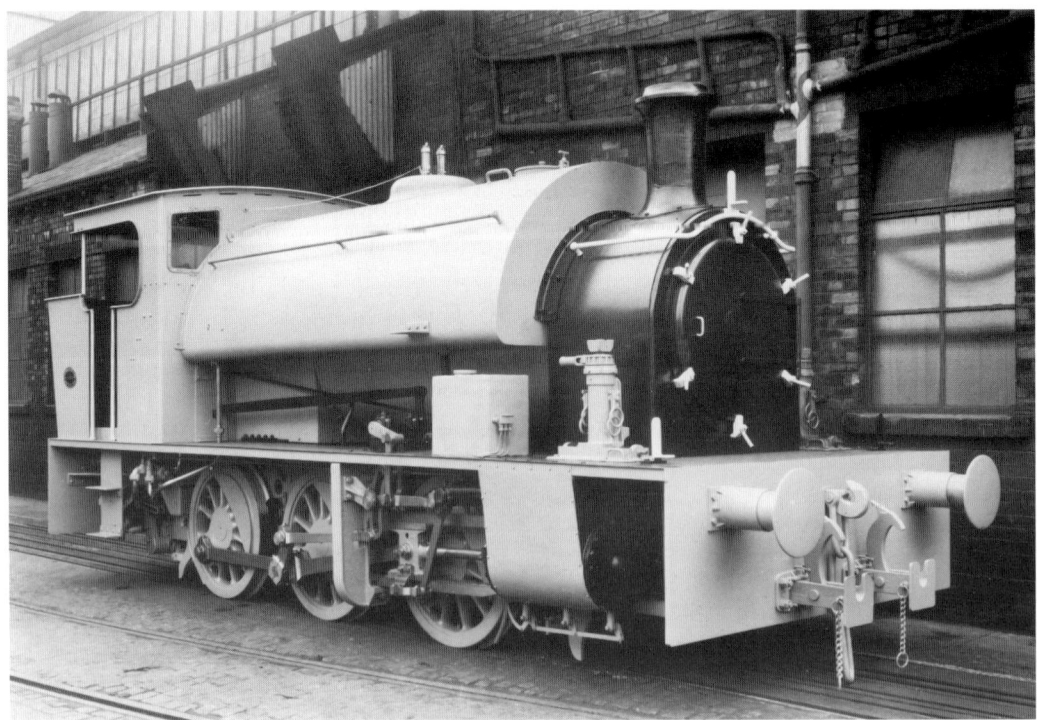

Two of these neat 16 x 24 in. 0–6–0 saddle tanks, HE 3774/5, were supplied in 1952 for the 5 ft. 6 in. gauge lines of the Madras Port Trust. They joined two earlier units of the same type, nos. 1507/8, built in 1926. Apart from the track gauge they were mechanically similar to the two British outside-cylinder 16 in. 0–6–0 saddle tanks supplied to Birch Coppice colliery and Tir John power station in 1929 and 1935 respectively. The two sets of narrow-gauge couplers were for handling metre-gauge wagons on the dual-gauge interlaced tracks in the docks.

Hunslet no. 3776 as built in September 1952 for Baggeridge colliery shows how little the design had changed over the intervening 420 locomotives. No. 3776 later went to Bickershaw colliery and is now Sir Robert Peel on the East Lancashire Railway.

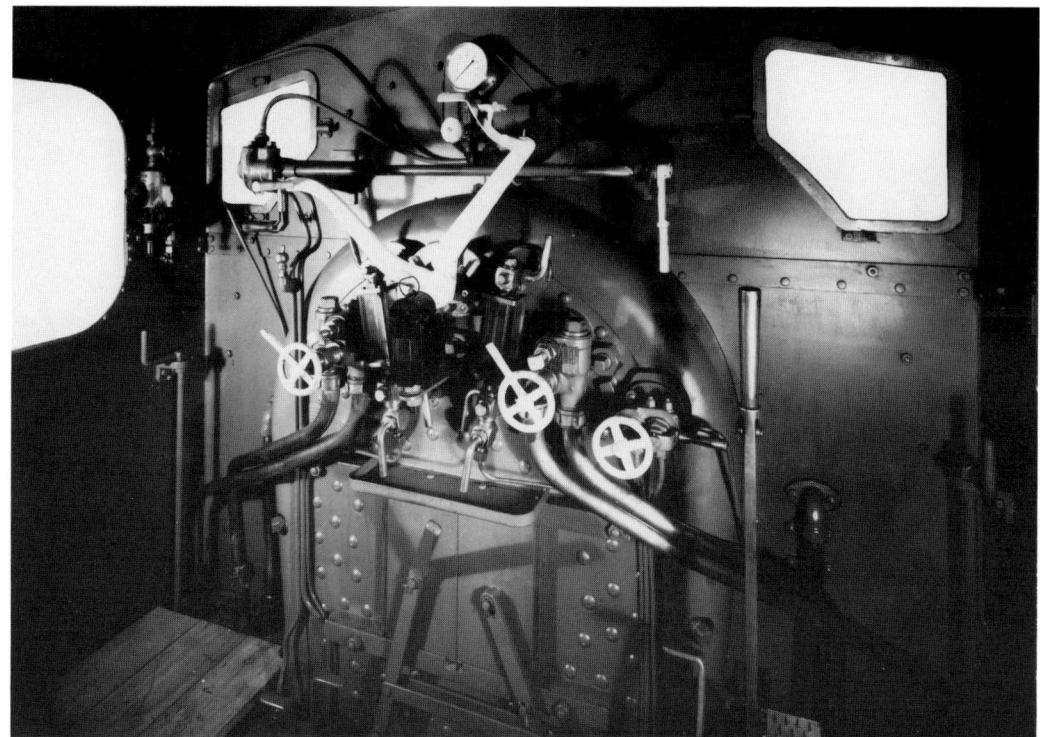

Cab interior of Hunslet Austerity no. 3847 – the 479th built, only six more to go – showing the simple controls. Reading from the left can be seen the left-hand injector spindle, sight feed lubricator, steam brake valve, regulator handle, the two combined steam and delivery valve handles, with gauge glasses between, and the combined sanding and blower valve. To the left of the reversing lever, which is in mid-gear, is the cylinder cock rod, while the two ashpan damper rods are to the left of the closed firedoor.

immediate, minute-by-minute contact with the steam locomotive. This is perhaps a suitable time to refer briefly to the extensive repair programme that was going on alongside the new construction. Most of the wartime-built Austerities were coming in for overhaul under a continuing Ministry of Defence contract. They all went back to military depots after rebuilding, but many were simply stored there unused until they were sold to industrial users, mainly the NCB. It was during this rebuilding that the opportunity was taken to strengthen the buffer beams. All new Austerities built after 1948 (beginning with Hunslet no. 3685) had two triangular gussets between the frame plate and the buffer beam, one above and one below the centre-line, whereas the original wartime locomotives had only one. The modification was brought in to counter the twisted buffer beams that had occurred in heavy industrial service. Every locomotive that was returned to Hunslet was modified in this way, with the result that apart from the LNER 'J94' locomotives, very few retained the single gusset.

Many of the repaired locomotives changed identities owing to the fact that on arrival at the works they were stripped down to the frame, which was then rolled outside on temporary wheels, while boiler, tanks, etc. went their various ways for attention; then it was just a case of marrying together the first set of parts that arrived back in the erecting shop. This could be done with minimum fitting, and shows the benefits of standardization in action. If the bunker happened to be in good condition it stayed on the frame, and if the maker's plate stayed there also, there was a good chance that at least the frame would retain its identity. In the main, though, it was a case of sending out a locomotive, *any* locomotive, with a maker's number that matched the Army serial number allocated to it, so as not to upset the accounting system.

Of the locomotives that were repaired at that time quite a number were of Avonside or Kerr Stuart origin. Particularly noteworthy were the Avonside-built Cadbury locomotives (No. 1 is preserved at Toddington), which came in one by one and returned to Bournville in crimson

The post-war repair backlog was considerable and went on for some years. A less-than-resplendent Cadbury Bournville No. 6 (Avonside 1921/23) awaits its turn for attention after having been pushed up the incline from the main line to the works by one of the Hunslet Lane 'Jinty' shunters in August 1950.

AUTHOR

This picture neatly encapsulates the steam–diesel transition. One of a series of over a hundred photographs taken between 1950 and 1952 to record every detail of construction of the 500 h.p. diesel no. 4000, it does indeed have no. 4000's frame in the foreground, being drilled for final riveting.

The background is kaleidoscopic, with the locomotive at top left presenting a history lesson in itself. In the final stages of a heavy repair, including authentic replacement of the tongued-and-grooved wooden boiler lagging, it is an 0–4–0 well tank locomotive, No. 10 Patricia from the Imperial Chemical Industries works at Burn Naze, Fleetwood. This unusual design, with its Stephenson's valve gear between the wheels and the locomotive frame, started in 1868 with James Cross & Co. of Sutton Engine Works, St Helens, who built early examples; was perpetuated by Edward Borrows & Co., also of St Helens; and was finally built under licence from Borrows by Kerr Stuart between 1915 and 1929. Patricia was Kerr Stuart no. 4431 built 1928.

At the bottom of the erecting shop can be seen a new shiny black 'Austerity', and behind that a 204 h.p. shunter destined for Algerian Railways. In the shadows to the left of the Austerity can just be discerned (on the original print) one of the pioneer LMS diesels (ex-No. 7054), still in its wartime War Department dark grey. The portable drilling machines survived for another thirty-five years.

lake paintwork with yellow and black lining and with the 'CADBURY BOURNVILLE' legend in gold leaf stretching the full length of the tank sides.

A particular favourite for repair was No. 10 *Patricia*, a Kerr Stuart-built 'Borrows' type 0–4–0WT from ICI at Burn Naze, Fleetwood. This was a design similar to that of *Windle* on the Middleton Railway, in which the Stephenson's valve gear was positioned between the

frames and the wheels. *Patricia* disgraced herself by demolishing the erecting shop doors while being driven by the charge-hand of the repair gang.

Most of the early batch of Swiss-built Snowdon Mountain Railway rack locomotives came and went, as did the Kerr Stuart 'Tattoo' class 0–4–2 saddle tank, Corris Railway No. 4, on its way to become *Edward Thomas* of the Talyllyn Railway.

1953 also produced twenty-eight new steam locomotives, but for the first time in over eighty years not one was for export, and only three types were involved. There were two standard 18-inch locomotives for Richard Thomas and Baldwin's Scunthorpe works, four 16-inch locomotives for the National Coal Board and twenty-two Austerities. The latter included the last fourteen (nos. 3790–3803) built for the Ministry of Defence. A brief return to steam locomotive exports was made in 1954 with three more of the Gold Coast Railway eight-coupled side tanks and two 2–6–2Ts for Sierra Leone. One of these Sierra Leone locomotives is now on the Welshpool and Llanfair Light Railway. Otherwise, the pattern was the same as the previous year, with two 16 in. locomotives and sixteen Austerities, all for the NCB. This discounts the twenty 0–6–0PTs, British Railways Nos. 9490–9 and 3400–9 (Hunslet nos. 3729–39) that finally emerged from the Meadowhall Works of the Yorkshire Engine Company between February 1954 and November 1956.

The Yorkshire Engine Company also produced ten Austerities in 1954 for the United Steel Company's Ore Mining Branch. Whether the decision to allow Yorkshire to build the Austerity design was part of the Western Region pannier tank deal I cannot say, but all the Austerity castings (cylinders, wheels, axleboxes, eccentrics, slide valves, chimney, etc.) were supplied by Hunslet. The wheels were of the 12-spoke cast-steel variety used on the Guest Keen and Baldwin locomotives, not the standard 14-spoke cast-iron type that was normally used.

The following year, in 1955, saw an Indian summer for steam locomotive exports. The obligatory Austerity was still much in evidence, with ten built for collieries in South Wales, Staffordshire, Lancashire, Yorkshire, Durham and Fife; but twelve locomotives were produced for export, including five more for Martin, Burn. These comprised two 'Eva' class for the Howrah–Amta, two 'Shada' class 2–6–4Ts for the Shahdara–Saharanpur and an 0–6–2T for the Arrah–Sasaram.

Mechanically the same as the fifty-seven 'African' 3 ft. 6 in. gauge 0–8–0 tank locomotives supplied between 1930 and 1954 were six 0–8–0 tank tender locomotives for Nigerian Railways in 1955. Hunslet nos. 3857–62, Nigerian Railways Nos. 53–6. The tender styling owed much to the British Railways standard tenders of the period.

This overhead view of no. 3857 gives a better impression of size than conventional views, particularly when compared with the diesel locomotive frame behind it. This frame was for HE 4679/55, a standard-gauge 153 h.p. shunter for Tilmanstone Colliery in Kent and of the same type as British Railways Nos. 11500–2. The Nigerian locomotives were 9 ft. 5³/₄ in. wide and 12 ft. 4 in. high. The tractive effort was 23,532 lb. at 75 per cent boiler pressure.

The Trujillo Railway, Peru, 2–8–0 tender engines were quite large considering the narrow track gauge (3 ft. 0 in.). This is HE 3413/47, the first post-war example; another – No. 32, HE 3808/56 – was shipped on 1 June 1956. 18¹/₂ x 26 in. cylinders, superheated, 88 tons 14³/₄ cwt. in full working order.

But the main attraction for 1955 was a batch of six 0–8–0 tender/tank locomotives for Nigerian Railways. Mechanically, the engine portion was the same as that of the tank locomotives previously supplied to Nigeria and the Gold Coast, but the side tanks and cab were restyled and a six-wheeled tender added. These tenders bore an uncanny resemblance to the standard British Railways tender, except that the water space was much greater in proportion to the coal space. These locomotives were intended for trip working, and, as against the earlier tank locomotives, the water capacity was increased from 1000 to 4800 gallons and the coal space from two to four tons. The locomotive weight was reduced slightly from 48 tons 8 cwt to 47 tons 2 cwt by the removal of the coal bunker, but the addition of the tender brought the all-up weight to 84 tons 12 cwt.

The final locomotive of 1955 was a repeat order for an 82³/₄ ton oil-burning metre-gauge 2–8–0 for the Peruvian Corporation's Guaqui–La Paz Railway. With coupled wheels of only 3 ft. 9 in. diameter, the 16 x 24 in. cylinders produced a tractive effort of 20,400 lb. The locomotive had a generally transatlantic appearance, with American-style stovepipe chimney, bell, buckeye couplers, pilot (i.e. cow-catcher), and a large sand box atop the boiler.

This was the beginning of the end for steam, and while diesel production continued to increase, the following year saw only twelve steam locomotives turned out. Ten of these were Austerities – nine for the National Coal Board and one for Guest, Keen & Baldwins' works at East Moors. Two locomotives were produced for export: the final Martin Burn 0–6–2T for the Arrah–Sasaram Railway and another 2–8–0 for the Trujillo Railway of the Peruvian Corporation. This was similar in appearance to the one built for the Guaqui, but ran on 3 ft. gauge. It was an oil-burning locomotive with an atomizing burner on the Laidlaw Drew system, and had larger, 18¹/₂ x 24 in.

cylinders, giving a tractive effort of 25,200 lb. The total weight of engine and tender was 88¾ tons.

1957 saw one 2 ft. gauge 0–6–2T for the sugar cane tramway of Natal Estates and one Austerity each for Lancashire and Somerset. Next year there was just one Austerity for Stewarts & Lloyd's ore field at Sewstern, and one standard 16 in. locomotive for Oxfordshire Ironstone.

The last big steam locomotives came in 1959, when two more 2–8–4Ts were despatched to Dholpur State Railways. Although mechanically similar to the Kerr Stuart design that had been supplied in 1951, these had full-length pannier tanks and greater coal and water capacity. To maintain a balanced axle load within the limitations imposed by the Indian Railway Inspectorate, the cab and bunker were made from aluminium sheet.

No more new steam locomotives appeared until 1962, when the last Austerity – or so it was thought at the time – went to Nailstone Colliery in Leicestershire. This was built from parts left over from a stock batch of thirty-six locomotives (Hunslet nos. 3816–51) laid down as far back as 1952. Also in 1962 the final two 0–6–2Ts were supplied to the Nepal Jaynagar–Janakpur Railway. The Snowdon Mountain Railway's No. 4 *Snowdon* was also rebuilt at this time; it had lain derelict outside Llanberis shed for many years and was far from complete when it was received at the Jack Lane works for rejuvenation. Apart from the frame plates and one cylinder, almost the whole of the locomotive sent back to serve the tourists was new.

By this time, the author had moved on from the design office to the greater flexibility of the Sales Department, and it was with much surprise that in 1963 the order was received for *the* final steam locomotives for British use built by a traditional British manufacturer on traditional commercial terms. The result was two new Austerities, nos. 3889 and 3890, which were sent off to Manvers Coal Preparation Plant and Cadeby Main Colliery respectively on 18 and 27 March 1964.

Hunslet 3873/4, built 1959, were 2–8–4 tanks for Dholpur State Railway, mechanically similar to the 1921 and 1951 built Kerr Stuart design but with pannier tanks and larger coal bunkers. To maintain the low axle load these two locomotives were unique in having aluminium cabs and bunkers. The gentleman standing to the left is Works Manager Albert Burns.

Nos. 3875 and 3876 were two 0–6–2 tank locomotives built in 1962 for the 2 ft. 6 in. gauge Nepal Jaynagar–Janakpur Railway. Painted gloss black and fully lined out in red and white, they carried brass name-plates Surya and Chandra when despatched. Surya is seen here in less formal dress, primed and filled ready for rubbing down, on final steam tests early in August 1962.

No. 4 Snowdon looks like a genuine 800 mm gauge Swiss-built locomotive from the first SLM batch supplied in 1896. It is in fact a virtually new locomotive built in 1961 from cannibalized remains that had stood at the back of Llanberis works since before the outbreak of war. Little more than one frame plate and one cylinder of the original locomotive survived – the rest was brand-new. The elliptical Hunslet plate on the side sheet proclaims its rebuilt status. All the remaining 1896 batch were rebuilt, not quite as drastically, in Leeds between 1956 and 1961, and new boilers were provided for the three 1922/3-built batch. KEITH R. CHESTER

The last three brand-new Austerity locomotives, nos. 3851, 3889 and 3890, were fitted with the Hunslet patent underfeed stoker and gas producer system, and a short description of this equipment is appropriate.

In its Autumn 1962 issue the magazine *Smokeless Air*, the journal of the National Society for Clean Air, acclaimed the Hunslet automatically fired shunting locomotive, saying, 'the effectiveness in reducing smoke to well below the dark smoke level has been confirmed to us by an experienced smoke inspector.'

There were four basic elements to the system: (*a*) the stoker itself, a reciprocating feeder actuated by a single-cylinder steam engine, (*b*) the gas producer system, (*c*) the jet system and (*d*) the blast-pipe arrangement.

The stoker was reasonably straightforward and arranged to take small coal down to half-inch singles. Early models had two feeders acting alternately, while later models had one large unit.

The gas producer system dealt with the problem of clinkering usually associated with the use of small coal. Steam was introduced into the ashpan from both the stoker engine exhaust and from a live steam pipe controllable by the driver.

To provide complete combustion a system of steam jets and ports admitted secondary air to the firebox over the fire and back up under the brick arch, while a four-jet blast-pipe of aerodynamically efficient design gave a better exhaust flow and reduced the back-pressure in the smokebox.

The result was a superb piece of kit that undoubtedly prolonged the life of steam locomotives used by the National Coal Board, but it was marred by the high cost of

Waterloo Main Colliery Railway, Leeds. The first Austerity to be fitted with a Hunslet underfeed stoker, HE 2876/43 Jess shows off its clean exhaust against the cooling towers of Skelton Grange power station on 27 March 1963. The stoker, with its associated secondary air system and multiple-jet blastpipe, was a good idea up to a point. It worked in theory and, when new and properly maintained, in practice, but it was a pain to keep in good order and provide with proper small coal. Unpopular with the drivers and fitters, its life in most cases was short. However, it kept to the Clean Air Act promises, satisfied the trade unions and prolonged the life of several dozen steam locomotives while providing a tidy income for its manufacturers. What else does one need for commercial success?

Hunslet 2868/43, newly rebuilt with underfeed stoker and renumbered 3883/63, leaves Kingham on the Worcester–Oxford line with the Western Region dynamometer car and test train on 23 April 1963. Still in works grey when the photograph was taken, the locomotive was subsequently sold to NCB Coal Products Division, Glasshoughton, as Coal Products No. 6 and later went to the Rutland Railway Museum. The large chimney cowl was made from fibreglass and covered the four-jet blast pipe which was an integral part of the stoker conversion except when the Giesl ejector was used.
P. H. WELLS

The underfeed stoker was also fitted to selected other NCB locomotives. This is the prototype Hunslet standard 15 x 20 in. locomotive, no. 1440 Airedale, originally built in 1923, rebuilt with an underfeed stoker in 1967 towards the end of the conversion programme and photographed at work at Ackton Hall colliery, Featherstone, on 25 August 1971. The stovepipe chimney identifies it as a stoker-fitted locomotive, while the cylinder of the auxiliary steam engine for the stoker can just be seen between the frames under the bunker.
J. A. PEDEN

maintenance. Additionally the traditional methods of fuelling colliery locomotives with whatever was available at the pit head played havoc with the stoker itself. Quite often drivers resorted to hand-firing, and in a lot of cases the equipment had a relatively short life.

Nevertheless, in addition to the three new locomotives, eighty further stokers were retrospectively fitted to locomotives of five different designs. Most of these were Austerities, but there were eleven Hunslet standard 15 x 20 in., four 16 x 20 in., one '50550' class 18 x 26 in. and seven Hudswell Clarke 17 x 22 in. 0–6–0 side tanks so equipped. Many of the locomotives came back to Hunslet for conversion, but several were dealt with at NCB collieries and workshops. Some of the conversions carried out on NCB property had Giesl ejectors (blast pipes) fitted instead of the Hunslet 'Kylpor' type. The term 'Kylpor' was derived from amalgamating 'Kylchap', the archetypal multi-jet blast pipe (itself short for Kylala–Chapelon), with the name of the Argentine locomotive engineer Livia Dante Porta who was Keith Alcock's friend and mentor and a confidant of John Alcock.

The first stoker conversion was on Hunslet Austerity no. 2876, a reconditioned locomotive sent away to Waterloo Main Colliery on 6 September 1962, and the whole programme was completed by the end of 1968.

Stokers were also fitted to nine of fourteen ex-Army Austerities that were purchased and rebuilt by Hunslet from 1961 to 1970. Most of the fourteen had been sold by 1963, but three, nos. 3885, 3892 and 3893, hung around for years without a taker. No. 3885 had been on loan to the Coventry Homefire smokeless fuel plant, had been involved in an argument with a British Railways diesel in the exchange sidings and was slightly the worse for wear. No. 3892 was a full rebuild with stoker and was in 'as new' condition, but no. 3893 had just been given a minor repair. In a final act of desperation, no. 3893 was cut up on the spot in 1970 and no. 3892 was sold to the Bahamas Locomotive Society for £1500. The following week, an urgent request was received from the National Coal Board at Walkden for a reconditioned Austerity! The best that could be done was to despatch the damaged no. 3885 to Walkden workshops with the thought that they could make one good locomotive out of two bad ones. It is worth pointing out that the invoice price of nos. 3889 and 3890 in 1964 was around £15,000 each.

In 1967, eight boilers were supplied to Hunslet Taylor Consolidated in Johannesburg for a batch of 2 ft. gauge

Table 9.1

The steam locomotive order book at the beginning of 1949	
3625–36	3 ft. 6 in. gauge 0–8–0 Nigerian Railways
3665	0–4–0T Cadbury No. 9
3667–9	2 ft. gauge 2–8–2, Barsi Light Railway, India
3670	2 ft. gauge 2–8–2 Eastern Province Cement Co.
3671–76	5 ft. 6 in. gauge 0–6–2T Calcutta Ports Commissioners
3677–83	3 ft. 6 in. 0–8–0T Gold Coast Railway (same as Nigeria)
3684	2 ft. 6 in. 0–6–2T *Shreebishnu* Nepal
3685–89	Austerity 0–6–0ST National Coal Board
3690	16 x 22 in. 0–6–0ST Peruvian Corporation
3692–3701	Austerity 0–6–0ST National Coal Board
3702–04	2 ft. gauge 0–4–2T Howrah–Amta Railway, India
3705–06	2 ft. 6 in. gauge 0–6–2T T. A. Martin, India
3707–08	2 ft. 6 in. gauge 2–8–4T Dholpur State Railway, India
3709–10	18 in. Standard 0–6–0ST RTB, Scunthorpe
3711–12	2 ft. 6 in. gauge 2–6–4T Shahdhara–Saharanpur Railway, India
3713–15	Standard 16 in. 0–6–0ST, National Coal Board
3716	Standard 16 in. 0–6–0ST Oxfordshire Ironstone Co.
3717–18	Modified Austerity 0–6–0ST GKB, Cardiff
3719	2 ft. 6 in. gauge 0–6–2T Baraset Basirhat Railway, India (cancelled)
3720–39	GWR '94XX' 0–6–0PT (subcontracted to Yorkshire Engine Company Ltd)
3740–49	14 x 20 in. 0–6–0T 3 ft. 6 in. gauge Sudan Government Railways
3750–55	5 ft. 6 in. gauge 0–6–2T Calcutta Ports Commissioners

'NGG16' Beyer–Garratts. Among the last boilers built at this time were two for the Isle of Man Railway's Beyer Peacock 2–4–0 tanks, one for the Festiniog Railway's *Linda* and the replacement boiler for the acclaimed *Russell*. There were also two double-ended boilers for the Festiniog Fairlies.

The first locomotive exported by the Hunslet Engine Company went to Java in the Dutch East Indies in 1866. It was an appropriate coincidence that on 27 August 1970 Hunslet received the order for the last steam locomotive to be built commercially (as opposed to being supplied for purely nostalgic and tourist purposes) in the western

Ten spare boilers were supplied by Hunslet to the London and North Eastern Railway for the former North British Railway class J37 0–6–0 goods locomotives in 1947.

Spare boilers were always big business. This is one of a batch for Indian Railways standard 'G' class passenger engines in 1954. The boiler mountings were picked out in white for photographic purposes only, but the stainless steel lagging straps were standard production quality. Even spare ashpans were provided.

The last Hunslet steam locomotive. HE 3902/71 stands at Trangkil sugar factory, Java, on 8 August 1984.

Flying Scotsman paid a brief visit to Hunslet during 1968 when it was repaired, retubed and fitted with a cow-catcher prior to its chequered visit to North America. In this photograph it is flanked on the left by one of the twenty-eight Ceylon Government Railways shunters and on the right by another 5 ft. 6 in. gauge locomotive, 235 h.p. HE 6984/68 for the port of Pasages in north-west Spain. The Festiniog Railway's vintage Leyland lorry has just brought the old boiler from Linda.

world, and this too was destined for Java. It was ordered by Robert Hudson (Raletrux) Ltd for the Transkil Sugar Factory near Surabaya in Indonesia (by then Java had become a part of the Republic of Indonesia), and was a small 9 x 15 in. saddle tank, mechanically the same as the 'Tamar' class built for the Cameroons in 1952; the design in fact dated back to 1909. No matter that it was a tiny little thing and antiquated to boot; it was a *steam* locomotive, and the enthusiasts greeted it with almost as much rapture as if it had been a reincarnation of *Flying Scotsman*.

Mention of *Flying Scotsman* is a reminder that this locomotive also passed through the works before its well-known and not uniformly successful visit to North America, as did the ex-LMS 'Jubilee' *Bahamas* and other preserved locomotives, but these were fill-in or friendship jobs as opposed to commercial ventures. For all practical purposes, the age of steam had come to an end. The diesel had triumphed.

CHAPTER 10

Diesel ascendancy, 1949–1962

While all the steam activity was going on, the railways and industries of the world were switching to diesel traction with ever-increasing momentum, so it seemed; demand was high up to the virtual elimination of steam in the late 1960s, and continued at a slower rate up to 1978. Not only that – the equipping of British coal mines with underground locomotives was proceeding at a rapid pace.

There were good years and not so good years as various national and international crises came and went, but in general the mood and structure of the company remained the same, and the diverse nature of its products allowed the *laissez-faire*, swings-and-roundabouts approach full rein.

Although the author's personal transition from school to industry had been quite dramatic, it was accepted, as things are at that time of life, as a reasonably commonplace happening with a trusting mind that assumed that everything would go right, because adults were assumed to know what they were doing; and go right things did. Which, in retrospect, is little short of miraculous.

The magnitude of the steam locomotive order book has already been mentioned. The diesel orders were no less

The Consett Iron Company, County Durham, purchased thirty Hunslet diesel locomotives between 1947 and 1959. The second one, no. 3580 – new on 27 April 1949 – is seen here with a train of molten slag ladles. By this time the pneumatic tipping equipment on the ladles was largely inoperative through lack of maintenance, and emptying was quite often achieved by the locomotive taking a run away from the train with a wire rope attached between the ladle and the locomotive drawhook. The snatch on the rope, taken round a strategically-placed pulley or two, was sufficient to start the mass of metal moving, but it was a practice that was frowned on and certainly was not kind to the drawgear or the locomotive frame.

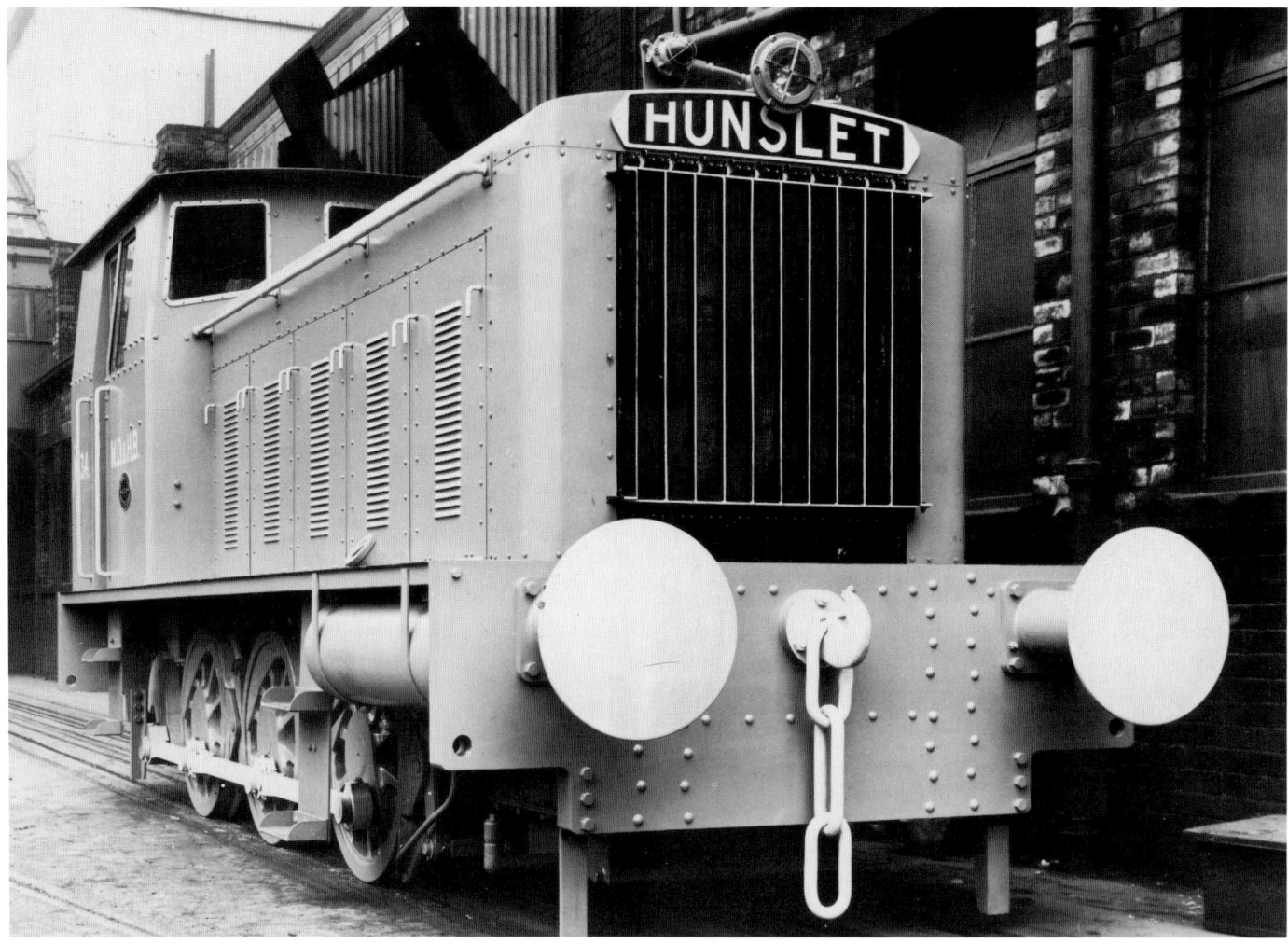

All dressed up in photographic grey with white details, Mersey Docks and Harbour Board No. 34 (Hunslet no. 4229) poses for the record. Note the flameproof head and tail lights above the 'HUNSLET' plate and the water-bath exhaust tank between the frames behind the buffer beam.

demanding, although at that time the only repair diesel work was that of the LMS foursome. These four original LMS diesels had come back to their birthplace after spending some years with the War Department. Three of them were to be rebuilt for further service. As *John Alcock*, no. 1697, the progenitor of them all, bore the name of its originator for a time, and found a retirement home, appropriately enough, on the preserved Middleton Railway. In 1949, however, it was being re-engined for further industrial use.

While the shop-floor equipment and the skills of the work force were the finest you would find anywhere, the administrative processes were all manual, with only mechanical typewriters and the odd stencil duplicator to ease the slog of handwritten ledger entries. There was one wet Copycat machine which produced photocopies of

customers' orders, etc., first as a negative and then as a positive, but this was by no means a fast process. Documentation was consequently kept to a workable minimum, and the whole process was driven and controlled by the drawing-office.

Just as the Austerity 0–6–0 saddle tank had established itself as the bread-and-butter standard steam locomotive, so had the 204 h.p. Gardner-engined 30-ton 0–6–0 locomotive become the surface diesel counterpart in terms of quantity production if not the equivalent in power. The design was based on that of the last three LMS diesels with progressive 'modernizations'. The Mersey Docks and Harbour Board and McAlpine examples had stabilized the design in a form that had resulted in the Belgian Congo, Algerian Railways, John Summers and Consett orders; and in 1949 repeat orders were in hand for all except the Belgian Congo and McAlpines, and with additional

The five sections of the six-speed gearbox used on the 500 h.p. shunters (nos. 4000–3) designed for the Peruvian Corporation.

Service Manager Albert Hancock stands alongside the finished 500 h.p. Peruvian locomotive gearbox.

examples for Pretoria Portland Cement and Freddies Gold Mines (both in South Africa), the Peruvian Corporation (Paita–Piura Railway) and Tanganyika Railways. All these had variations in detail, and covered track gauges of metre, 3 ft. 6 in. and 4 ft. 8½ in., but nevertheless the basic appearance remained more or less the same.

The second Mersey Docks 204 h.p. locomotive (no. 4229), which was completed in January 1951, showed a slight design change in that it imitated to a degree the largest of the LMS diesels in having a slight rake to the cab sides and shaped cab front and rear windows rather than the traditional and almost universal 'portholes'. But this was only cosmetic, and the larger 300 and 500 h.p. designs already in production had a much more up-to-date appearance.

Talk of the 'three hundreds' and 'five hundreds' was usually accompanied by expressions of reverence, and parts for these mystic beasts were given the highest priority. Diesel-mechanical locomotives, shunting or otherwise, larger than 204 h.p. were still very rare even in 1949, and transmitting the torque of the slow-speed marine-type diesel engines of the day to the wheels of a shunting locomotive was not a simple process. The higher the power, the slower the engine and the bigger the clutch. The 500 h.p. clutches were *big* big.

The 500 h.p. examples had in fact come first in terms of conception. Shortly after the end of the war, the Peruvian Corporation had ordered two 0–8–0s and one 0–6–0 for the Paita–Piura and Southern Railways, respectively, and these had been allocated Hunslet numbers 3478–3480. Before design work had commenced, however, these three prestigious locomotives were given new numbers 4000–4002, thereby starting a new series exclusively for diesel locomotives. A second 0–6–0, no. 4003, was also put in hand, to be built for stock.

Before going to Peru the prototype 500 h.p. diesel locomotive underwent extensive trials on the Midland main line north of Leeds to Lancaster and on the West Leeds local branch lines. Permission to run on British Railways lines was easily obtained by local arrangement in those days. Here, no. 4000 runs through Guiseley station heading towards Yeadon on 20 October 1951, contrasting sharply with the nineteenth-century Midland Railway footbridge and the clerestory-roofed North Eastern Railway dynamometer car, designed by Worsdell, built in 1906 and now in the National Railway Museum.

The driving cab of the 500 h.p. locomotive. Hand brake wheel in the centre; all other controls, i.e. gear selector, gear change, forward and reverse, air brake and sanding duplicated at each driving position. No danger of any of these 'coming off in your hand'

The four locomotives were identical except that the six-wheelers were 3 ft. 9 in. shorter and four tons lighter than the eight-wheelers, the weights in working order being 51 and 55 tons respectively. A Colchester-built Davy Paxman 12 RPHXL turbo-charged diesel engine developing 500 h.p. at 1375 r.p.m. drove through a massive Hunslet friction clutch to a Hunslet air-operated gearbox giving six speeds forward and reverse up to a maximum speed of 33 m.p.h. Starting tractive effort was 35,400 lb. (cp. the LMS Stanier class 8F 2–8–0 freight locomotive, which developed 32,438 lb.), and a train of 600 tons could be hauled at a steady speed of 20 m.p.h. on the level.

Before it was shipped to Peru, no. 4000 was fitted with British-style buffing and drawgear and underwent extensive trials in revenue freight service on the London Midland Region of British Railways.

Between July and October 1951, four weeks of heavy coal-train shunting at Stourton Down Yard (on the former Midland Railway southern approach to Leeds) were followed by five weeks in regular freight traffic between Stourton, Guiseley and Yeadon, and culminated in just over a week of dynamometer car tests. These tests were covered by a comprehensive report issued by the North Eastern Region Mechanical and Electrical Engineer's Department at Darlington. The dynamometer car used was the former North Eastern Railway clerestory-roofed vehicle built in 1906, later used behind *Mallard* in the record-breaking 126 m.p.h. run before the war, and now exhibited in the National Railway Museum.

On one occasion the 3.30 p.m. class F train from Lancaster Ladies Walk Sidings to Stourton Up Sidings at Leeds (67 miles) was deliberately overloaded to 67 empties plus dynamometer car and brake van, a trailing load of 511 tons as against the normal class 8F steam locomotive's load of 320 tons. On another occasion the locomotive produced a sustained drawbar pull of 39,200 lb. (17.5 tons). This was only just within the capacity of the dynamometer car, which was 45 years old at the time of the tests and had not been not designed to cope with pulls of over 18 tons.

After proving itself no. 4000 was refitted with its buckeye couplers and cowcatchers and shipped with no. 4001 to Paita on 24 March 1952. The six-wheeler, no. 4002, followed to Callao on the Peruvian Central Railway on 19 September 1952.

Hughie Wilkinson, who had supervised the testing of no. 4000 on the Midland main line, accompanied it to Peru in a similar manner to his predecessor Albert Hancock with the pre-war Guaqui–La Paz 0–6–0. Some time after his return he found himself demonstrating the stock 0–6–0 500 h.p. no. 4003 first at the BR Stourton yard, where the Southern Region Norwood Junction shunter no. 11001 had also worked briefly in 1952, and then on the National Coal Board Manchester Collieries' Walkden system. The locomotive then returned to Leeds for repainting and the fitting of a coke stove in the cab before going on 19 May 1954 to the John Summers steelworks at Shotton, where it worked on incoming iron ore trains until the shipment of ore through Bidston Dock ceased in the mid-seventies. In later main-line steam days, the ore trains were hauled by British Railways standard class 9 2–10–0s to Shotwick sidings, where no. 4003 took over: quite often 'on the fly' when the load was particularly heavy – usually well over 1000 tons. The coke stove was a mandatory requirement on all the John Summers diesels, both for cab heating and so that the drivers could maintain their traditional steam-locomotive 'billy' of strong stewed tea. Smoke pouring out of a chimney at the back of a diesel locomotive could be quite disconcerting for the uninitiated onlooker.

The 300 h.p. locomotives were equally impressive. Two standard-gauge four-wheelers, nos. 4010/11, with 4 ft. 0 in. diameter wheels and weighing 40 tons, were delivered to the Consett Iron Company at the end of 1950, followed in April 1951 by a third, no. 4261, for the Barrow Iron Works. The fourth was no. 4211, a 3 ft. 6 in. gauge six-wheeler with 3 ft. 9 in. wheels and weighing 37½ tons for the Roan Antelope Copper Mines in Rhodesia (later Northern Rhodesia, then Zambia) and followed in November of the same year. All four were fitted with Crossley five-cylinder two-stroke scavenge-pump engines thumping along at 900 r.p.m. Two more four-wheelers went to Consett in 1953, but these had the four-stroke Mirrlees, Bickerton & Day J5 engine instead of the rather indifferently-performing Crossley, as did five later six-wheelers, no. 4550 for the United Sulphuric Acid Corporation in 1954, no. 4551, built as a demonstrator but sold to NCB South Wales for the Aberaman Phurnacite plant in December 1961, and three, nos. 5392–4, for the Consett Iron Company. The last four mentioned also had a Brockhouse/Hunslet torque converter in addition to the four-speed gearbox, making them in effect diesel-hydro-mechanical, although the manufacturer preferred to refer to them as diesel-hydraulics. Consett also built themselves two 0–6–0s from spares, the author's original erecting

A photograph taken by Hunslet's commissioning engineer Hughie Wilkinson shows Hunslet no. 4000 on its initial run in Peru – a location far removed in distance and style from its previous Midland main line test runs, on which Hughie also officiated.

Possessing the aesthetics of a brick the four-wheeled 300 h.p. locomotives for the Consett Iron Company and Barrow Ironworks nevertheless had a purposeful air about them and a useful starting tractive effort of 22,400 lb. HE 4010 in works grey, November 1950.

A record photograph intended to show the controls of 45 h.p. mines locomotive no. 4255 destined for the Ashanti Goldfield in the Gold Coast, but interesting in that it is one of the few photographs to show the utilitarian works shunter, no. 1850 (on the left in this picture). Also evident is a Ministry of Supply 'Austerity' saddle tank fresh from overhaul and awaiting a visit from the painters. The dark shape behind the 'Austerity' is ex-LMS 0–6–0 diesel No. 7053 awaiting a decision on its fate. The time is June 1951.

shop foreman, George Cowell, having gone to the steelworks as locomotive engineer – a job he retained until the works closed, when he became engineer to the Beamish museum. At the time of checking this manuscript I spoke to George, who was still taking an active interest in locomotives on the Tanfield Railway at the ripe old age of 84. Here was a direct link with George McArd, mentioned in chapter 4, for, as an apprentice, the young George Cowell had erected the Armstrong-Whitworth diesels designed by McArd. Such was the interchange of personnel between the private locomotive builders.

No. 4551 was a bit of a plaything and was used as a mobile test-bed for experiments with gear-change mechanisms, engine exhaust brakes, etc., and was provided with a near-centre cab. It was the first Hunslet locomotive to be so fitted, although not the first to be despatched to a customer. A Perkins P4 engine was also fitted to drive the auxiliary equipment, making the total power available approximately 350 h.p. For some time no. 4551 toured the country, sometimes with 204 h.p. diesel-hydraulic no. 5511 in tow. Oxfordshire Ironstone was one port of call, the National Coal Board in Northumberland another. No. 4551 ran for some time on British Railways trains in Durham coupled to the later 1940-built LNER dynamometer car.

But we are getting ahead of the situation in 1949. Alongside the 204 h.p., 300 h.p. and 500 h.p. locomotives already described were a batch of six very neat 72 h.p. 0–4–0 chain-drive estate (or plantation) locomotives. These were nos. 4004–9, ordered by Robert Hudson Ltd. Each had a Meadows 4DT420 engine and a simple canopy with Perspex front and rear screens and canvas side

Hunslet 44 ton 325 h.p. 0–6–0 shunter no. 4551/56 on test in County Durham in April 1958. Fitted with a Mirrlees J5 diesel engine and used as a mobile test bed for transmission and braking experiments, this locomotive travelled widely – the Oxfordshire Ironstone Company and the Bowes Railway being but two locations – before being sold 'shop soiled' to the NCB's Aberaman Phurnacite smokeless fuel plant in South Wales in December 1961. The dynamometer car is the 1951 reincarnation of the LNER Gresley car that was authorized in 1937 but was destroyed by fire while under construction. Note the reflection of the other track in the highly-finished crimson lower panels of the car.

The first Hunslet diesels for the Sudan Gezira Board, nos. 4341–4, appeared in 1953. This is no. 4344 in works grey. 120 h.p., 18 tons, 2 ft. 0 in. gauge. With its McLaren M6 Mk I engine and four-speed Hunslet gearbox it was capable of hauling 150 tons at 20 m.p.h. over distances in excess of a hundred miles at a time.

curtains. Two of them, nos. 4004/5, were of 2 ft. 0 in. gauge for Borneo; no. 4007, also of this gauge, went to South African Portland Cement. Another two were of 2 ft. 6 in. gauge: no. 4006 for Palestine Portland Cement and no. 4008 for Roan Antelope Copper Mines. No. 4009 was an 80 cm gauge machine for Port Louis, Mauritius. All of them went away between July 1949 and January 1950.

The 300 h.p. and 500 h.p. locomotives had set new standards for driver comfort and view from the cab, with their pedestal driving seats, duplicated controls, air-operated gear-change and fully-enclosed cab with large windows all round. Visibility forward over the engine compartment was much facilitated by sloping the top corners of the bonnet inwards at 30 degrees to the vertical. This modernizing treatment was being applied progressively to the other surface locomotive designs, notably the 153 h.p. and 204 h.p. models; but some orders that were already in progress emerged visibly similar to their pre-war antecedents. These included four Ruston-engined 132 h.p. 2 ft. 6 in. gauge 0–6–0s (nos. 4156–9), based on the African Manganese units but intended for the Kelani Valley line in Ceylon; another 1 ft. 6 in. gauge 0–4–4–0 'Carnegie' for the Woolwich Arsenal; a 2 ft. 0 in. gauge

0–4–4–0 (no. 4509) for Doornkop Sugar Estates; and some 102 h.p. and 153 h.p. standard-gauge 0–4–0s for C. A. Parsons, London and Thames Haven Oil Wharves, and Portsmouth Dockyard.

Brief reference was made in chapter 6 to the Kerr Stuart diesel locomotive supplied to the Sudan Gezira Light Railway in 1929. The Sudan Gezira Board was responsible after World War II for development of the cotton plantations centred around Barakat in Blue Nile Province about 150 km south of Khartoum. The 60 cm gauge Gezira Light Railway expanded rapidly to become a system with hundreds of miles of track connecting plantations and ginning factories. In January 1953 Hunslet supplied four 18 ton 0–8–0 diesel-mechanical locomotives for main-line use. The McLaren M6 diesel engine developed 120 h.p. and was started by a Morris petrol engine. The four-speed Hunslet gearbox gave a top speed of 20 m.p.h. on the level with a load of 150 tons. The design was to be produced at intervals, with only detail differences, over a period of 34 years.

The second-generation Gardner 6L3 and 8L3-powered shunters were developed as a 'design pair' and incorporated the maximum number of interchangeable

The first two Hunslet shunters for British Railways at work at Ipswich Docks in 1956. No. 11500 (Hunslet no. 4625/54) is one of three 153 h.p. 0–4–0 machines supplied specifically for the quayside lines, while No. 11136 was the first of twenty-four 204 h.p. 0–6–0s (Hunslet nos. 4866–73 and 4996–5011) for the Eastern Region of British Railways. The cow-catchers and aprons on No. 11500 were deliberately styled to be in keeping with the ex-Great Eastern Railway four-wheeled tram engines. A sort of 'diesel Toby' if you like.

Two standard 204 h.p. diesel shunters against a backdrop of the new Avenue carbonization plant just south of Chesterfield in May 1957. Five of these locomotives were supplied in 1955/6 to the builders of the plant, Woodall-Duckham Ltd, for handing over to the National Coal Board, imaginatively named Avenue 1 to Avenue 5. A degree of nostalgia for these locomotives may be forgiven; for, with the almost identical British Railways and steelworks examples, they were the author's last major design involvement before military service intervened in June 1955.

parts while still allowing the customer to have a choice of detail fittings. The first of the 153 h.p. units with the improved cab and gear-change, no. 4539, was in fact different in being a 2 ft. 6 in. gauge 0–6–0 for Umfolozi Co-operative Sugar Planters, sent away on 30 November 1953. The same design team, of which the author was a member, concurrently produced the drawings for the British Railways Eastern Region 0–4–0 and 0–6–0 shunters (BR nos. 11500–2 and 11136–59 respectively) and for the industrial versions that went to Tilmanstone Colliery (153 h.p.) and to the NCB, the Consett Iron Company, John Summers, etc. (204 h.p.).

Consequently there were very few basic differences: for example the Tilmanstone 153 h.p. locomotive differed visibly from the BR Ipswich four-wheelers only in that the latter were fitted with electric marker lights and had cow-catchers and side aprons for street running, copied unashamedly from the Great Eastern Railway steam tram engines, using the picture on the cover of *Model Railway News* for September 1939 as a guide. Similarly the five 204 h.p. 0–6–0s supplied to the new Avenue smokeless fuel plant at Chesterfield differed from the BR 0–6–0s only in not having the marker lights and vacuum brake. Three examples of a flashproof version were supplied for use on the metre-gauge Iraqi Republican Railways tracks at Basra oil refinery in 1955.

The last two batches of British Railways shunters, D2574–D2585 for the Scottish Region in 1958/59 and D2586–D2618 for the North Eastern Region in 1959/61, had a further improved cab profile, larger fuel tanks and 3 ft. 9 in. diameter wheels. All other Hunslet 204 h.p. locomotives, except for six four-wheelers for the Consett Iron Company and the CEGB, had 3 ft. 4 in. wheels. A final stock batch of three standard locomotives was put in hand in 1959, and two of these were delivered to John Summers in May 1960. The diesel-mechanical shunter had had its day, however, and no. 5673 remained unsold until September 1963 when it was despatched unpainted in response to an urgent telephone request from a contractor, Smiths Enterprises (Glamorgan) Ltd, performing work on slag disposal at Richard Thomas and Baldwin's Spencer Steelworks, Llanwern. A total of 158 Gardner 8L3-engined Hunslet 204 h.p. diesel-mechanical locomotives had been built for eight different countries and three track gauges over a period of 23 years. Fifteen basically similar locomotives had been supplied in the period 1954–7 (nos. 4528–37 and 4697–4701) to New Zealand Government Railways, but these were fitted with the National Gas and Oil Engine Company's 250 h.p. engine; and there was a solitary McLaren-powered example (no. 5436) for South Africa in 1958.

For twelve years after the cessation of hostilities the locomotive business thrived, and in 1957 one hundred and five diesel locomotives were produced by the Hunslet Engine Company. Steam locomotive production numbered only three examples that year, but the overall turnover was very good indeed. There was, however, a sense of change in the air. Customers were now regularly demanding larger locomotives than the Gardner-engined units. While the 300 h.p. and 500 h.p. locomotives had been superb examples of the locomotive designer's art, the gearboxes had been heavy, complex and costly, with their multitude of high-quality nickel–chrome steel gears, clutches and shafts; and the Crossley and Mirrlees engines were not without their problems. The slow running speed, 900 r.p.m., of these two engines, compared with the 1200 r.p.m. of the Gardner, demanded a gearbox over twice the size to compensate for the compound effect of lower speed combined with higher power.

The National M4AA6 engine, developed from the 250 h.p. unit employed in the New Zealand locomotives, now offered 264 h.p. at 1700 r.p.m., and the higher speed meant that the 204 h.p. gearbox could more than cope with the increased power without detriment. Consequently in 1956 a decision was made by the Hunslet Board to put in hand for stock a batch of four locomotives based on the 30-ton 204 h.p. frame, gearbox, wheels and running gear but weighing 32 tons and incorporating the National M4AA6 engine. Of the four locomotives two, nos. 5238/9, were to be straight diesel-mechanical machines, while nos. 5240/1 were to have a Brockhouse torque converter in addition to the four-speed gearbox.

No. 5240 was rushed through first, and was despatched on 14 August 1957 for extended trials at Yorkshire Main Colliery, where in fact it remained for the rest of its days. There was no real urgency for the other three stock units, and the order book was still healthy. The result was that events, shortly to be described, overtook them and rendered them obsolescent before their time. No. 5241 was eventually sold in 1961 to Caltex for an oil refinery in Turkey, and a package deal was negotiated in 1962 with the National Coal Board, East Midlands No. 4 Area, under which no. 5238 went to Pleasley Colliery, no. 5239 to Blackwell 'A' Winning Colliery and a stock 153 h.p. 0–4–0 (no. 5623) to Newstead Colliery.

D2574 (Hunslet no. 5456/58) was the first of twelve 204 h.p. locomotives for Scottish Region. Mechanically similar to the Eastern Region examples except for larger driving wheels (3 ft. 9 in. diameter) in cast steel instead of 3 ft. 4 in. diameter cast iron and a larger cab with better visibility. Photographed at St Rollox on 13 September 1958.

The last batch of Hunslet 204 h.p. shunters for British Railways were nos. 5635–67 supplied to the North Eastern Region between 1959 and 1961. They worked from a number of depots throughout the region, from Percy Main in the north to Ardsley in the south. Here D2614 (Hunslet no. 5663/61) is seen at Goole docks on 21 March 1962.

A classic piece of Hunslet mix-and-match to produce a 'special' from well-proven components. HE 4689/55 was one of two (HE 4754/63 was the other) 2 ft. 0 in. gauge 230 h.p. Bo–Bo locomotives supplies to Illovo Sugar Estates in Natal. Effectively a marriage of the upperworks of the standard-gauge 204 h.p. shunter and the bogies of the pre-war Avonside and Hunslet geared steam locomotive. The rear bonnet housed a 204 h.p. Gardner 8L3 engine to provide traction. In front of the cab was a 26 h.p. 2LW engine driving compressor, fan and other auxiliaries, together with the radiator.

Built like a battleship. The 637 h.p. 3 ft. 6 in. gauge 0–8–0 shunter no. 4835 supplied in June 1957 to Mufulira Copper Mines in Northern Rhodesia (now Zambia). Driven half-way across Africa from docking at Beira in Mozambique through Salisbury (now Harare) and Bulawayo, over the Victoria Falls and up through Lusaka to Ndola – 1000 miles at a maximum speed of 25 m.p.h. – by another of Hunslet's unsung heroes, Arthur Johnson.

An aerial view of the Hunslet Engine Works from the south-east. This and the next photograph were taken at 11.30 a.m. on 21 October 1958. The extension to the boiler shop into an office frontage on Jack Lane can be seen, together with the footbridge connection to the main office block. Building of the new Drawing Office above the stores (which themselves grew out of the first Manning Wardle boiler shop) is well advanced (left of centre) and all but one building of the old quadrangular Railway Foundry (centre right) has been demolished. Compare this view with the 1949 one in chapter 8.

The Hunslet Engine Works from the north-east, also on 21 October 1958. The re-roofed south building of Shepherd & Todd's Railway Foundry and the boundary wall and entrance arch from Pearson Street can be seen in the centre. There are still quite a number of old terraced houses to be demolished.

Brief reference has already been made to the demonstration 350 h.p. locomotive no. 4551, and even more briefly to its 204 h.p. counterpart no. 5511. Before the last drawings for the National-engined 264 h.p. batch (nos. 5238–41) had been issued, the decision was made to address another customer requirement, that of further improvements in the visibility from the cab in both directions – hardly a surprising prerequisite in a shunting locomotive – and also to concentrate on diesel-hydraulic transmission for the larger surface locomotives.

The commercial impetus for this decision came from a National Coal Board (Durham Division) order for two 270 h.p. 32-ton 0–6–0 locomotives, one for Fishburn Colliery (no. 5341) and one for Sherburn Hill Colliery (no. 5342), both supplied on 21 December 1958.

The two most formidable obstacles to harnessing the internal combustion engine for railway purposes were, first, how to apply the engine's power to the rails without snatching or stalling on starting, and, second, how to move up the speed range without jerking. The difficulty lies in the fact that, whereas a steam engine produces its maximum torque at almost zero speed, and is therefore ideally suited to accelerating a load from rest, the internal combustion engine has a torque curve that slopes the other way, and will not function at all below a certain speed. Anyone who remembers early car driving lessons using a manual gearbox, especially one without synchromesh, will understand the problem. But steam locomotive men were not car drivers, and the blend of brute strength, manual dexterity and sensitive timing demanded by the early diesel-mechanicals was a rare commodity.

A fluid coupling between engine and clutch helped to solve the starting problem. The down-side was the loss of braking assistance from the diesel engine, and Hunslet disliked fluid couplings for this reason. There also remained the need to be able to go up through the gears. The Brockhouse single-stage torque converter, fitted between clutch and gearbox on nos. 5240 and 5241 was in effect an up-market fluid coupling. It gave a speed range of up to 8 m.p.h. in each gear, which effectively rendered the first and third gears redundant. Consequently, from no. 5341 onwards, Hunslet hydraulic transmissions only had two-speed gearboxes.

As with the earlier stock batch, this new design retained the traditional 204 h.p. frame and running gear, with the 9 ft. 0 in. wheelbase, 3 ft. 4 in. wheels with cast-iron centres, jackshaft drive, and coupling-rods with integrally-machined oil siphons. The superstructure changed considerably, however, with the driving cab moved as far towards the engine as the Hunslet friction clutch would allow, in effect straddling the driving shaft between clutch and gearbox. Some attempt to lower the bonnet line was made, which led to slightly larger windows, but the side-entry doors to the cab were retained. Pushing the cab hard up against the clutch had its down side, however, in that the control desk could not be accommodated in its normal position. The whole cab was turned round, and the controls consequently faced the gearbox end, which was somewhat confusing to a generation steeped in the tradition that the radiator end of a diesel locomotive was usually the front end.

These two Durham Division locomotives again had a National engine, but of a later type, the NM6T. This was not a happy choice. Two further locomotives, ostensibly identical to nos. 5341/2, put down to stock shortly afterwards, eventually appeared with the old reliable Gardner 8L3 of 204 h.p. One of these, no. 5510, went to Cuba on 18 January 1960, and the second, the previously-mentioned no. 5511, to Caerphilly Tar Plant as an ex-demonstration unit at the end of July the following year. A variation on the same theme was two metre gauge examples, nos. 5618 and 5670, which had the same National engine but were cut down in width and height to clear a particularly difficult tunnel in the Basque mountains. These went to the Orconera Iron Ore Company in Northern Spain on 12 June 1960; but when Orconera bought two further locomotives, nos. 6600 and 6698, in 1965 and 1967 they specified Rolls-Royce engines (not available at the time of their first order) and also ordered two Rolls-Royce engines to replace the Nationals. One consolation in 1957 was that no one else was really having much luck with anything other than the Gardner engine, and another was that the large orders placed by British Railways for 204 h.p. diesel-mechanical shunters bought a certain amount of time.

This centre, or near-centre, cab design was in reality an interim stage between the pre-war designs and the final Hunslet standard large surface shunting locomotive, although that was not anticipated at the time. Again we had the design pair concept, a six-wheeler and a four-wheeler. All told, four of the 0–4–0s were built and thirty-one 0–6–0s authorized, the last of which was not assembled. The Madras Port Trust ordered four 5 ft. 6 in. gauge examples arranged for operation as tandem pairs. In

To meet the requirements of firms with limited rail traffic, Hunslet developed the Yardmaster, a 13-ton 71 h.p. chain-driven diesel-hydraulic locomotive with a Dorman 4LB engine and Brockhouse torque converter. It could be driven from the usual position inside the cab or by someone standing outside on the long shunter's platform. One-man operation was therefore possible, with consequent cost savings. With the reshaping of British Railways and the closure of the Gas Works and other small rail-connected concerns the market disappeared and only a handful were built. HE 5176/57 is seen at the Yorkshire Copper Works, Stourton, shortly after delivery.

these the National engine was replaced by the more reliable although rather more expensive Paxman 6RPH unit, but the two-speed hydraulic transmission was retained. All four left for India on 17 March 1960. Two 3 ft. 6 in. gauge examples (nos. 5624/5), which went to the Tsumeb Corporation in South West Africa (now Namibia) in September 1960, brought yet another change of engine. This time it was the Cummins NRT6 of 230 h.p. – a happier choice than hitherto but still not quite powerful enough for most envisaged applications.

After several years of enjoying an almost Savile Row-style market – one did not 'buy' one of their engines, rather did one request an 'allocation' (and like Bugatti engines they had the reputation of being hand-carved from the solid) – Gardners bowed to the inevitable and uprated their

larger engines to become the 6L3B and 8L3B, developing 195 and 260 h.p. respectively at 1300 r.p.m. This resulted in a further pair of stock 260 h.p. locomotives, nos. 5590/1 for the home market, followed by five for Thailand State Railways, one UK demonstrator (no. 6230), two more batches of four 0–6–0s (nos. 6286–9 and 6691–4) and two stock 0–4–0s (nos. 6262/3), all Gardner-powered. The first two 0–6–0s had the Hunslet two-speed hydraulic transmission; the remaining 0–6–0s and the 0–4–0s the new 'Hunslet Twin' transmission.

On the two-speed diesel-hydraulics there was still the bugbear of changing between first and second gears, an operation difficult to achieve without disconcerting noises. The 'Hunslet Twin' circumvented the problem by providing each gear with its own torque converter: the two

3 ft. 6 in. gauge 230 h.p. 0–6–0 no. 5624/60 destined for the Tsumeb Corporation in South West Africa (now Namibia). Cummins NRT6 diesel engine for traction purposes with a Gardner 2LW auxiliary engine driving the fan, compressor and vacuum exhauster. Hunslet two-speed hydraulic transmission.

converters drove a common output shaft through simple gear trains of different ratios and a unidirectional device. As the locomotive approached the maximum speed attainable in the first ratio, the associated converter began to freewheel and the drive was taken up by the second converter. The arrangement was simple in theory but needed development. Neither time nor quantity was on Hunslet's side, for the transmissions produced in significant numbers for earth-moving and off-highway vehicles in the United States were beginning to take over in railway locomotive applications also.

Thus these were very much 'end of range' locomotives. 'Don't buy in anything that can be made in house' had long been the Hunslet dictum, and this policy had served the company well for as long as diesel engines remained slow-running and transmissions required tailoring to individual needs. Market forces were to change all that. Quick-running Cummins and Rolls-Royce diesel engines (1800 and later 2100 r.p.m.) were coming on to the market in

increasing numbers. Rolls-Royce had acquired the American Twin Disc Clutch Company torque converter licence, and had purchased the Sentinel factory in Shrewsbury whence they introduced their Design Award-winning diesel shunter range in 1959. One could buy three Rolls-Royce 300 h.p. diesel engines for the price of two Gardner 8L3Bs, and at least four Twin Disc torque converters for the cost of one Hunslet Twin transmission. The economics and marketability of the 'traditional' product no longer added up. Sentinel monopolized the market for quite a while, and the stock Hunslet units were slow to dispose of; indeed no. 6694 was never built and no. 6691 was used as a hire locomotive and within the works until it was sold off to Woolley colliery ten years after completion.

This was not the end of the world; there was an answer. More of that later, for so far in this chapter we have not mentioned the other lucrative locomotive markets.

The National Coal Board's requirement for flameproof underground locomotives had remained pretty steady in the period under review, with all three basic models, 65 h.p., 70 h.p. and 100 h.p., much in demand. These, it will be remembered, had a single driving position at the rear, but by 1957 the Mines Inspectorate was requiring that new locomotives weighing more than $9\frac{1}{2}$ tons should have either a central driving position, which was not really practicable with the diesel, in view of the restrictions on height and width, or a driving position at each end. Out of this requirement developed the 66 h.p. (Gardner) and 76 h.p. (Meadows) 10-ton 0–4–0 and 100 h.p. (Gardner) 16-ton 0–6–0 'double-ended' locomotives. To facilitate transportation in the cage to pit bottom or intermediate roadway level, the driving wells at each end were detachable. The author made most of the drawings for these new locomotives, of which the frames, or bedplates, were intricate one-piece castings in high-grade iron, incorporating not only the horn blocks but also the fuel tanks, sand-boxes, engine carrier, auxiliary water tanks and the lower section of the gearbox.

A long-wheelbase 0–4–0 with a 170 h.p. Cummins horizontal engine and measuring 7 ft. 0 in. wide and 3 ft. 3 in. high was produced for the 1953 American Mining Exhibition at the Cleveland Show from 11 to 14 May, but it failed to break the United States' monopoly of electric locomotives underground and returned to Leeds unsold. At the other end of the scale were the 15 h.p. 'Tiny Tim' and 23 h.p. 'Tiger Tim' classes. The first of these had an Enfield horizontally-opposed two-stroke engine, and the latter the Ailsa Craig RFS2 engine that was still being used in relatively large numbers in the 21 h.p. 'Hudson Hunslet' $3\frac{1}{2}$-tonner. Customers for these were legion – gold mines in Ghana, phosphate mines in Morocco, sulphur mines in Greece, uranium mines in India, coal and metal mines in Canada, and so on.

The Irish Peat Development Authority (Bord na Mona) provided another steady income for Hunslet for 25 years

The final development of the traditional Gardner engined mines locomotive was the double-ended version of the earlier 65 and 70 h.p. 0–4–0 and 100 h.p. 0–6–0. This photograph shows HE 5516/59, a 66 h.p. 0–4–0 for Glyncorrwg colliery high up in the Rhondda Valley. The 100 h.p. version had the same profile but was longer to accommodate the larger engine and additional pair of wheels. Just short of a hundred double-enders were built in total.

The 170 h.p. mines locomotive no. 4538, designed for the American market and exhibited at the Cleveland Show, under construction early in 1953. The exhaust gas conditioner, with its four large double flame traps, can be seen in the centre of the front section flanked by the two radiator and fan assemblies. The horizontal Cummins diesel engine, antecedent of the present-day Class 158 DMU engines, can just be seen to the left of the three oil-bath air cleaners. Not a lot of space in a locomotive only 3 ft. 3 in. high. In the background, left, can be seen the frame and bunker of War Department 'Austerity' locomotive No. 75274, in for repair to order no. 58429, and the boiler for a new 16 x 22 in. standard locomotive. Behind the 170 h.p. locomotive is the frame for no. 4522, a 102 h.p. 0–4–0 diesel-mechanical shunter for C. A. Parsons at Newcastle. Top right of the photograph is ex-LMS No. 7053, brought in for possible re-engining but later scrapped as beyond economical repair.

Dignity and Impudence. The 170 h.p. mining locomotive compared with a 15 h.p. 'Tiny Tim' destined for Canada. 'Tiny Tim' had a 23 h.p. bigger brother, 'Tiger Tim', and large numbers of each size were built both in Leeds and in South Africa. The 170 h.p. machine was to be a one-off, rejected by the American trade unions and defeated by the 'Buy American' policy.

Mechanically the 1950s revamp of the 20 h.p. 'Hudson Hunslet' was little changed from the legions of pre-war examples. Christened 21 h.p. for identification purposes, the engine casing was restyled and a seated driving position provided. This is no. 4479, one of a batch of four supplied to ICI in 1953 for the 2 ft. 6 in. gauge railway at the Ardeer Works of its Nobel Division, originally Nobel's Explosives Co. Ltd.

from 1961. The Authority's Chief Mechanical Engineer Mr W. Green, an Englishman from Kent, had designed a locomotive especially for the unique conditions experienced in the vast peat bogs, and a prototype had been built in Bord na Mona's Derrygreenagh works. In 1961 tenders were sought from twenty locomotive builders throughout the UK and western Europe, and against this competition Hunslet gained the first production order for twenty-five 'Wagonmaster' locomotives. A further twenty-eight were built in 1964–5, twenty-five more in 1971–2, twenty-three in 1977 and another twenty-five in 1979–81.

The Wagonmaster was a 9.3 ton, 85 h.p., 3 ft. 0 in. gauge locomotive powered by a Ford diesel engine – type 590E in the first two batches, 2713E in the remainder. The coupled wheels were driven from a jackshaft, which was driven from the engine through a Vulcan Sinclair fluid coupling, plate clutch, five-speed Ford gearbox, cardan shaft and special heavy reverse gearbox of Hunslet design. This seemingly elaborate arrangement provided five speeds in forward and reverse, allowing a comparatively small engine to give both a high starting tractive effort and a useful top

speed for medium distances. Much thought was given by the Bord na Mona engineers to adhesion, and specially-dried sand was used so as to give the best-possible adhesion on the short steep gradients that were characteristic of terrain in which the mechanics of peat extraction can cause abrupt changes in level.

During the first half of the fifties, the combined steam and diesel locomotive output from Hunslet hovered within a whisker of 100 locomotives a year, peaking at 118 in 1953. The mix was roughly three diesel to one steam in numbers, but in reality the steam locomotives were of a more uniform nature in terms of size and value whereas the diesels varied from under 3 tons to over 50 tons in weight. The order book and work-load values combined were consequently nearer two to one in favour of the diesel locomotive.

As steam locomotive production tailed off after 1955, diesel production initially rose to maintain the work-load at a steady level. It was realized, however, that the post-war boom was at an end, and that some diversification would be necessary if momentum was to be maintained. This is a subject that will be discussed later; but owing to

A total of 126 Wagonmasters were built between 1962 and 1981 for the 3 ft. 0 in. gauge lines of Bord na Mona, the Irish Peat Board. One of the later examples, HE 8934/80, heads back across the bog at Blackwater, Co. Offaly, with an empty train after delivering a load of peat for the never-ending appetite of Shannonbridge power station's boilers on 5 December 1996.   COLIN BOOCOCK

the long lead times prevalent at that time (up to two years), and the fulfilment of orders for British Railways, John Summers, the Consett Iron Company, the National Coal Board, not to mention the huge export market, another problem manifested itself. The traditionalists in the company were slow to realize that no matter how technically superior the locomotives with slow- and medium-speed engines, with their heavy in-house transmissions, might appear to be, the fact was that they were just not selling against the sleek new Rolls-Royce Sentinel. In the years 1959–62 Sentinel had despatched over 120 standard-gauge shunting locomotives to British industrial users; sixty-two of these were for Dorman Long alone. The same four-year period had seen the equivalent Hunslet figures drop to 250 in total – an average of just

over sixty a year, of which only fifty-three were standard-gauge units for the home market. Most of these fifty-three had been sold, but not yet built, before the appearance of the first Sentinel. Other manufacturers were also moving over to the Rolls-Royce engine and Twin Disc transmission; indeed Andrew Barclay had been the first to use an imported Twin Disc, coupled to a National diesel engine, in 1957, and the Yorkshire Engine Company had been using the Rolls-Royce engine in diesel-electrics since 1955. From being UK market leaders in 1955 Hunslet had dropped to fifth place by 1960, and while the steam locomotive underfeed stoker conversion programme was to cushion the blow for a while, something more drastic had to be done.

CHAPTER II

Standard locomotives

A Nelsonian interpretation by the Sales Department at Jack Lane of an internal instruction from John Alcock that 'Hunslet transmissions shall always be offered except when Sales [the Sales Department] consider that the chance of an order will be seriously jeopardised' contrived to allow an order to be placed with Hunslet in mid-1963 by the National Coal Board East Midlands No. 7 Area comprising two locomotives for a new coal preparation plant that was to be built at Snibston colliery in Coalville,

Leicestershire. The chance was taken to design a rather angular but nevertheless stylish six-wheeled locomotive having a Rolls-Royce C8SFL supercharged diesel engine. This developed 311 h.p. at 1800 r.p.m., and was connected via a CF11500 series Twin Disc torque converter and a Hardy Spicer propeller shaft to a new design of Hunslet single-speed reverse and final drive gearbox mounted on the trailing axle. Roller-bearing axleboxes and Davies & Metcalfe brake equipment were fitted as standard.

Part of the new drawing office built above the stores, shortly after moving in. This is in effect an upper storey built on top of what was the original Manning Wardle boiler shop. The date is 28 November 1962 but the Mavitta draughting machines on which the author had trained in the forties were to last for several more years.

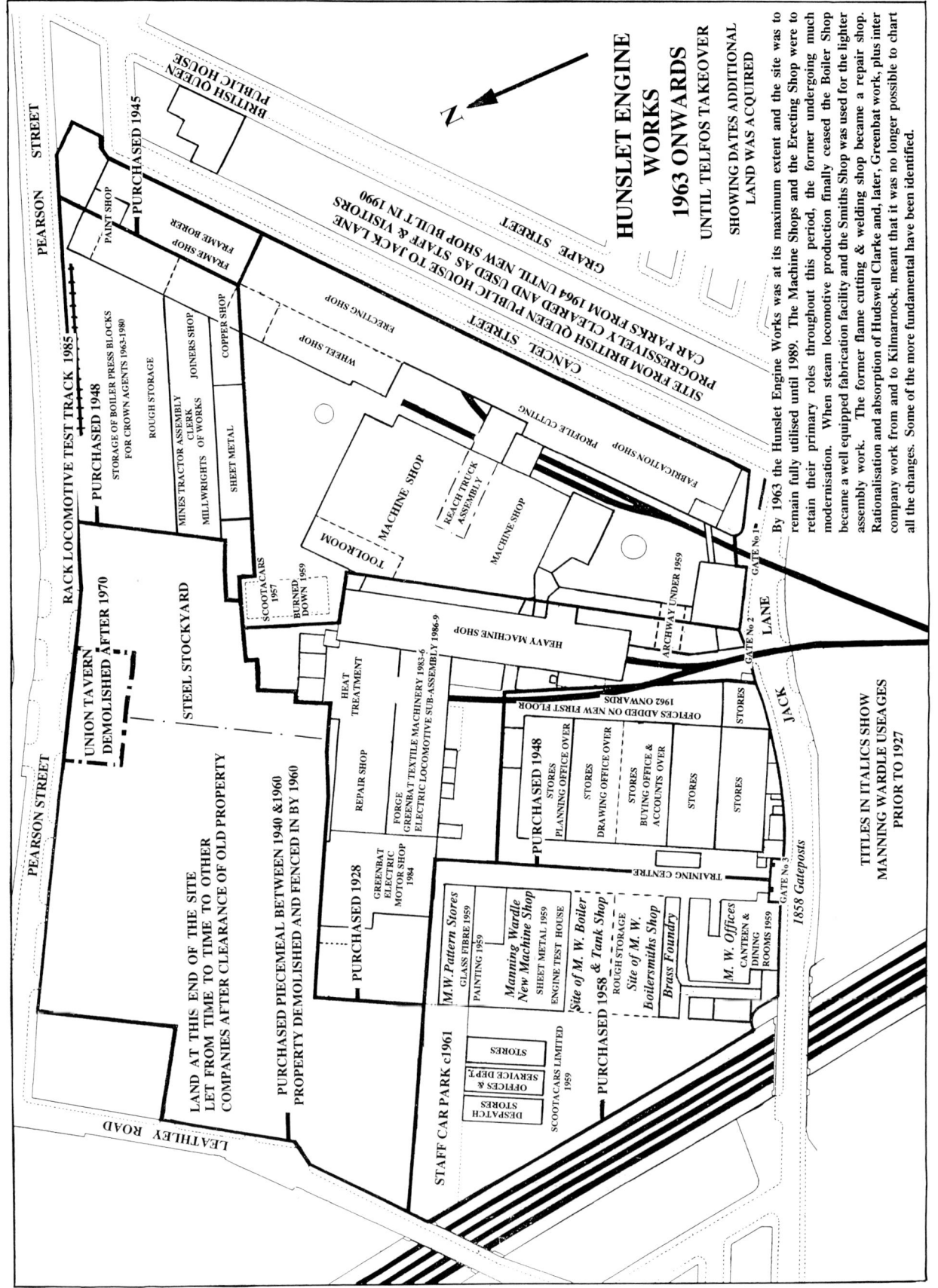

CHAPTER 11

HUNSLET ENGINE WORKS 1963 ONWARDS

UNTIL TELFOS TAKEOVER

SHOWING DATES ADDITIONAL LAND WAS ACQUIRED

By 1963 the Hunslet Engine Works was at its maximum extent and the site was to remain fully utilised until 1989. The Machine Shops and the Erecting Shop were to retain their primary roles throughout this period, the former undergoing much modernisation. When steam locomotive production finally ceased the Boiler Shop became a well equipped fabrication facility and the Smiths Shop was used for the lighter assembly work. The former flame cutting & welding shop became a repair shop. Rationalisation and absorption of Hudswell Clarke and, later, Greenbat work, plus inter company work from and to Kilmarnock, meant that it was no longer possible to chart all the changes. Some of the more fundamental have been identified.

TITLES IN ITALICS SHOW MANNING WARDLE USEAGES PRIOR TO 1927

This was the first market-led Hunslet design of diesel locomotive, and the drawing office was prevailed upon to produce something that embodied most of the virtues and as few as possible of the vices of the Rolls-Royce Sentinel and other competitors' products. They responded magnificently, and, moreover, the design provided basic sub-assemblies that would allow the future building of locomotives within a power range of 200–800 h.p., weights of 35–80 tons, track gauges from 3 ft. 6 in. to 5 ft. 6 in., and of 0–4–0, 0–6–0 and 0–8–0 wheel arrangements.

The two Snibston locomotives, nos. 6294/5, weighed 40 tons and were carried on six 3 ft. 9 in. diameter wheels, and were despatched during January 1965. A near-centre cab with rear entry doors and an unobstructed, relatively high floor gave excellent visibility in both directions, while convenient steps and platforms provided safe access and allowed the shunter and pointsman to ride within the profile of the locomotive at any of the four corners. The paint scheme adopted was a striking two-tone green livery – dark Brunswick green at top and bottom with a central band of mid Brunswick green, the three bands separated by $1^{1}/_{2}$ in. wide vermilion lining. This scheme was carried further with carmine and crimson lake with yellow lining for NCB North Yorkshire, and grey and yellow with silver for Shell-Mex and BP.

While nos. 6294/5 were at the design stage, a request came from NCB Northumberland and Durham Division for proposals covering 35-ton 0–4–0 and 55-ton 0–6–0 locomotives, and from NCB Coal Products Division came an enquiry for two large locomotives to work the steeply-graded branch (1 in 49) from Lythalls Lane sidings on the British Railways Coventry–Nuneaton line to the projected Coventry Homefire smokeless fuel plant at Keresley $1^{3}/_{4}$ miles away.

The Hunslet Engine Company celebrated its centenary year (1964) with an open day at the Works, the conveniently coincidental installation of Patriarch John Alcock as President of the Institute of Locomotive Engineers and the publication of Tom Rolt's rose-tinted *A Hunslet Hundred*. All pretty upbeat stuff, but enacted

The first Hunslet diesel-hydraulic shunter with Twin Disc transmission, HE 6294/64 for the Snibston coal preparation plant, is shown here with the equipment compartment doors removed to demonstrate the simple layout and good accessibility. Eight-cylinder Rolls-Royce diesel engine, radiator and transmission flexibly mounted in the front compartment; gearbox accessible from rear of cab. Fuel tank with direct-reading contents gauge and with air reservoirs attached forms the rear sub-assembly. Stylish layout without being over-fussy; cheap without being nasty. 311 h.p., 40 tons weight.

Very much in transition. The Hunslet erecting shop on 26 November 1964. The two prototype Rolls-Royce-engined, Twin Disc transmission shunters nos. 6294 and 6295 can be seen, one half-way down each row. Three 'Austerities' and a Hudswell Clarke 17 in. steam locomotive (frame in centre, wheels left foreground) are being converted to automatic stoking, and the last 'traditional' Gardner-engined diesel shunters are timorously facing the future. These are nos. 6286/65 and 6287/65 for Elsecar and Wath Main collieries (heading the right and left rows respectively), no. 6263/64 for the Exeter gas works (centre front) and no. 6257/64 for Thailand State Railways (back right). Just discernible in the far distance on the original print are two Wagonmaster locomotives crated up for shipment to the Bord na Mona in Ireland and a 66 h.p. mines locomotive for Blidworth colliery.

The frame for the first 55 ton 311 h.p. locomotive, no. 6611, assembled on 8 March 1965. The rigidity of the 3 in. thick frame plates will be noted. Rivets were still used for assembly, but the 6 in. thick buffer beams were bolted on, since it was considered that the mass of metal would have prematurely and unevenly cooled the rivets if they had been used at this point.

against a background of Sentinel competition, a recession and the loss of Empire business which had seen the huge North British Locomotive Company out of business late in 1962. The looming shadow of Beeching rationalization would render the Swindon-built type 1 0–6–0 shunters, for which Hunslet built the gearboxes in 1963, redundant while they were less than five years old; and these would join other British Railways shunters to flood the second-hand industrial market.

Only thirty Hunslet locomotives were produced in 1964, but the NCB Northumberland and Durham Division and Coal Products Division orders materialized to produce eight 311 h.p. six-wheelers and two 60-ton 776 h.p. eight-wheelers respectively, which, when added to the five Thailand State Railways 240 h.p. locomotives, a repeat order for twenty-eight Bord Na Mona 'Wagonmasters', several 66 h.p. mining locomotives and others, brought business back to normal for 1965. The Sentinel had been caught off guard, but it had taken five long years.

Inherent in the Northumberland and Durham 'standard specification' was the proviso that the resultant locomotive should be able to be built subsequently by any competent industrial locomotive manufacturer through competitive tender. The use of standard 3 ft. 9 in. diameter wheels was mandatory, as was the incorporation of certain 'proprietary' items. These were the Rolls-Royce engine, Twin Disc transmission, Hunslet type 350 final drive gearbox and Hunslet control desk. In practice, all future Northumberland and Durham 'standard' locomotives were ordered from either Hunslet or Andrew Barclay, right up to the last one, which came from Barclay, in 1982.

The eight 0–6–0 locomotives, Hunslet nos. 6611–18, were allocated four to Springwell Bank Foot locomotive sheds, whence they worked the Bowes Railway; one to Westoe Colliery in South Shields; and three to Wearmouth Colliery, Sunderland. Apart from the 15 tons extra weight they differed little in appearance from the Snibston prototypes except that a six-inch increase in wheelbase, to 9 ft. 6 in., and an increase in overall length of 10 inches, allowed for a cross-over walkway at each end of the locomotive rather than just at the rear as on nos. 6294/5.

The Coventry locomotives had two Cummins NT400 diesel engines of 388 h.p. each at 2100 r.p.m., two Twin Disc CF11500 torque converters and a Hunslet type 350 gearbox on each of the two intermediate axles. All wheels were coupled by rods. In effect these were the equivalent of two six-wheel locomotives mounted back-to-back sharing a single cab on a common rigid frame.

The 35-ton 0–4–0 with Rolls-Royce C6SFL engine (233 h.p. at 1800 r.p.m.) came shortly afterwards. The first two, nos. 6675/6, went to Bentinck and Silksworth collieries respectively. It now only needed the insertion of a smaller locomotive in the range – this came on 10 May 1967 with the authorization of a batch of four 25-ton 0–4–0 179/240 h.p. locomotives with 3 ft. 4 in. wheels for building to stock – to produce a series which, by careful adjustment of weight, and a variety of engines and detail options, could meet most conceivable customer requirements over the next twenty years.

The documentation that was issued by the company at the time described the types available as 'a fully integrated series incorporating the maximum number of standard components and available with either Rolls-Royce, General Motors or Cummins diesel engines as standard options with other types fitted at customer request. The hydraulic transmission [it went on] is normally of British Twin Disc Type, with Voith as an alternative.' This standard range covered all powers from 179 h.p. up to 800 h.p. and weights from 25 tons up to 80 tons. There were five basic variants, as follows:

Light 0–4–0 locomotive	Weight 25–30 tons	179–300 h.p.
Heavy 0–4–0 locomotive	Weight 35–45 tons	250–450 h.p.
Light 0–6–0 locomotive	Weight 36–45 tons	300–575 h.p.
Heavy 0–6–0 locomotive	Weight 48–60 tons	325–750 h.p.
0–8–0 locomotive	Weight 50–80 tons	500–800 h.p.

Notwithstanding the excellent recovery in 1965, the following two years were disappointing for a number of reasons, mainly political. The widespread growth of decolonization in general and the reshaping of the African political map in particular had a detrimental effect on a number of traditional markets. The Yorkshire Engine Company had ceased building in 1965, Rustons finished in 1966, Fowlers lasted until 1968, when their goodwill was purchased by Andrew Barclay; and effectively by the mid-1960s it was left to Sentinel (and their associates Thomas Hill), Hunslet, Hudswell Clarke, Andrew Barclay and English Electric to meet the mainstream industrial locomotive requirement of the United Kingdom, together with any export opportunities that might arise.

In the wake of the Durham and Coventry deliveries it became the normal practice to put through stock batches of parts for the standard range of locomotives, and, depending on perceived requirements, these could either be limited to common parts or they could be agreed as to wheel arrangement; they could even be settled at a full

The Snibston design developed into the 55-ton Northumberland and Durham standard unit. The wheelbase was longer, 9 ft. 6 in. instead of 9 ft. 0 in., and the wheel centres were cast steel rather than cast iron. Access across both front and rear of the locomotive was provided from now on. The locomotive pictured here is HE 6613/65.

The 0–4–0 233 h.p. version used all the major sub-assemblies of the 0–6–0 but with the six-cylinder Rolls-Royce engine. This one is HE 6676/67, supplied to Silksworth Colliery.

Evolution not revolution. The bogies and frame are almost exactly pure 1930s Avonside geared steam locomotive design, but the cab, superstructure, Caterpillar diesel engine and Twin Disc transmission are classic late 1960s. Hunslet no. 7066, a 250 h.p. 25 ton Bo–Bo supplied in 1972 to the Compagnie Générale de Sucreries du Zaïre for the 60 cm gauge sugar cane railway.

The basic Twin Disc design appeared in a number of variations and liveries. This is HE 7041/71, a 350 h.p. flashproof unit, in a night-time scene at the Salt End, Hull, refinery of BP Chemicals Ltd.

final specification. Consequently any casual observer trying to understand the chronology of works numbers would be in difficulties, since the orders did not always match what was available in stock, and quite often a new batch would be started against a specific order.

The difficult trading conditions throughout 1966 and 1967, and on into 1968, meant that stock batches laid down in 1965 were slow to clear. An example of this is provided by batch J11, which was intended to cover four 40- to 45-ton 311 h.p. Rolls-Royce-engined 0–6–0s identical to the original Snibston twins. The first of these, no. 6683, went reasonably quickly in February 1967 to join its brethren at Snibston; nos. 6684/5 were delivered to NCB Castleford Area in March 1968; while no. 6686 was fitted with a Cummins 308 h.p. engine and sent as a flashproof refinery locomotive to the Slovnaft petrochemical plant in Bratislava, Czechoslovakia as late as January 1970.

Ten Rolls-Royce-engined 268 h.p. 3 ft. 6 in. gauge 0–6–0s (nos. 6665–74) were however supplied during 1966 to the Nigerian Ports Authority against an order placed by the Crown Agents. These had Voith transmissions and German-style 'steeple' cabs giving sufficient height to allow the cab access doors to be below the large rear windscreens. At least two of them were subsequently destroyed in the Biafra war when less than a year old.

The design traditionalists tried a final hand on 29 March 1966, when, against all the commercial odds, a programme amendment was issued: 'It has been decided to build a batch of four Hunslet Twin Drive standard 260 h.p. locomotives nos. 6691 to 6694. Delivery to commence week 43 (1966) at the rate of one every three weeks.' The fate of these four locomotives has already been mentioned.

So it was that the end of 1966 saw almost twenty stock locomotives complete, or nearly complete, without any

HE 6665/66 was the first of a batch of ten 3 ft. 6 in. gauge 268 h.p. locomotives supplied to the Nigerian Port Authority. The driving cab was of 'steeple' pattern, very Germanic, and the Rolls-Royce C8S diesel engine was matched to a German Voith hydraulic transmission.

immediate buyers in view, and these were to move but slowly over the next two to three years.

The continuing underground mining locomotive business and the steam locomotive automatic stoker conversion programme had in the first instance enabled Hunslet to maintain full contact with the National Coal Board, and, in the second, the steam revival had delayed the purchase of more than a handful of new diesel locomotives for colliery surface shunting, thereby hurting the competition more than it hurt Hunslet. At that time the Coal Board had a combined surface and underground fleet of around 2500 locomotives, of which the average railway enthusiast was completely unaware. The surface shunter hiatus had resulted in only eight of the first 175 Sentinel locomotives being sold to NCB, and it was mainly the steelworks business that had been temporarily appropriated by Sentinel from the likes of Hunslet.

With a new and acceptable range of locomotives, mostly available at short delivery, the major steelworks were targeted in an intensive Hunslet sales campaign in the run-up to Harold Wilson's Labour government's intended nationalization of the steel industry. One of two stock 534 h.p. locomotives, no. 6689, with its newly-developed Rolls-Royce DV8T diesel engine, was demonstrated at steelworks in South Wales and Yorkshire and at Hams Hall power station in the Midlands.

Nor were other customers neglected. Orders were received from Ceylon Government Railways for twenty-eight 5 ft 6 in gauge DV8T-powered 0–6–0 shunter/trip locomotives, 0–6–0 and 0–4–0 locomotives for the National Coal Board, refinery locomotives for Shell-Mex and BP at Hamble and for BP at the Isle of Grain; while the smaller 25-ton 0–4–0 machine sold to the Admiralty and to chemical plants in Roumania and Chile. The first Cummins-powered 388 h.p. 50-ton 0–6–0 (no. 6973) was sold to the CEGB at Carrington Power Station near Manchester. Business was on the up again, and it only needed the steelworks back into the fold to put the icing on the cake.

Over the years Spain had been a useful market for Hunslet, but by 1968 the emergence of indigenous suppliers was making the importation of complete products of any kind into the country virtually impossible. Hunslet concluded a deal with Babcock & Wilcox CA whereby Babcock & Wilcox built Hunslet locomotives under licence at Bilbao. Two 776 h.p. 0–8–0 diesels were built for the Port of Gijon-Musel, and these were followed by a handful of 0–6–0s for other customers. All of these had gearboxes supplied from Leeds.

Timing was everything, and a decision was taken in early June 1969 to hold an open day at the works, to which prospective customers were invited. Consequently the twenty-sixth of June saw just short of a hundred guests from the mining, petroleum, steel-making and manufacturing industries assemble at Jack Lane. To quote the press release of the day: 'The basic theme was industrial diesel locomotives but the items on show presented a vivid picture of the varied range of products produced at the well known Leeds works.' On the locomotive test track was to be seen a standard 388 h.p. industrial shunting locomotive – this one in the striking marigold livery of the Central Electricity Generating Board – together with a glossy black 530 h.p. 5 ft. 6 in. gauge unit for Ceylon Government Railways. The latter was the twentieth of a £750,000 order for 28 locomotives, production of which was running ahead of schedule. Its commodious cab allowed parties of six or so to ride together without becoming detached from their guide.

Battery locomotives for a gold mine undertaking in India, and diesel locomotives for nickel mining in Canada, salt mining in India and bauxite mining in Greece, served as a reminder of Hunslet's forty years' experience in the production of high-quality compact locomotives. Rubber-tyred mining equipment was represented by a white-painted MT.60 personnel carrier for Canada and a standard MT.25 mines tractor – the latter specially painted yellow for the BBC2 television series covering the excavations at Silbury Hill in Wiltshire.

In an advanced stage of construction in the Erecting Shop was the first of a batch of giant tractors designed to handle the Boeing 747 'jumbo jet'. Weighing 70 tons and powered by a 635 h.p. diesel engine, this rubber-tyred monster was destined for Heathrow Airport.

For those interested in mechanical handling, the Hunslet 4-Way Travel side-loading reach truck, which was by then well established, gave a demonstration of its high manoeuvrability and ability to travel down narrow aisles impassable by the more conventional fork-lift trucks of the day.

Several reminders of the century-old Hunslet tradition were to be seen in the variety of steam locomotive spares that were still being produced for overseas railways. There were parts for Burma, Benguela, Ghana, Portugal and many others, to say nothing of new boilers for the Festiniog

Fishing for the big one: Hunslet's Northern England and Scotland sales agent John Leybourne, on the running plate left, shows Ravenscraig steelworks Chief Engineer Bob Jamieson around the CEGB locomotive while other interested parties look on. 'We have another just like this inside,' says John; 'what colour would you like?' Painted yellow, sister locomotive HE 6974/69 duly went to Ravenscraig. Several others followed over the years. YORKSHIRE POST

Keith Alcock, right, describes the glossy black Ceylon Government Railways 530 h.p. Rolls-Royce-powered shunter no. 694 (HE 7150/69) to an admiring audience. This the twentieth of a batch of twenty-eight. Note the footwear of the day, competing with the finish on the locomotive. YORKSHIRE POST

Railway and the Welsh Highland Railway and the overhaul of a former British Railways main-line steam locomotive. These contrasted sharply with precision gear-cutting facilities, developments in container handling, aircraft towing tractors, and so on, but served also to demonstrate the solidarity and versatility of this closely co-ordinated group of companies.

A tour of the machine shops provided evidence of the faith that the Hunslet management were able to place in the future. Engineering was a competitive business, both on the home front and abroad. No effort had been spared in recent years to find and obtain the finest machine tools for the particular job in hand. Visitors could see gearboxes being machined on an optically-controlled horizontal boring machine of the type used by leading diesel engine manufacturers on cylinder blocks calling for the highest

precision, and the recently installed bevel gear cutter was an outstanding example of machine tool design.

While all the invited guests had been carefully selected for their future business potential, one small party from north of the border was watched with particular care. The Colvilles group of steelworks in Scotland (Clyde Iron, Clydebridge, Dalzell, Glengarnock, Lanarkshire Steel and Ravenscraig) were operating a fleet of over sixty Ruston & Hornsby 165 h.p. diesel-electric locomotives, some of which operated in tandem pairs. Trains were getting heavier, and the older Rustons were beginning to wear out. Replacement of the Rustons with a mix of 50 ton 0–6–0 and 35 ton 0–4–0 designs with standardized common parts was an obvious solution. The positioning of the CEGB locomotive, no. 6973, at the top of the yard was no accident. Its follow-on, no. 6974, as the last of a batch of four, was unsold and almost complete in the erecting shop,

'Up the 'Craig'. Three of the Hunslet 0–6–0 50-ton shunters in a panoramic view of Ravenscraig steelworks, Motherwell. From left, nos. DH63, DH62 and DH64, HE 7188/70, 7190/70 and 7061/71 respectively. No. 7188 had a Rolls-Royce 400 h.p. C8T engine for comparison purposes; all the rest had Cummins power.

with other batches of 0–6–0 and 0–4–0 locomotives not far behind. Barclays already had a demonstration locomotive at Ravenscraig, but it was about 5 tons lighter and less powerful than the Hunslet machine.

The strategy worked. Shortly afterwards, no. 6974 was despatched by road, ostensibly on hire to Ravenscraig works, with no owner's identification but painted in a style specifically requested by Colvilles. It stayed, to be included early in the new year as part of an order for three 0–6–0 locomotive for Ravenscraig and two 0–4–0 locomotives, one for Lanarkshire Steel at Flemington and the other for Dalzell Works in Motherwell. These five locomotives were the start of a complete replacement programme at all the Colvilles works. Eight 0–6–0 and twenty-one 0–4–0 locomotives were supplied from Leeds at intervals over the next five years, and several others were built by Barclays using the standard locomotive arrangement after the Scottish company had become part of the Hunslet Group in 1972.

Coincidental with preparations for the open day came locomotive orders from Ceylon, Egypt and Northern Ireland.

The Ceylon order was for a further two locomotives similar to the twenty-eight already building but this time for the Ceylon Cement Corporation's works at Puttalayanam. From Egypt the requirement was for thirteen 340 h.p. 75 cm gauge 0–6–0 coupling-rod drive locomotives, arranged to operate either singly or in tandem pairs. A Rolls-Royce C8TFL engine drove through a Twin Disc CF11500 torque converter to a frame-mounted jackshaft-driven Hunslet final drive and reverse gearbox. Ordered under the auspices of the Egyptian Delta Light Railway, itself already moribund, they were despatched between September and December 1970 to operate on the sugar estate railways of the Egyptian Sugar Corporation.

Northern Ireland Railways required three main-line locomotives to meet their share of the agreement with Córas Iompair Éireann for the joint Belfast–Dublin 'Enterprise' express service. The competition was from

Northern Ireland Railways No. 101 (later Eagle*), HE 7197/70, stands on a short length of 5 ft. 3 in. gauge track at Doncaster Works in preparation for the handing-over ceremony.*

Built to very restricted dimensions for use serving the electric arc furnaces of the British Steel Corporation's Templeborough rolling mills, No. 36, HE 7001/71, handles an incoming train of scrap shortly after delivery. Seven of this type were built in all; three are now used at Rover's Longbridge works.

The frame for the Templeborough locomotives was unique. Six-inch-thick frame plates and footplates, and eight-inch-thick buffer beams, machined and welded to produce one 25 ton indivisible unit.

America, General Motors, but there were strong nationalistic pressures to buy from mainland Britain. The few British main-line builders were disdainful of an order for only three special locomotives, and the result was a classic compromise. Hunslet took contractual and design responsibility, and with GEC support took the former English Electric-built 1350 h.p. diesel-electric locomotive design for Portuguese Railways (itself based on the successful British Railways type 1, later class 20, shunter/trip unit) and transformed it into an 80 m.p.h. twin-cab express locomotive to operate one at each end of a rake of British Railways-built Mark II style coaches. (The third of these special locomotives was an operating spare.) All design and drawing work was done at Leeds, as was manufacture of the bogies. GEC provided the traction package, while fabrication of the superstructure and final assembly took place at BREL, Doncaster. They carried Hunslet numbers 7197–9 and became NIR Nos. 101–3 named *Eagle*, *Falcon* and *Kestrel*. This was probably the first example since nationalization of a private/public-sector collaboration deal; others were to follow.

The June 1969 open day brought other rewards. By early 1970 work was in hand for two other steelworks, at Rotherham and Ebbw Vale. The Rotherham locomotives were for the former Steel Peach and Tozer melting shop at Templeborough and were to replace earlier Yorkshire Engine Company 44 ton 0–4–0 diesel-electrics which had been handling the hot ingot casting car trains since 1950. Modernization of the plant with new electric arc furnaces had increased the output to a stage at which more powerful locomotives were required. There was a snag. While the tonnages had increased dramatically owing to the new furnaces, the railway infrastructure of the shop had remained the same, with its limited clearances and tight curves. The new locomotives were to be physically no larger than the existing Yorkshire Engine Company four-wheelers, but must provide 45,472 lb. tractive effort, almost twice that of the Yorkshires, must have an axle load less than 25 tonnes and must be able to negotiate 90 ft. reverse curves. Additional escape doors were required for the crew and protective heat shields were needed at front and rear. GEC Traction declined to bid, saying that the task was impossible. ('Rubbish!' countered the Hunslet Sales Manager.) Thomas Hill provided an effete twin tandem master and slave chain-drive unit with one cab, while Hunslet's proposed solution was virtually carved from the solid. The resulting 67 ton locomotive was powered by a Paxman 403 h.p. 8RPHL diesel engine. It had a heavy-duty

Hunslet final drive and reverse gearbox, six-inch-thick main frame plates and buffer beams 8 in. thick. Four locomotives were built and received with acclaim.

Carrying Hunslet works numbers 7001–4 (BSC Rotherham Works nos. 36–9), these entered service between 5 March and 12 May 1971. Three more of the same type were built, no. 7357 in 1973 and nos. 8805 and 8902 following in 1978. In 1990 three of the seven were transferred to the Rover motor car works at Longbridge.

The Ebbw Vale locomotives were based on the 60-ton 0–8–0 units for the Coventry Homefire plant but increased in weight to 80 tons. They used a rubber chevron suspension unit instead of steel springs, based on the unit initially developed for the Rotherham six-wheelers, the aim being to minimize the down-time due to the breakage of springs. Unlike most similar arrangements for cardan-shaft drive locomotives this suspension was placed above the conventional axleboxes and was directly interchangeable with the steel springs. The interchangeability was never widely practised, but the rubber chevrons did not suffer the same rate of failure as their steel forerunners. The majority of the locomotives supplied to the ex-Colvilles works in Scotland were similarly fitted.

Three locomotives, nos. 7063–5, went to Ebbw Vale in July 1971, and a fourth, no. 7200, in April 1972. They were put to work on the heavy hot-metal ladle traffic from the blast furnaces to the rolling mills.

The next five years were to see a constant stream of standard shunting locomotives, mainly put through in stock batches, for use both at home and overseas. What attracted most attention, however, were the ten locomotives for the British Steel Corporation's new showpiece steelworks at Scunthorpe. Named Anchor, the new works was intended to co-ordinate, and gradually supersede, the three existing Scunthorpe works: these were the former John Lysaght's Normanby Park Works, the Richard Thomas and Baldwin's Redbourn Works, and the English Steel Corporation's Appleby Frodingham Works. The project from a transport aspect was led from the start by Appleby Frodingham-based engineers who had long experience of the very successful Yorkshire Engine Company twin Rolls-Royce-engined 400 h.p. 'Janus' locomotive. The new scheme required larger locomotives to augment the Janus fleet, and while Hunslet would have preferred to supply diesel-hydraulic locomotives based on the Coventry and Ebbw Vale examples it was realized that the deep-rooted diesel-electric culture at Scunthorpe

Similar in appearance to the earlier Coventry Homefire locomotives but twenty tons heavier, four of these 80-ton 776 h.p. locomotives were supplied to British Steel's Ebbw Vale works. Here No. 173, HE 7200/72, hauls a train of slag ladles up the gradient from the steelmaking plant in 1973. These locomotives were in effect two standard 0–6–0 locomotives back-to-back on a single rigid frame and sharing a common cab. Each of two Cummins NT400 engines provided 388 h.p.

would be difficult to counter. The Janus design was, and still is, remarkably trouble-free.

Another trouble-free design of the time, and in the higher power bracket, was the Steel Company of Wales's Brush-built 750 h.p. Bo–Bo operating at Margam and Abbey works. Why not, it was argued at Jack Lane, use two of the largest Rolls-Royce engines as fitted to the recently-built Ceylon locomotives and marry them to the Brush traction equipment and running gear to produce a 90 ton 1124 h.p. twin-engined diesel-electric. This combination would produce a powerful centre-cab locomotive using previously-tested components, thereby satisfying the Scunthorpe engineers' preference for electric traction, which would look modern and forward-thinking and yet pose minimum risk of failure in service. The quotation was

submitted and accepted, and the ten locomotives nos. 7281–90 were driven under their own power to Scunthorpe between 31 August 1972 and 12 March 1973. An eleventh locomotive, no. 7473, was ordered especially for the hot-metal transfer traffic between Normanby Park Works and the Anchor project and followed on 24 April 1977. Also ordered for Normanby Park were four very elegant but not tremendously successful 75 ton 0–6–0 diesel-electrics using one of the Anchor locomotives' power units in a lengthened Rotherham-type frame. The cab arrangement and the whole of the forward superstructure were identical to the corresponding items on the Anchor Bo–Bo locomotives. These four, nos. 7400/1, 7472 and 7499, were supplied between 18 April 1975 and 28 May 1977. Two of the standard Bo–Bo traction motors were mounted side-by-side just forward of the trailing buffer beam and drove via a mixing gearbox and cardan shaft to a final-drive gearbox on the centre axle. Coupling-rods with roller bearings then drove to the leading and trailing wheels. It was a good engineering solution – perhaps a little too good. The high torque produced by the motors, combined with the effects of backlash in the gears, allowed the drive to 'wind up' in such a way that power was produced at starting and slow speed far beyond the design expectancy. The movement of wheel centres within tyres and of axles within wheel centres, and the breaking of crankpins, were common faults. Normanby Park works closed prematurely in the

early 1980s, and the four 562 h.p. locomotives saw very little use.

During the construction of all these 'new age' steelworks locomotives the manufacturers themselves were undergoing change. Rolls-Royce had ceased production of the Sentinel locomotive at Shrewsbury in 1968, transferring all their locomotive work to Thomas Hill & Co. at Kilnhurst near Rotherham, by this time a wholly-owned Rolls-Royce subsidiary. The English Steel Corporation, which owned many steelworks and iron ore mines, had wound up its Yorkshire Engine Company subsidiary shortly before this, selling its designs and goodwill to Rolls-Royce. The last Fowler locomotive had emerged in January 1968 with the

For the brand-new Anchor steelworks at Scunthorpe Hunslet produced ten 92 tonne 1124 h.p. shunting and transfer locomotives in 1972/3. An eleventh example followed in 1977 for hot metal transfer between Anchor and the former John Lysaght's works at Normanby Park. This view shows the first five locomotives under construction on 22 September 1972. On the right is the frame for a 55-ton 0–6–0 for the National Coal Board Staffordshire Area, while lost amongst the clutter are two 29 h.p. diesel-mechanicals for Colombia, a 144 h.p. tunnelling locomotive for the Madrid Metro, a 65 h.p. mining locomotive for south Wales and a 100 h.p. mining locomotive for the north-east.

Hunslet's magnum opus. British Steel Corporation No. 73, HE 7287/73, hauls two Distington torpedo ladles full of molten metal, a trailing load of about 1200 tonnes, in the early days of Anchor steelworks activity.

Against the background of the 'three queens' – Scunthorpe blast furnaces Victoria, Mary and Elizabeth – is assembled the most expensive single-purchase steelworks fleet made prior to 1973 (and probably since). 920 tonnes of diesel-electric power with a combined haulage capacity of 73,000 tonnes.

Designed to haul trains to the coal face seven miles under the sea at Easington colliery in County Durham, this 216 h.p. locomotive had two Gardner 6LW diesel engines each driving a Twin Disc hydraulic transmission. Continuous welded rails were used, to a track gauge of 3 ft. 0 in., and gradients outbye the coal face were as steep as 1 in 15 against the load. Inbye speeds could reach 25 to 30 m.p.h. HE 7099/73.

drawings and goodwill being purchased later in the year by Andrew Barclay, Sons & Co. Ltd at Kilmarnock. English Electric acquired the Ruston and Hornsby business in 1968 and transferred locomotive work to the Vulcan Foundry, although only English Electric and, later, GEC designs were to come out of Vulcan until locomotive building ceased there in 1980. Brush were only occasional participants in the shunting locomotive business, concentrating on one standard design of 350 h.p. 0–6–0 diesel-electric. Five of them went to the Tyne and Wear Metro, but they sold principally to Tanzanian, Nigerian and Malaysian Railways.

The success of the new range of Hunslet locomotives with Twin Disc transmissions, the ever-present mining locomotive business (although the latter was cyclic, being quieter than normal between 1965 and 1972), a massive spare parts business and the diversification previously referred to had sustained Hunslet through some difficult political years during which four of its major competitors had perished. In 1967 Rae Fryers and John Alcock had

discussed a possible Hunslet and Barclay merger with Andrew Barclay's managing director, Stanley Kewney, but nothing came of it at that time. Barclays were busy with twenty-two 690 h.p. twin-engined eight-wheelers for East African Railways and Harbours for which Hunslet supplied the frame plates. The two companies had a reasonable, if competitive, working relationship, and in December 1970 Hunslet acquired a block of unissued Barclay preference shares. This just happened to coincide with the Colville re-fleeting exercise, which Hunslet had won much to the dis-appointment of the Kilmarnock work-force. Amicable discussions continued: Barclays built the bogies for the Hunslet 1124 h.p. Bo–Bo locomotives, and Hunslet provided the gearboxes for fifteen 302 h.p. shunters for East African Railways and Harbours, an order that Barclays received during 1971. Only six locomotives of any kind had been turned out of Kilmarnock in 1970, and only three in 1971. During the latter half of 1971 and most of 1972, while the Scunthorpe locomotives were being built, regular management and project meetings were held between the

Works Director David Gawthorpe discusses the machining of a locomotive gearbox on the new Scharmann horizontal boring machine in the Gun Shop early in 1967. A very advanced machine for its day, it was similar to those used for machining cylinder heads on the Rolls-Royce diesel engines.

By June 1968, after standing empty for some time, the old diesel drawing office had become a machine shop for the apprentice training school. The average intake at that time was twenty apprentices at the end of each school year.

two companies, on neutral ground in a hotel at Temple Sowerby on the A66 trans-Pennine road roughly half-way between Leeds and Kilmarnock. Ultimately the Barclay directors accepted Hunslet's offer for the remaining share capital, and in August 1972 Andrew Barclay, Sons & Co. Ltd lost its independent status and became a member of the Hunslet Group of Companies, although continuing to trade under its own name. Rationalization of resources took place. Hunslet eventually closed down their own foundry, which had traded under the name of Hill Brothers, and concentrated foundry work at Kilmarnock, while Leeds supplied gearboxes, frames and other fabrications in return. A number of overspill locomotives from Leeds were later to be subcontracted in their entirety to Kilmarnock.

Across Jack Lane Hunslet's long-standing competitive neighbour, Hudswell Clarke, was also having a lean time. The works had never really been modernized; and, while the diesel shunters supplied to the Mersey Docks and Harbour Board and the Ford Motor Company were superb, and Hudswells' 100 h.p. mining locomotive was streets ahead of its Hunslet equivalent, the meagre work-load and the old-fashioned practices did not impress the finance company that had ultimately come to own the business. Concurrently with the Barclay agreement, Hunslet acquired the locomotive interests of Hudswell Clarke, including all drawings and the transfer of those staff who wished to cross the road. The name Hudswell Clarke was retained for some time as a sales company, supplying spare parts and any repeats of the Hudswell 100 h.p. mines locomotive, which was offered to customers as an optional alternative to the host company's product. In one month, with very little outlay and without external financial help or debt, Hunslet acquired access not only to the businesses of Hudswell Clarke and Andrew Barclay but also to those of John Fowler and the North British Locomotive Company, with any residual spares business and traditional customer bases that all this might bring.

Coincidental with this risk-free enlargement of the Group, the opportunity was taken to modernize the blacksmiths' shop and replace most of the pneumatic hammers with a 500 ton forging press. This press was custom-designed and -built jointly by Hunslet and the nearby press specialists Henry Berry and Company Ltd. Not only did it speed up the production of forgings and improve their accuracy but it was much quieter and did not produce the vibration that had always been one of the side-effects of the hammers it displaced. The latest type of ultrasonic flaw detection equipment was also purchased at this time.

On 15 February 1973 over two hundred customers and potential customers joined in a tour of the works prior to hand-over of one of the Scunthorpe 1124 h.p. locomotives. Other surface shunting locomotives were to be seen in various stages of construction for Belgium, Chile, Italy and Malawi, in addition to further examples for the home market.

The acquisition of Hudswell Clarke coincided with increased mining locomotive business from the National Coal Board, predictable enough if it is remembered that very large numbers of locomotives had been supplied since 1949, and these early examples were reaching the end of their lives. Demand now was either for larger locomotives, i.e. Hudswell and Hunslet 100 h.p. or small 28 h.p. machines. There was also a new trend. Conveyors were increasingly being used to take coal from the coal face to the pit bottom, and sometimes even to the surface; locomotives were tending towards the transport of men and equipment (supports, coal cutters, etc.) rather than for coal haulage. At Easington colliery in County Durham, the coal face had progressed some seven miles under the North Sea and the amount of time lost by miners travelling to and from their place of work was becoming untenable. Hudswell Clarke, prior to the take-over by Hunslet, had supplied some tandem pairs of 100 h.p. 0–6–0 locomotives regeared to the 'high' speed of 18 m.p.h., and these were put to work at Easington on purely man-riding duties.

Soon after the inauguration of this pilot 'high-speed' scheme, the Chief Engineer of the National Coal Board Durham Division gave a paper in the Mechanics' Institute in Newcastle and commented on the tendency of the tandem locomotives to interact adversely with each other, probably owing to the large overhang, and precluding any further attempts to gain a higher speed. Over tea and sandwiches following the paper, those of us with an interest in such matters produced a few thumbnail sketches on table-napkins; and thus was the high-speed man-riding bogie locomotive conceived.

The result was a 216 h.p. fully-flameproof B–B locomotive fitted with two Gardner 6LW engines each with a Twin Disc DFR 10,000 torque converter driving into a centrally-mounted mixing and reverse gearbox and then by cardan shaft to final-drive gearboxes on each axle. The length was 7600 mm, the width 1370 mm and the height 1830 mm; and the locomotive weighed 21 tons. Gradients

This 75-ton 562 h.p. 0–6–0, one of four, was in effect half of one of the Anchor locomotives mounted on a rigid frame developed from the earlier 67-ton Rotherham locomotives.

Apart from single-figure quantities of 25 and 50 h.p. locomotives, Hudswell Clarke concentrated their mining activities on the Gardner-engined 68 h.p. 0–4–0 and 100 h.p. 0–6–0, at first in single-ended and finally in double-ended versions. The 100 h.p. version was a superb machine, and well over three hundred were built before Hunslet bought the business in 1972. Another twenty-three were built by Hunslet, the first, HE 8850, being delivered to the National Coal Board on 25 March 1982. This is Hudswell Clarke DM1352 before delivery to Allerton Bywater colliery in 1967.

at Easington were as steep as 1 in 17 from pit bottom down to the face, and the locomotives were to run on 3 ft. 0 in. gauge track with specially-prepared long welded rails. Each train consisted of seven specially-designed lightweight articulated gondola-type cars. A total of 168 men could be carried in one train, the train's laden weight being 26 tonnes. Speeds of up to 30 m.p.h. were expected. Four units of this initial design were produced, nos. 7099/100 for Easington in 1973 and nos. 7309/10 for a similar scheme at Ellington and Lynemouth collieries in Northumberland during 1974. Later examples of the class were uprated to 300 h.p.

The first 90 h.p. flameproof locomotive, no. 7387, was produced at this time, a particularly narrow four-wheeled, cardan shaft-driven unit designed especially for the Polish coal mines around Katowice but later to find favour also with the National Coal Board.

There were difficulties. An Aslef overtime ban from 12 December 1973, a three-day working week imposed from 31 December to 9 March and the month-long miners' strike from 10 February 1974 brought down the Conservative government led by Edward Heath on 28 February 1974 and did little to help manufacturing industry. Delivery promises were hard to keep. Not surprisingly, business was becoming harder to find, and as late as July 1976 the Company felt it necessary to remind its agents world-wide of the 'roaring inflation' that was being experienced, not only in Britain but elsewhere, and of the inability to quote firm prices without adding a considerable surcharge.

The year 1975 was the 150th anniversary of the opening of the Stockton and Darlington Railway. An Austerity saddle-tank built by Hunslet took part in the Cavalcade at Shildon on 31 August, while a month earlier a Hunslet

In typical latter-day National Coal Board condition, 562 h.p. 0–6–0 HE 7396/74 brings loaded 24-ton mineral wagons from the colliery to the exchange sidings at Bagworth on the former Midland Railway Leicester to Burton-on-Trent line.
ALAN ETHERINGTON

562 h.p. diesel shunter had been exhibited at the Railway Technical Centre in Derby alongside the first production HST InterCity 125 power car. Over two hundred guests attended Derby, travelling between there and St Pancras on the prototype High Speed Train. The railways of Britain were changing more radically and more rapidly than at any time in the previous 150 years.

Also during 1975 was signed a contract for a further twelve diesel-mechanical locomotives, nos. 7500–11, for the Sudan Gezira Board, bringing this customer's Hunslet fleet up to 77 locomotives. An order for two 65 ton 776 h.p. 0–8–0 flashproof refinery locomotives, nos. 7402/3, for the Czech Slovnaft petrochemical plant at Bratislava was completed at the end of the year; and these two events led very nicely into 'Export Year', launched from 1 June 1976 as a result of a combined initiative by the Confederation of British Industry, the Trades Unions Congress, the Association of British Chambers of Commerce, the Institute of Export, the Committee on Invisible Exports, the British Overseas Trade Board and its Advisory Council. A press notice from the Central Office of Information contained the following paragraph on Hunslet:

Also taking part in Export Year is Hunslet (Holdings) Ltd of Leeds, a group which exports an average of more than 50% of its output of specialized locomotives, fork lift trucks, mining equipment and Textile Presses. The Hunslet Engine Company which makes the locomotives is presently building units for use in the Sudan, Indonesia, New Guinea and Zambia. The firm prides itself on being one of the few prepared to design and manufacture locomotives in small quantities to meet the specific needs of the customer.

To underline this statement, orders for mining locomotives were received that year from Peru and Nigeria, together with a third repeat order for twenty-three Wagonmaster units, nos. 8529–51, for Bord na Mona in Eire, bringing their fleet of Hunslet locomotives to 101, with more to follow in the next few years.

'Exporting is fun,' someone said, and so it was – sometimes. It was also necessary in order to survive. 1976 had ended on an upbeat note, but there was more hard work ahead; more change, technical, commercial and political; no time for complacency.

CHAPTER 12

Diversions

'When a firm of reputation sets out to make a unique vehicle to meet a specific need, it is fair to approach the product with an open mind and proper appreciation of its purpose. Scootacars Ltd is a subsidiary of Hunslet (Holdings) Ltd, whose largest unit is the Hunslet Engine concern which has been making locomotives for over 100 years [actually 94] at Jack Lane, Leeds. It is there that the Scootacar is made; and if anyone thinks that locomotive manufacture nowadays is a matter of tons of metal and rule of thumb, he should see the little diesel 'Tiny Tim', a fireproof [*sic*] locomotive about the size of a packing case which hauls coal trains down the mines.'

So eulogized the road test correspondent of *The Motor Cycle* magazine in the 18 September 1958 issue, having spent a day three-wheeling in and around the City of Leeds. We shall return to this early 'go to work in an egg'

theme shortly; but first let us look at the basis of spin-offs from locomotive manufacture into less likely but nonetheless logical trajectories.

Diversification, or, more correctly, direction by government into the manufacture of armaments, tanks, ships, aeroplanes, etc. during times of armed conflict had often been the lot of locomotive and other railway workshops, and quite well equipped they were for these tasks. As has been demonstrated, Hunslet was no stranger to other products in these circumstances. In peacetime, however, there was very little movement away from locomotives apart from helping out with the odd pressing on or off, and manufacture, of gears and similar items for neighbouring factories that were not blessed with suitable heavy equipment. Much sheet metal work, mainly spiral mash conveyors in stainless steel, was however produced for the local Tetley brewery.

The first Scootacar had simple styling and is shown here against the roundabout in the Hunslet works yard. Later 'de luxe' versions were more fussy and became dated more quickly, in much the same manner as the Vauxhall Victor of the same period.

The Hattersley Pickard presses were assembled and tested alongside the locomotives on a specially prepared floor at the lower end of the erecting shop. This late hydraulically powered model is on the customer's premises, however, having been works tested, dismantled, transported and reassembled. The location is not known: possibly one of Courtaulds' factories.

By the mid-1950s the steam locomotive was definitely on the way out, and while orders for diesel locomotives were rising more or less in proportion to the fall-off in steam production, the workshop balance was not quite right. While the diesel locomotive gearboxes provided additional machining, in the way of gears, and some fabrication work for the welders, the diesel engine took away the major need for pressure vessel, cylinder and valve gear expertise. There was surplus heavy machining and foundry capacity, and fitting skills to match.

F. Hattersley Pickard & Co. Ltd had produced pressing and finishing machines for the woollen and worsted trades over a period of 65 years from the Excel works in Great Wilson Street. This lay almost half-way, in a straight line, between the site of Murray's Round Foundry 400 yards away and the original Midland station at Hunslet Lane, itself only a quarter of a mile from the Railway Foundry site. By 1957 Hunslet had acquired the business, closed the Great Wilson Street works and transferred all

manufacture, design and sales to Jack Lane. Hattersley Pickard henceforward traded as a wholly-owned subsidiary of Hunslet (Holdings) Ltd., the public company formed in 1958.

The automatic flat-plate presses were huge and took two basic forms, both of which were to be continually developed and improved over the next 25 years and built alongside the locomotives.

Type XW was intended for high-quality finishing of woollen and worsted piece goods up to 64 inches wide at varying outputs up to 750 yards per hour, while type CR coped with cotton, rayons and other light piece goods at up to 1500 yards per hour. The iron platens, with steam heating coils cast *in situ*, and the rams and cylinder units were ideally suited to the Hunslet shop practice, as were the large steel pillars connecting the top and bottom beds.

The pillars on the largest multi-plate presses were machined from forgings, some as large as 18 inches in diameter and 8 feet long. Assembly and testing of each

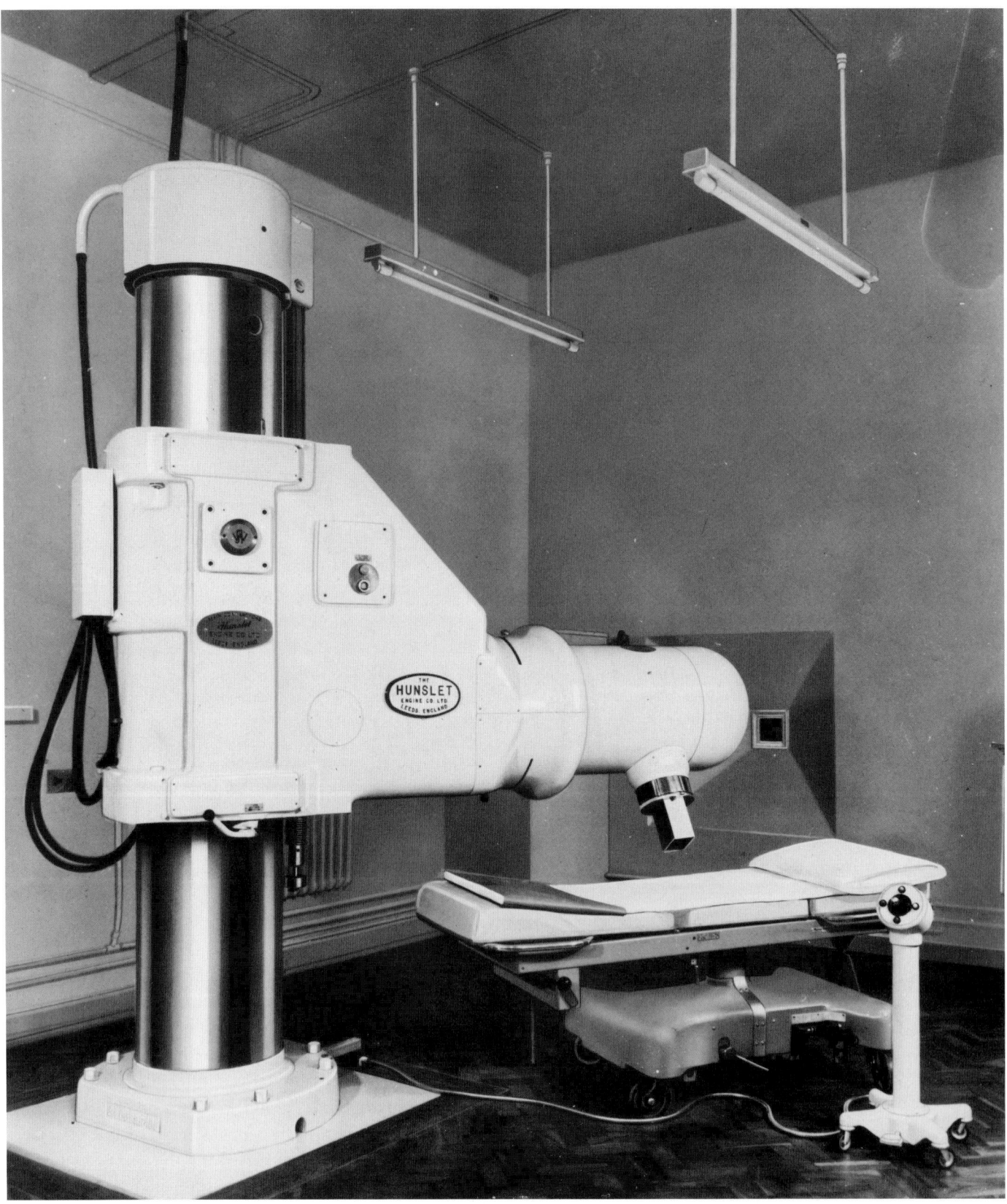

This first 'Hunslet' Radio Cobalt Therapy Unit was developed in collaboration with a Leeds teaching hospital to be used both for medical treatment by radiotherapy and for research. For medical purposes the treatment could only be given from above.

press was a major operation, followed, inevitably, by stripping down into convenient sections for transportation. The main market for Hattersley Pickard presses in later years was the Far East, including Japan and Taiwan; while British customers included such household names as Crombie (of overcoat fame) and British Celanese. France and Romania were also good customers.

Ancillary to the presses was a range of low- and medium-pressure industrial steam heating boilers designed for solid fuel, bagasse or oil firing, manufactured on behalf of another Leeds firm, Clayton-Chambers Ltd.

One of the Hattersley Pickard presses for Taiwan had its moment of glory during shipment from London in 1973. The considerable difficulties involved with this order,

combined with a fast-expiring letter of credit, resulted in an eleventh-hour rush to get the unit on board ship and loaded as the gang-plank was being raised. The crew of the *Bendorran*, aware of the difficulties that had been overcome in getting this equipment safely on its way, manned the rail and sang the British and Taiwanese national anthems as the ship moved away from the jetty.

A major diversification in 1958, and one which at first glance was again totally foreign to the capability of a locomotive works, was the development of radioactive cobalt therapy equipment for medical treatment and research. The first such dual-purpose unit was designed at Hunslet in collaboration with a Leeds teaching hospital, and over the next few years many cobalt units of both the

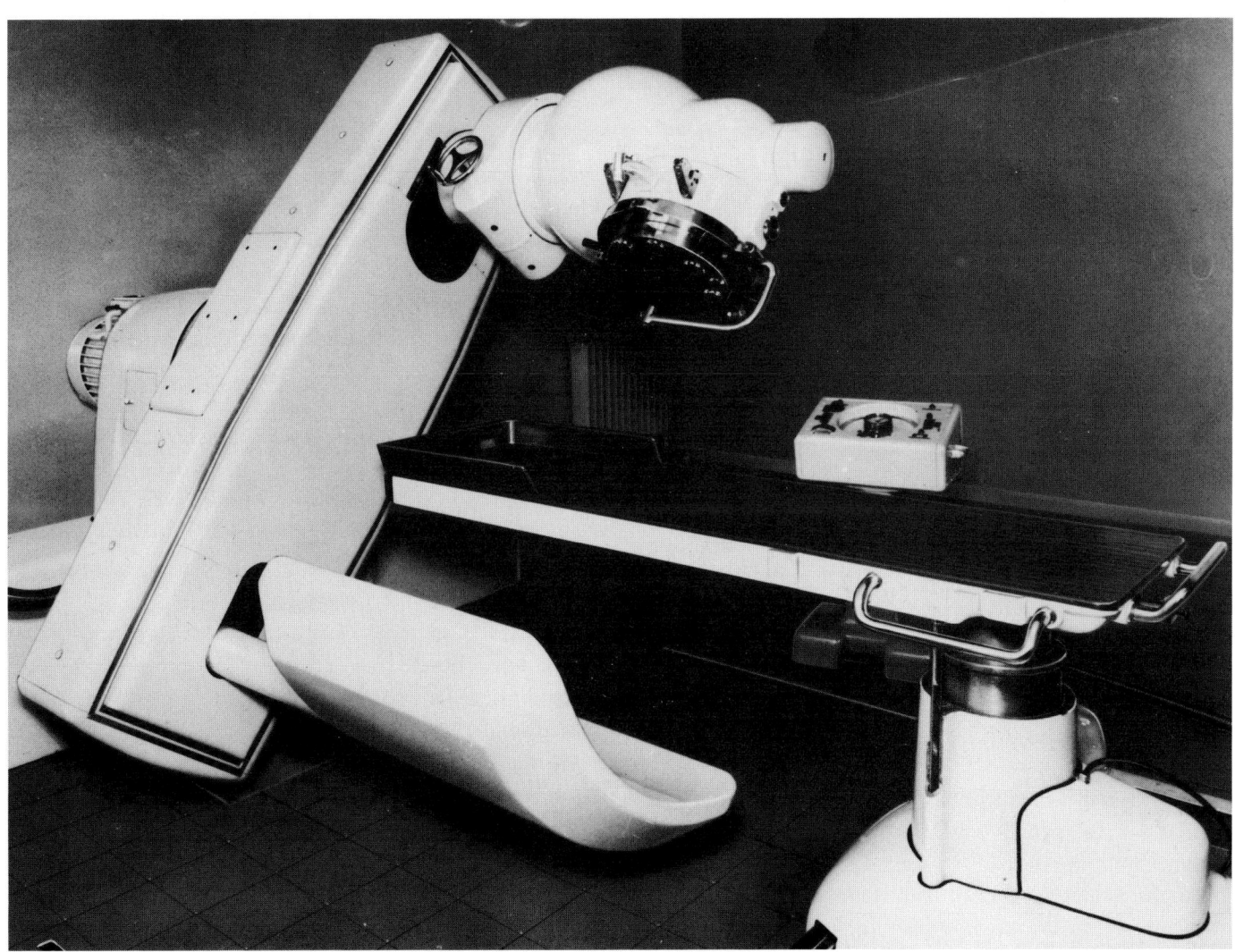

This later Rotating Therapy Unit enabled the fullest use to be made of all moving-field techniques while retaining the ability to be used for stationary work. The speed of rotation was infinitely variable from 0.2 to 2.0 r.p.m. and an arc movement with automatic reverse could also be set in steps of 10 degrees up to a maximum of 360 degrees.

stationary and the moving-field (rotational) type were produced for the country's leading hospitals and infirmaries.

The radioactive source housing, although relatively small in size, was constructed in heavy metal and weighed several tons. It was here that the locomotive experience came to be of great value, for the ability to fabricate and machine such items, and then to produce the means of accelerating and decelerating large masses with precision, was not readily available elsewhere. As the treatment of previously fatal diseases become more widespread and accessible to patients, and as the scientific complexity later overshadowed the engineering expertise, the business was sold to Fairey Engineering of Stockport. However, Hunslet Precision Engineering Ltd, which had been set up as a subsidiary of Hunslet (Holdings) Ltd to market and administer the cobalt units, continued to operate as the sales unit for gearboxes, flameproofing and other specialist items produced at Jack Lane for a further twenty years or so.

The Hattersley Pickard presses and the cobalt units were both very complex examples of heavy precision engineering well suited to the skills and facilities of a modern locomotive works. They were produced in sufficient numbers to balance the considerable but fluctuating demand for locomotives. And there were other opportunities. With the extension of the works described in chapter 8, the original joiners' and pattern shop became vacant. This was a self-contained part brick, part timber building with no heavy craneage, and therefore not suited to anything other than lightweight work.

This was at a time when car ownership was becoming the aim of, if not every individual, at least every household; and, while Japanese cars had yet to make an impact, mainland European vehicles were already commonplace. In an attempt to provide more economical transportation a number of small unorthodox vehicles had come on the market, notably with Heinkel, Isetta, Messerschmitt and Trojan 'bubble-cars' and the Bond and Reliant three-wheelers.

In 1953 one Henry Brown, a young Leeds engineer, produced a relatively conventional small car called the Rodley, after the district in west Leeds where Thomas Smith and Joseph Booth had their huge crane and excavator works and where Brown had been producing his Castra washing-machine. The project failed, not least for lack of interest and lack of financial support from the massive Rubery Owen conglomerate, which had acquired

the Castra business; and in 1955 Brown looked around for another backer. He found one in Rae Fryers, at that time director and general manager of the Hunslet Engine Co. Ltd, who offered to provide facilities for development of Brown's ideas. By one of those strange coincidences, another Brown, David J. this time, was to come along at around the same time with other good ideas, of which more later.

The pattern shop became the home for the new small car venture, but not before Rae Fryers had influenced the design away from a cheap (but not too cheerful) four-wheeler, of which sixty-five in all had been built, towards a robust three-wheeler. Selling at £235.10s.3d. plus £60.19s.9d. purchase tax (reversing gear £10 extra including Purchase Tax, spare wheel and carrier £4.15s. extra including Purchase Tax) and with annual road tax at £5, the Scootacar provided comfortable accommodation for the driver plus one adult passenger or two children.

The aim had been to provide an economical, easy-to-park small car with a driving position high enough to provide safe, all-round vision. The method of styling was typically Hunslet: Rae Fryers, who was a tall man (well over six feet), sat on a box against the shop wall while dummy controls were adjusted in front of him, and a chalk line was then drawn on the wall to indicate leg- and headroom requirements. A full-size wood and Plasticine model was then made, and developed to provide the desired overall styling.

The robust welded box-section chassis with sheet-steel floor had three interchangeable wheels with independent suspension, the rear wheel being driven by a 197 c.c. air-cooled Villiers reversible engine unit. Lockheed hydraulic brakes were standard. A major feature of the design was the full-height large, car-type door, which allowed easy access from the kerb side without acrobatics. The maximum speed was around 45 m.p.h., the fuel consumption was 80 m.p.g. at 30 m.p.h., and the range was around 220 miles on a tank of two-stroke petrol. The body was a one-piece lamination composed of polyester resin bonded together with glass fibre laid up cold, which after curing hardened into a tough high-tensile plastic quite advanced for its day. A *de luxe* version with slightly restyled bodywork was introduced later, and a sporty Anzani twin-cylinder engine could be provided as an option to provide a very nippy runabout indeed. Space required between other vehicles for kerb-side parking was six feet.

The converted pattern shop was soon too small for series production, and two bays of the recently acquired additional Manning Wardle property were used to set up a new assembly line, while three Nissen huts served as stores, servicing bay and offices. A fire destroyed the timber portion of the original shop around the time of the transfer.

Production of the Scootacar continued for a number of years into the early 1960s, by which time the ubiquitous Mini had taken the small car market and 'bubble-cars' had gone out of fashion. The Scootacar tooling and drawings were sold to India, but it is not known whether any were ever made on the subcontinent.

A spin-off from the Scootacar product line was the ability to produce speciality glass-fibre items, and this resulted in the production of a range of items including floral pattern domestic trays, warehouse doors, ducting and piping for chemical works and even *flèches* (spires) for Mormon churches. Consequently the choice of fibreglass for the chimney cowl on the underfeed stoker-fitted Austerity steam locomotives was more logical than many might have thought.

As in a number of other instances, Hunslet was ahead of popular thinking in the use of plastics for locomotive components and vehicle cabs, yet suggestions at the time that the new range of locomotives with Twin Disc transmissions should have GRP cabs were laughed out of court, which was a pity.

David J. Brown came to Hunslet with two design innovations looking for products in which to incorporate them. One was the idea of using two standard agricultural tractor units comprising engine, transmission and drive axle, complete with road wheels, and mounting them in tandem with a swan-neck joint. This gave a rugged four-wheel drive tractor with steering undertaken by hydraulic rams acting on the swan neck. The turning circle was only 26 ft. 6 in. wall-to-wall. A prototype, the ST125, was built, and extensively tried and tested, eventually going to South Africa as a sugar cane loader. No more were ever built, but David Brown went on to form DJB Engineering at Peterlee, County Durham, where he developed the Artix swan-necked load haul dump truck. He later sold out to Caterpillar and with the proceeds bought the Bedford truck plant at Dunstable, which he renamed AWD (all-wheel drive).

Brown's other idea took off in quite a big way at Hunslet. It was for an all-wheel-drive tractor with slew steering.

This was developed into a 25 h.p. flameproof underground machine only 3 ft. 5 in. high, 3 ft. 0 in. wide and 8 ft. 11$\frac{1}{2}$ in. long with four 3 ft. 0 in. diameter Power Grip rubber tyres on a 3 ft. 2 in. wheelbase. This diminutive machine had a carrying capacity, evenly distributed, of 27,000 lb. and a drawbar pull of 4200 lb. The weight in working order was 5500 lb.

The MT25, as it was known (the designation denoted Mines Tractor, 25 h.p.), obtained a 'Buxton' certificate issued by the Ministry of Power, and was flameproofed to a standard similar to that adopted on the Hunslet mining locomotives. Through a torque converter, a three-cylinder Perkins 3–152 engine drove an epicyclic forward and reverse gearbox, thence through reduction and transfer gearing, with sintered metal multi-plate clutches, and finally to a worm drive in each wheel. Steering was of the slew type, controlled by the multi-plate clutches and brake bands on both sides, giving a turning radius of 5 ft. 6 in.

A wide range of ancillary equipment was provided, including trailers, winches, power shovels, angle dozers, transfer stations, etc., all built in house; and the MT25 quickly became the mechanical equivalent of the pit pony, taking materials to the working face and generally assisting with mine maintenance.

Over two hundred were supplied to the National Coal Board. Others went to Coal India and to Polish salt mines. One of them was exhibited in Moscow and later specially prepared for the BBC documentary series during 1969 on the excavation of Silbury Hill, that strangely menacing Neolithic mound that crouches by the A4 in Wiltshire. The mound is a considerable size, and it had been hoped that a tomb or a sacrifice would be found at the bottom; but in the event, although we now know a lot more about *how* it was built, the reasons for doing so remain tantalizingly inconclusive. A number of specially adapted units were purchased by the Ministry of Defence (Navy Department) for use as air tugs in Royal Navy aircraft carriers, and proved their worth in the Falklands war.

Developing the theme of rubber-tyred underground transportation there followed the MT30 and MT60 general-purpose vehicles, four-wheel drive, four-wheel steered units of 30 and 60 h.p., which could carry up to 26 men or 3 tons of materials, and a larger tractor, the MT40, for moving heavy powered supports and coal cutters from pit bottom to the coal face. A rough-terrain haulage unit, the MT90, was built but not put into series production, though the prototype was sold to a Canadian metal mine.

The MT25 flameproof underground mining tractor was produced in large numbers for the National Coal Board, Coal India, Polish salt mines and others. This surface installation at Bevercotes colliery in Nottingham shows a complete stockyard transfer station with tractor, trailers and hydraulic hoist all manufactured complete by Hunslet.

This illustration is typical of the environment in which the MT25 tractors usually operated. This is Dawdon colliery in County Durham on 28 October 1964.

A number of these modified MT25s were supplied to the Royal Navy for handling aircraft in the confined hangars below decks on aircraft carriers.

Several of the MT60s also went to Canada, principally to the coal mines around Crows Nest Pass in the Rocky Mountains; and others, by contrast, found work in the copper mines of Zambia.

With the exception of the Hattersley Pickard presses, all these diversifications had been developed within Hunslet. In 1962 came a completely new range of products. Two retired Army officers, Cols. Hartmann and Whitfield (brother of the actress June Whitfield), had built up a company in Maidenhead called Materials Handling (Great Britain) Limited. It built side-loading fork-lift trucks and reach trucks to designs licensed from Albert Irion GmbH of Stuttgart. Hunslet purchased MHE (as it came to be called for convenience) and moved manufacture to Jack Lane, where it followed on quite nicely from the Scootacar.

The side-loaders were diesel-driven, and although the range was advertised as being up to 50,000 lb. capacity there were two basic sizes, a 4 ton capacity ('Kestrel') and an 8 ton capacity ('Falcon'), with Ford 592E and 590E four- and six-cylinder engines respectively.

The reach truck was a four-way travel battery-operated machine with a unique three-castor wheel arrangement – two motorized and steerable, one of them free rather in the manner of a present-day supermarket trolley. The maximum capacity was 2 tons, and the truck was known as the 'Lizard' from its ability to sidle along in any direction. It could operate down aisles of just over 2 ft. in its narrowest form. Longer platforms and forks for aisles up to 4 ft. wide were available.

The diesel side-loaders sold well enough over 15 years or so, mainly to timber merchants and steel stockholders. The major competitor was Lancer, of Leighton Buzzard, a company started by two brothers who had also been part of the original Maidenhead-based MHE. To Lancer the

From left to right in this 17 July 1972 view are MT25 and MT40 tractors, MT30 and MT60 personnel carriers and the MT90 tractor. The figures represent the horsepower of their diesel engines. All sold in reasonable numbers, to NCB and others for underground use, with the exception of the MT90, which was a one-off, eventually going to the McIntyre Porcupine mine in the Crows Nest Pass, Canada.

The Hunslet 'Lizard' four-way-travel reach truck was a semi-bespoke machine with exceptional manoeuvrability. This two ton lift example is shown handling empty spools for man-made fibres at Courtaulds.

side-loader was the core business: to Hunslet it was at most a useful gap-filler, and when Lancer and the front-loading fork-lift truck manufacturer Boss joined forces to become Lancer-Boss, the hassle of heavy competition was not really worth the effort. The intellectual property rights were sold to the British Crane and Excavator Corporation, and Materials Handling Equipment (GB) Ltd concentrated on refining the Lizard four-way-travel reach truck, which it manufactured and marketed through to the Telfos take-over in 1987. The Lizard was unique in its narrow-aisle capability combined with four-directional travel and lifts of over 20 ft. were achievable with some variants. Several hundred were produced, and its customers were widespread. Rolls-Royce used them for stacking car bodies, Courtaulds for handling long spools of fibre, and ICI purchased an expensive and exotic specimen constructed almost entirely from stainless steel. In 1976 a 3 ton model, the Iguana, was introduced using four wheels, two free-running and two motored, and with a number of attachments specially designed for the textile industry. Again Courtaulds was a major customer, using Iguanas in their plants at Skelmersdale, Leigh and Derby.

The Lizard itself was uprated to 3 tons capacity in 1985 with new electronic controls, improved steering and modular construction.

When Boeing introduced the 747 Jumbo Jet at the end of the 1960s, airports and airlines alike were faced with a revolution in ground handling methods. Not only was the passenger volume a problem, but how to move aircraft weighing up to a million pounds (approximately 500 tons) all-up weight was also of great concern. The International Air Transport Association (IATA) drew up a standard specification for the largest-ever aircraft towing tractor, against which the major airlines went out to tender.

The British Overseas Airways Corporation (BOAC) identified a need for seven of these tugs, with the result that a Hunslet internal instruction dated 18 April 1968 said: 'It has been decided to put in hand 7 Aircraft Towing Tractors suitable for the Boeing 747. These are allocated numbers T1200 to T1206. Test date for the first unit week 12/1969, second week 36/1969 and the balance at two per month from October 1969.'

The result was the Hunslet ATT77, 30 ft. long, 10 ft. 0 in. wide and 5 ft. 2 in. high, with four 5 ft. 0 in. diameter pneumatic tyres on a 16 ft. 0 in. wheelbase. The all-up weight was 70 tonnes and the 12-cylinder 635 h.p. Cummins turbocharged vee engine and General Motors–Allison six-speed Torqmatic Power Shift hydraulic transmission provided a tractive effort of 77,200 lb. at stall and a maximum speed of 20 m.p.h. The basic structure was a locomotive-style main frame 23 ft. 0 in. long and 10 ft. 0 in. wide comprising two longitudinals and four transverse members each 4 ft. 6 in. deep by 4 in. thick. Driving cabs at each end were normally within the 5 ft. 2 in. overall height, but could be raised hydraulically to 6 ft. 8 in. when not in the vicinity of the aircraft fuselage.

Only four of the seven were built. One went to New York JFK and three to London Heathrow. One of the Heathrow three was driven from Leeds to Hull on completion whence it went by ferry to Amsterdam to feature in an exhibition of airport equipment at Schiphol. The three sets of parts remaining on stock were gradually disposed of as spares.

The reasons for the manufacturing shortfall were various. The IATA specification was full of 'nice to have' features and consequently made for an expensive product. In true British fashion BOAC had taken the IATA specification as mandatory rather than advisory, which meant that their chosen manufacturer, in good faith, produced probably the best machine but at a price that was more than other airlines were paying for lower-specification but probably adequate units. Another factor that did not help was the BALPA dispute in 1970, which considerably delayed the entry into service of BOAC's 747 fleet. By the time the customer was able to accept the Hunslet ATT77s the other airlines had operational experience not only of Boeing 747s but also of other manufacturers' towing tractors, and were beginning to pool resources. Hunslet sold the intellectual property rights to Reliance Mercury and took no further part in the aircraft handing market until Reliance Mercury built a small number of very similar but battery-operated vehicles, which were driven by Greenbat-designed, Hunslet-built traction motors (this was after the Hunslet takeover of Greenbat in 1980). With the 1980 takeover came a whole catalogue of further non-railway products, but these are not relevant to this chapter.

The diverse products outlined in this chapter all had their own production lines, albeit in the same factory, and were self-supporting through their own subsidiary trading companies. More varied were the items covered by the 'GM' order book originally conceived in 1961 to cover general items of a 'non-locomotive' nature although it provided an accounting mechanism for locomotive and railcar gearboxes supplied to the other manufacturers, e.g.

Effectively a 77-ton locomotive on four large rubber-tyred road wheels, the squat lines of the ATT77 aircraft towing tractor are well demonstrated in this works view of the first unit in British Overseas Airways Corporation livery, taken on 28 January 1970. BOAC and British European Airways (BEA) were to merge soon afterwards to become the state-run British Airways.

The ATT77 axles incorporated disc brakes and spherical constant-velocity joints to allow four-wheel drive and four-wheel steering. They were much larger than anything of a similar nature previously manufactured. Design and construction was by a partnership between Hunslet Holdings plc and Kirkstall Forge (later GKN Axles Division). The class 5 4–6–0 in the background provides a comparison in scale.

Flameproof equipment, i.e. exhaust gas conditioners (sometimes called 'scrubbers'), flame traps, etc. developed for the mining locomotives were adapted and sold for a number of diesel engine applications in hazardous areas. Sometimes a complete 'drop-in' power unit would be produced bespoke for the job. This is a 90 h.p. 'power pack' supplied in October 1978 to specialist mining machinery manufacturers British Jeffery Diamond (BJD) Ltd of Wakefield.

During the 1970s several manufacturers attempted to introduce viable underground transportation systems in place of conventional locomotive haulage. Becorit, UMM and, later, GMT and others all produced monorail or trapped-rail systems using either rope haulage or diesel traction. Somewhat perversely, Hunslet designed and produced the flameproof power units for some of these systems, and in some cases built complete 'locomotives' under subcontract. This is a Becorit-designed, Hunslet-built 25 h.p. trapped-rail 'Road Railer' on the short-lived surface line at the NCB's Birch Coppice Training Centre. On this line, a senior National Coal Board Engineer of the time recalls, 'Ellistown and Desford men learned to drive these abominations and fitters learned to change the [neoprene] tyres — frequently.'

A. R. ETHERINGTON COLLECTION

–241–

Andrew Barclay, Clayton Equipment, Baguley–Drewry, Kirloskar (India) and including the 59 (including three spare) for the British Railways D95XX series Swindon-built 650 h.p. diesel-hydraulics.

'GM' items covered anything not otherwise classified, from complete vulcanizing presses for motor car tyre manufacture to heavy-duty stores racking for use in conjunction with the reach trucks.

By far the greater proportion of 'GM' orders covered flameproofing equipment of one kind or another to be incorporated in other peoples' varied products. Hunslet was pre-eminent in exhaust conditioning and flameproofing, and some of the applications were competitive with their own activity. In this competitive area came complete power packs for monorail and trapped-rail locomotives manufactured by Becorit (GB) Ltd and Underground Mining Machinery Ltd. Some of the Becorit

locomotives were manufactured complete at Jack Lane. H. Stephens and Sons Ltd of Gloucester built several rubber-tyred tractors around Hunslet power packs for Navy armaments depots around the country. Ten road/rail Trackmobiles were built in Holland by Nederhorst Staalfabriken for Iraqi State Railways in 1975, and these had Hunslet power packs on account of their intended use in the refineries at Basra. The prototype for these had been one built in 1967 by Strachan and Henshaw in Bristol, in conjunction with Hunslet, for the BP refinery on the Isle of Grain.

In the early 1970s a number of diesel Land Rovers were flameproofed and marketed by Hunslet for 'inspection duties in areas of oil refineries otherwise inaccessible by internal combustion engined vehicles'. The inlet and exhaust manifolds were completely changed. After passing through the air filter, inlet air was passed through an

The flameproof headlamps, air cylinder and air horn can be seen in this photograph of a Land Rover modified by Hunslet for use at the Esso refinery at Fawley near Southampton.

The author stands in front of the partially completed North Eastern Region six-foot and shoulder ballast cleaning machine (HE 5617/61) on 6 March 1961

The completed ballast cleaning machine on display at the plant exhibition organized by British Railways' North Eastern Civil Engineers Department at Ponteland in June 1961.

The High Commissioner for Western Australia, Mr G. C. Wayne, and two colleagues with John Alcock, Peter Alcock and the author on 24 July 1962. HE 6056/62, the drainage trenching machine for BR North Eastern Region provides the backdrop in the Hunslet works yard. Mr Wayne is on the extreme right, standing next to John Alcock; Peter Alcock extreme left, the author third from left.

automatic shut-down valve and then through a Hunslet flame trap and manifold to the cylinder head. The exhaust was protected by a water-cooled manifold and exhaust pipe, thence to a Hunslet exhaust gas conditioner fitted under the rear of the vehicle. On the outlet side of the conditioner the gases passed through a standard Hunslet flame trap before emission to the atmosphere. All the original electrical equipment was removed and replaced by a flameproof switch box, dynamo and directly-lit flameproof head and tail lamps. An engine-mounted compressor fed two air cylinders, one mounted above the front bumper and the other behind the front seats, to provide air for engine starting, windscreen wipers and horn.

The major fork-lift truck manufacturers of the day – Coventry Climax, Hyster, Conveyancer and others – were also major buyers of Hunslet flameproof power packs and exhaust gas conditioners. Other applications for diesel vehicles in hazardous areas or confined spaces also received Hunslet's attention.

There were a number of items of railway-related non-locomotive work worthy of special mention. Among these were a ballast-cleaning machine, a trenching machine and a rail broaching machine, all for British Railways North Eastern Region and supplied in 1961/2.

The ballast-cleaning machine, HE 5617/61, was a 65 ft. 2 in. long rail-borne vehicle weighing 65 tons and designed to clean railway track ballast to provide efficient drainage. When working it could excavate, clean and return ballast in the shoulder and the six-foot at operating speeds of up to 600 yards per hour. The shoulder is the section from the outer sleeper ends to the edge of the ballast, while the six-foot is the part between two adjacent tracks. The theory then put forward by the North Eastern Region management of the time was that it was not necessary always to clean under the track, since the provision of thoroughly clean ballast at both ends of the sleepers permitted drainage of rain water throughout the road bed to cleanse the centre section. The machine was self-propelled and self-contained, and was able to travel from its stabling point to the job site under its own power. All this, it must be said, long before the arrival of Messrs Plasser & Theurer on these shores.

The trenching machine, HE 6056/62, was similar in size but was not self-propelled. Mounted on a custom-built underframe in the style of a well wagon, it could dig drainage trenches eighteen inches wide by six feet deep from rail level either side of the track at a speed of 150 yards per hour. Propulsion was by hydraulic winch, hauling on a wire rope suitably anchored to the track bed.

The broaching machine was not given a number in the locomotive series, and was a static device using hydraulic cylinders to power broaches to provide perfectly-formed fishplate holes in rail ends, thereby reducing the risk of stress fractures caused by poorly-fitting bolts. Hydraulically-powered rail pre-stressing equipment for long welded-rail installation provided another useful source of income for the company.

There is no doubt that the existence of a comprehensive yet straight-forward Drawing Office led system of document control and recording ensured that the multiplicity of both locomotive designs and the products decribed in this chapter progressed smoothly through the works. Futhermore the policy throughout the privately owned years of promoting by merit from within ensured continuity of purpose without stifling initiative and innovation.

Harold Dean retired in 1949 after twenty five years as Chief Draughtsman and was succeeded in turn by Harry Pybus, Philip Garnett and Len Cliff, all having many years of locomotive design experience and each having understudied the previous incumbent. In the mid eighties the growth of computerisation and complexities of modern management led to a change of style to Design Manager, a role which Roger Burnley fulfilled until shortly after the Telfos take-over. Roger had played a major part in design of the Anchor Project diesel electric locomotives.

Supporting the Chiefs were a number of Section Leaders, together with design and detail draughtsmen. Worthy of particular mention is Colin Broadhurst who started at Hunslet on the same day as the author and was ultimately responsible for the production drawings of most diesel locomotives from 1958 to the end. Both Colin, who died in 1993, and the author were schooled by Len Cliff and the relationship brings back many happy memories. Len is standing centre front in the photograph on page 160 with Colin just behind his right shoulder.

CHAPTER 13

A decade of change, 1977–1987

'Opportunities fall in your lap if you place your lap where opportunities fall.' A hackneyed and cynical statement, perhaps, but a truism that had well served those firms in the railway industry that had survived between the wars and after the BR modernization and rationalization programmes. By the end of 1976 the heady days of widespread dieselization by British Rail, British Steel and the National Coal Board had just about come to an end. In particular, the requirement for new shunting locomotives on British Railways had ceased, and was not only shrinking rapidly in respect of industrial units but then was aggravated by the increasing availability of second-hand machines. From now on, any new shunters for the British market would, in the main, be specialized designs for military, petrochemical and other safety-critical applications.

The Hunslet results for the year ending August 1976, announced early in 1977, showed a turnover increased to £6.2 million with pre-tax profits just short of £1 million – quite remarkable at the time – and enthusiasm was high. The batch of 0–8–0 locomotives for the Sudan Gezira Board was on the shop floor, together with shunters for Patent Shaft Steelworks, Associated Octel and the Central Electricity Generating Board, and a number of underground mines locomotives. In March 1977, the press were notified of a contract from Kenya Railways for 35 large shunters placed in conjunction with 1200 railway wagons, the latter to be built at the Ashford (Kent) works of British Rail Engineering Ltd. Coincidentally with the Kenya contract came orders for a 450 h.p. shunter for the Esso refinery at Fawley near Southampton and two 56 tonne tandem operation 562 h.p. diesel-hydraulics from the West German company Polysius for the new Ssan Yong cement works in South Korea. As if that seemed insufficient, the company launched its revolutionary flameproof rack-and-pinion locomotive during the international mining machinery exhibition held from 10 to 15 October 1977 at the National Exhibition Centre in Birmingham. Castle Bromwich 1932 all over again. To round off the year came an order in November from Bord na Mona (the Irish Turf

Board) for yet another twenty-five Wagonmaster locomotives.

The Kenya contract was of major significance and required careful management from initial tender to final completion to maintain the Hunslet policy of minimum exposure to risk. It also fully proved the value of the contacts with British Embassy and High Commission officers that had been encouraged by the Leeds office of the British Overseas Trade Board.

Worth in total just over £40 million (say £200 million in 1995 terms), of which the locomotives accounted for £7.5 million, the contract was large by the standards of the day. In a deal brokered by the British High Commission in Nairobi and financed by a line of credit through Lloyds Bank, the contract was negotiated by BRE-Metro Ltd and Hunslet Holdings plc acting in concert. BRE-Metro took the commercial lead for the whole, and technical responsibility for the wagons, while Hunslet was technically responsible for the locomotives and had a back-to-back commercial arrangement with BRE-Metro. BRE-Metro Ltd was at that time a sales organization formed to co-ordinate the export activities of the state-owned British Railways workshops (British Rail Engineering Ltd) and the privately-owned Laird Group company Metro-Cammell Ltd.

The 53 tonne 0–8–0 shunter trip locomotives were fitted with Shrewsbury-built Rolls-Royce 562 h.p. DV8T vee-eight engines derated to produce 525 h.p. at an altitude of 5250 ft above sea level in Nairobi. The drive was taken from the engine through propeller shafts first to a frame-mounted German Voith L2R3 automatic reversing hydraulic transmission and thence to a Hunslet final-drive gearbox on the third axle. (The North British experiment in building Voith transmissions under licence had failed long ago.) The four axles were connected by rods mounted on roller-bearing crankpins. A very exacting specification called for the maximum tractive effort with precisely balanced axle loads determined by complex bridge stress calculations. Perpetuating Cecil Rhodes's never (yet) to be

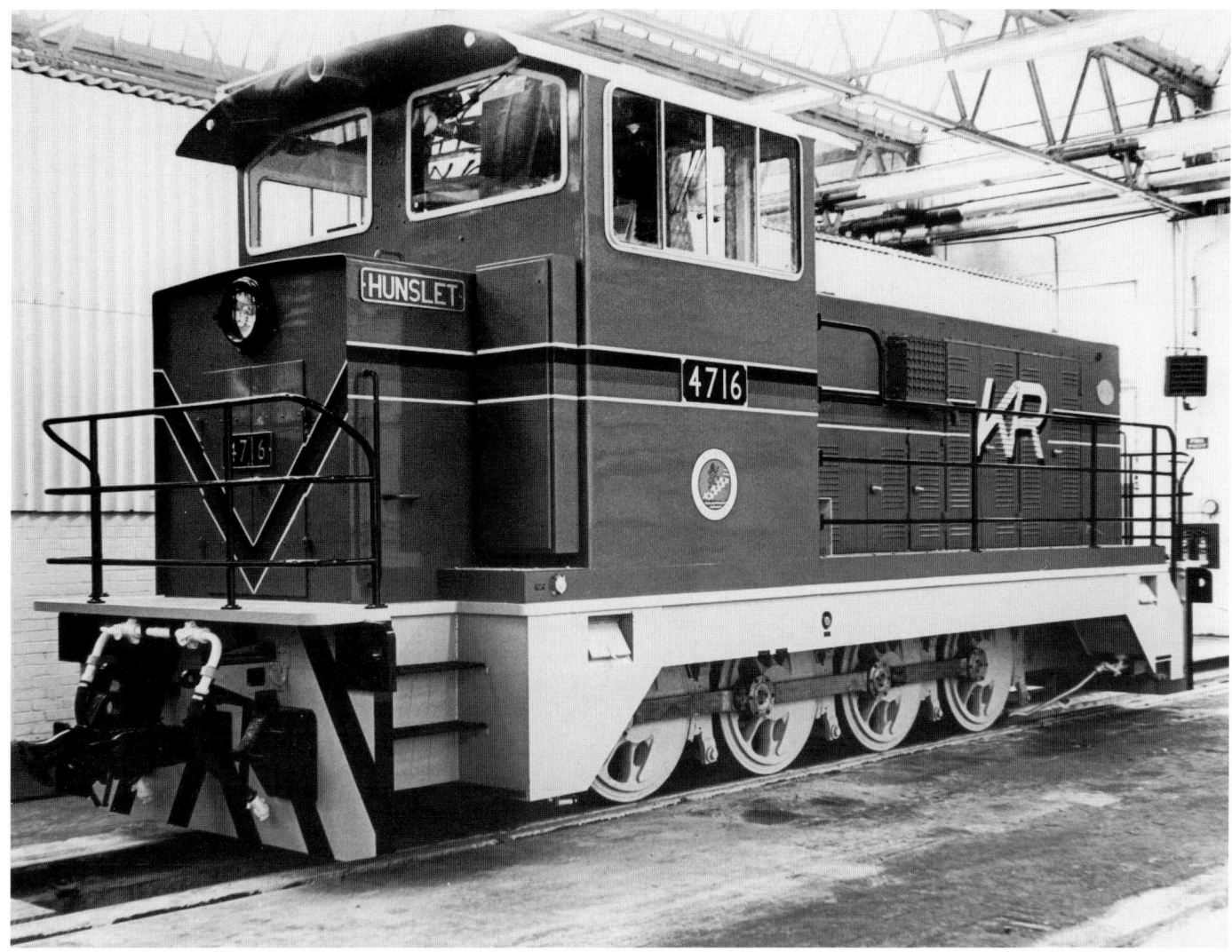

With one exception the Kenya class 47 locomotives were shipped in grey primer for finishing at Nairobi. The first Swindon-built example was painted, for publicity purposes, not without political and morale-boosting undertones, in blue and grey with the Kenyan national flag colours in a broad waistband. The makers' plate read 'BUILT BREL SWINDON' but gave the Hunslet number 9016, built 1979.

realized Cape-to-Cairo dream, the East African Railways and Harbours standard practice of easy gauge convertibility from metre gauge to the 3 ft. 6 in. 'Cape' gauge was followed in the design of brake gear, wheel centres and tyres; though the East African Community, and in consequence EAR&H also, had finally broken up into Kenya, Uganda and Tanzania during the course of the contract negotiations. Designated class 47 by Kenya Railways and numbered 4701–35 the locomotives took Hunslet numbers 9001–35.

Delivery was urgently required and in a smaller way the wartime Austerity steam locomotive practice of partnership was adopted. Hunslet produced all drawings and documentation, purchased all material and

manufactured all gearboxes, running gear and wheelsets together with fifteen frame assemblies. Twenty locomotives were erected in the BREL ex-Great Western works at Swindon, nine at Hunslet and six at Kilmarnock. The Swindon-built class 47s were the last new locomotives to leave the famous 'A' shop, the previous new build being the fifty-six class 95XX BR diesel-hydraulics for which Hunslet had also provided the gearboxes a dozen or so years earlier.

The Esso locomotive, a 45 tonne flashproof unit with a two-stroke General Motors Detroit 12V71 engine was also assembled at Kilmarnock under similar arrangements. To confuse the enthusiast and historian, however, this and the

Identified as a Barclay-assembled machine by having only one handrail flanking the headlamp, Hunslet no. 9011 shunts a heavy train at the Changamwe marshalling yard just outside Mombasa early in 1980. AUTHOR

Hall of the Kings. Of the Kenya Railways order for thirty-five Hunslet locomotives, twenty were built by British Rail Engineering Ltd at the Swindon works of the former Great Western Railway; wheels, gearboxes and other components were supplied from Leeds. This is the famous 'A' shop early in 1979, with seven locomotives in various stages of construction.

six Scottish-built class 47s carried 'HUNSLET BUILT LEEDS' maker's plates, as did other, later, subcontracted overspills from Leeds.

This practice of partnership did however enable the Hunslet Group to meet other commitments, for example the National Coal Board, Bord na Mona and South Korea orders, so as to maintain a production balance and maximize potential at both Leeds and Kilmarnock with minimum risk and optimum customer satisfaction.

In 1977 coal mining was still big business, and the NEC exhibition was a lavish affair. As has already been recorded, mining locomotives were also relatively big business to Hunslet, and the large stand space was taken more because the customers expected a presence than by way of advertisement.

The centrepiece of the display was the prototype 91 h.p. fully flameproof underground rack-and-pinion locomotive, which had already been proved on a special test track developed, with the co-operation of NCB North Yorkshire Area Chief Engineer Mr Jack Smith and his staff, at Ledston Luck Colliery eight miles from Leeds. Eleven units

of the type were already on order for five Yorkshire pits – Kellingley, Wheldale and Fryston in North Yorkshire Area, and Maltby and Manton in South Yorkshire – in contracts worth £600,000.

On an adjacent stand was a linear motor powered man-riding train of very futuristic appearance exhibited by a newcomer in the mining transportation business, Gyro Mining Transport (GMT for short). Quite understandably, with the absorption by Hunslet of Hudswell Clarke in 1972 and the earlier demise of Ruston & Hornsby, the National Coal Board was becoming nervous of a monopoly situation and subsequently encouraged GMT to compete in the rack locomotive programme. GMT we shall deal with later, for here was to be a twist in what has so far been a relatively straightforward tale.

Meanwhile, though, the Hunslet rack locomotive had been in the design and development stage since two pre-production prototypes, nos. 7455/6, were authorized early in 1974 against a requirement for Maltby colliery that was not to materialize until 1979. However, this stock building was increased on 14 November 1974 to eight locomotives

The prototype 91 h.p. flameproof underground rack-and-pinion locomotive undergoes testing on the specially prepared circuit at Ledston Luck colliery early in 1977. The mine car is weighted up with concrete; the gradient post (1 in 4) says it all.   NATIONAL COAL BOARD

Demonstrating the natural state for most mining locomotives after only a few weeks underground, a Hunslet 91 h.p. rack locomotive rests for a while at Kellingley colliery some time in the early 1980s.   NATIONAL COAL BOARD

with the addition of nos. 7470/1 and 7488–91. No. 7492 was added shortly afterwards, with the number 7493 being used for a spare transmission assembly. The original intention was that the four lowest-numbered locomotives should be built first, naturally, and completed to run, initially, as adhesion-only locomotives for Maltby, with the facility for the rack/adhesion change-over equipment to be fitted subsequently; but the Maltby operation was delayed, and no. 7488 was built first and extensively tested at Ledston Luck before going to the NEC.

Hunslet's first commercial application of the rack principle was in effect also the world's first steep-gradient underground locomotive. The rack traction principle permitted uninterrupted haulage on inclines impassable to conventional locomotives. The maximum gradient permitted underground for a conventional adhesion locomotive was (and still is) 1 in 15; but the rack locomotive was designed to operate regularly on 1 in 6 and the prototype was extensively tested on 1 in 4.

The rack principle did not necessitate extensive modifications to existing trackwork, since the locomotive used existing track and could haul conventional cars. The rack sections only needed to be added on sections where the gradient was too steep for normal adhesion traction. The Hunslet 91 h.p. unit had a hydraulic drive which allowed power to be transmitted to the four driving wheels, in adhesion mode, or to the two rack pinions when it was required to ascend or descend a steep gradient. Change-over was by means of heavy-duty clutches mounted outboard on each wheel. Selection of adhesion or rack drive also selected the appropriate braking and safety systems for each mode. In adhesion mode, the pinions rotated freely, while in rack mode it was the road wheels that were free to rotate. This allowed entry to and exit from rack sections without the danger of derailment or damage by a tooth-on-tooth engagement. Short spring-loaded, sleeper-mounted rack entry sections were provided for easy engagement of the pinions at speeds of up to 12 km/h.

Thus the locomotive could run uninterrupted for the whole of the underground journey, travelling at speeds of up to 10 m.p.h. on normal track and at 2 m.p.h., with a full load, on the short rack sections. The rack was profile-cut from thick steel plate, and the track was supplied in modular panels complete with steel sleepers. Turnouts, also incorporating the rack plate, could be supplied when required. These track components were subsequently marketed from Kilmarnock under the name Barclay Racktrack.

By August 1978, plans were well advanced for the new Selby Coalfield (planned to produce 10 million tonnes a year), this necessitating diversion of the British Rail East Coast Main Line from north of Shaftholme Junction to Colton Junction. The National Coal Board issued an enquiry in August for eight rack locomotives for its main spinal roadway. Hunslet offered its new 150 h.p. loco-motive, then under development, but lost out to GMT, who built seven locomotives against this requirement. Before the GMT locomotives were delivered, however, six Hunslet 91 h.p. rack locomotives went into service, in 1981/2, at the Gascoigne Wood end of the Selby development; these were to be followed over the next ten years by six 150 h.p. and six 64 h.p. Hunslets as the Riccall, Stillingfleet and Wistow mines, which were all interconnected with Gascoigne Wood, developed.

This is not to say that the NCB had stopped buying conventional locomotives. The world energy crisis had focused attention on coal once again – this was before North Sea oil – and the mood was bullish. In the years from 1977 to the disastrous 1984 miners' strike, 162 flameproof underground locomotives left Jack Lane for collieries throughout the country. In addition to the thirty-seven 91 h.p. rack locomotives, these comprised sixty-one 28 h.p., thirty-six 100 h.p., sixteen 90 h.p. and twelve 300 h.p. locomotives. These last were larger and more powerful versions of the 216 h.p. B–B high-speed man-riding locomotives previously supplied to Easington, and they too went to the North East coastal pits. Towards the end of this period, the Industrial Railway Society published a booklet detailing all the National Coal Board's flameproof locomotives, and an analysis of this, demonstrating Hunslet's predominance in the market was used by Hunslet in its publicity. Since 1939 British coal mines had purchased approximately 2250 flameproof underground

locomotives, and in 1983 just over half were still in service. Sixty-three per cent of them were built in Leeds. The business in spare parts from this vast fleet was tremendous. 1984 was yet to come.

Sixty per cent of the 162 mining locomotive orders after 1977 came during the currency of the Kenya Railways contract and the other commitments outlined at the beginning of this chapter. The order book was overfull. Sixteen of the 90 h.p. locomotives (nos. 8517–20, 8566–70, 8585–9 and 8801–2) and eight 300 h.p. locomotives (nos. 8509–16) were subcontracted in their entirety to Kilmarnock, as were a 60 h.p. flameproof narrow-gauge locomotive for the Royal Naval Armaments Depot at Dean Hill and a 240 h.p. 5 ft. 6 in. gauge shunter for North Bengal Paper in Bangladesh.

Five very profitable years had seen the turnover at Hunslet rise from £4.83 million in 1975 to £10.79 million in 1979, and with no overdraft and a full order book the company was cash-rich. Not so two miles away in West Leeds. Greenwood and Batley was a well-respected general engineering company with its roots in weaponry for the Crimean War. Its Albion Works, fronting on to Armley Road and overlooking the Midland Railway lines just to the north-west of Holbeck low-level station, where they ran parallel to the North Eastern Leeds and Thirsk tracks, had employed over two thousand workers in 1900. By the late 1960s the company had been absorbed into the Fairbairn–Lawson group of companies. Bad trading figures in 1978 and 1979 resulted in the group's going into receivership in April 1980. On Friday 9 May 1980, Greenbat (as it had by then become known) ceased trading and all 480 employees were made redundant; though 150 of them were immediately re-employed the following day when the business, factory and goodwill included, was acquired by Hunslet Holdings for £1.65 million cash on the nail.

It was the Greenbat range of electric-powered mining and coke car locomotives, with its associated traction motor manufacturing facility, that interested Hunslet, as it had done for some time, as an obvious extension to the diesel range available from Jack Lane. But the Official Receiver was adamant. 'Take the lot, for cash,' said he, 'or nothing at all.' Peter Alcock, on behalf of Hunslet, took the lot.

So what was 'the lot' that now came under the overall administration of Jack Lane? Perhaps 10 acres of the former 20-acre site was still operational, although very much under-utilized. There was machinery and

A number of the later standard 28 h.p. flameproof mining locomotives were built as tandem pairs for man-riding duties. The 'cab' portions were enlarged to provide for a 'second man'. These are Hunslet nos. 9047/8 of 1981 for Blidworth colliery in Nottinghamshire. The design had originated over twenty years earlier in a range of 38, 40 and 60 h.p. low-cost diesel-hydraulics that were intended for the Canadian market but also sold to mines, plantations and other users throughout the world.

stock-in-trade for a multiplicity of products – mechanical handling equipment, cold forging machinery, circular looms for polypropylene sack manufacture, paper sack machinery, die-handlers, furnace chargers, steam turbines and other items in a similar vein, all with some sales potential. On the debit side there was a change of culture, a tinge of acrimony even, fuelled no doubt by the circumstances surrounding the receivership. The dog-eat-dog, 'me first' attitude in business was beginning its apprenticeship.

More importantly, to return to the positive aspects of the acquisition, there was the battery and trolley locomotive expertise and an existing contract for thirty-four unit trains to re-equip the Post Office Railway (the 2 ft. 0 in. gauge tube from Whitechapel to Paddington) in London. This contract, for which Hunslet had tendered in competition with Greenbat, was still in the pre-production stage and was renegotiated to reflect the changing circumstances.

Some fabrication work, including frame components, was undertaken at Jack Lane, but most of the trains were assembled at Armley Road. The customer took delivery of the units between October 1980 and March 1982.

The non-locomotive work continued at Albion Works for the time being, mostly finishing off existing orders for circular looms and sack machinery and the sale of stock-built items such as cold forging machines (for making screws, etc.) and baling presses for compacting paper, cardboard and other waste products. The employees who had been retained stayed for the most part on the Greenbat premises.

Between 1927 and 1980 Greenwood and Batley had produced 1367 electric locomotives, from the small battery-powered 5 h.p. 'trammers' to some large 150 h.p. radio-controlled trolley locomotives weighing 20 tons for Mount Isa Mines in Queensland, Australia. The total included sixty-five specialist coke car locomotives for use in steelworks and smokeless fuel plants, an application in which Greenbat was the market leader. Greenbat's locomotive business was export-led, only 25 per cent being for the home market; and the countries to which locomotives had been exported, forty in all, also featured in the Hunslet export map. There was thus an opportunity for rationalization of overseas selling agencies over the coming months.

Most of the designs were individually suited to customer requirements, although there were some 'standard' types. They could all be described as 'robust', with virtually indestructible home-built traction motors. But, paralleling the Hunslet situation with slow-speed engines 15 years earlier, they were expensive, unexciting and ripe for updating if they were to meet the competition from the other makers, particularly Clayton Equipment at Hatton near Derby. Sixteen locomotives were built – three for

Thailand, eight for the Philippines, four for India and one flameproof 14-tonner for the National Coal Board – before Albion Works was 'mothballed' in 1984. Two of the Philippines locomotives were shipped fully assembled; the other export units were shipped in kit form.

Meanwhile in Jack Lane, the Kenya contract had effectively been completed by the end of 1979, although the last Swindon- and Kilmarnock-built examples were not shipped until 16 January and 20 February 1980, the Scottish one being delayed by a shipping dispute. Mining locomotive orders still predominated: twenty-five of them were built in 1980, together with eighteen of the batch of twenty-five Bord na Mona locomotives (nos. 8922–46). There were four flameproof 28 h.p. machines for the Royal Ordnance Factory at Bishopton (more were to follow), and standard six-wheeled shunters for CEGB's Carrington power station, BSC at Workington, the Oxcroft open-cast coal site and the Lombarde Falck steelworks in Italy. Two

The two locomotives built for Esso in 1981 were the last to be delivered to any customer by rail on their own wheels. This one is no. 8998, a 60 ton 450 h.p. unit for the Milford Haven refinery seen on 22 February, having just crossed Jack Lane on test prior to its despatch the following day. No. 8999, a 45-tonner, went to Fawley a month later. There were destined to be only four more standard-gauge Hunslet diesel shunters built new for British industrial customers. The track across Jack Lane was lifted three years later, only to be reinstated (on a new alignment) in 1990.

smart 25-ton 240 h.p. four-wheelers were supplied to the Royal Naval Armaments Depots at Frater near Gosport. All in all, a good business year, and the company reports for August 1980 and August 1981 showed operating profits of £1.975 million and £2.672 million on turnover figures of £10.474 million and £14.389 million respectively.

The following year, 1981, showed a similar performance and a similar product mix, with 37 locomotives for the National Coal Board, the balance of the Bord na Mona contract, two more flashproof 450 h.p. standard-gauge shunters for Esso Petroleum (one each to the Fawley and Milford Haven refineries) and a 252 h.p. 0–4–0 to Steetley Magnasite Ltd at their Hartlepool works. Three ex-British Railways 650 h.p. Swindon-built shunters, the ones for which Hunslet had provided the gearboxes in 1963–4, came to the works from British Steel at Corby towards the end of the year to be altered to 5 ft. 6 in. gauge on behalf of a Spanish civil engineering contractor.

The falling-off in orders for new shunting locomotives for UK industrial use was dramatic but not unexpected, as noted in the opening paragraphs of this chapter. Two new 300 h.p. 0–4–0s and a rebuilt 0–4–0 were delivered to British Nuclear Fuels Ltd at Sellafield in 1982, followed by a 400 h.p. 0–6–0 in 1983, all in connection with the THORP nuclear waste recycling project, and it was to be five years before another, and indeed the final, new 4 ft. 8½ in. gauge shunter was produced at Jack Lane for use in the British Isles.

There were other markets, however, and these were receiving a greater degree of attention than hitherto. Nevertheless, the situation demanded close scrutiny. Russell Wear, in his book *Barclay 150*, published in 1990, describes how Barclays also only produced three locomotives in 1982 and two in 1983, and how discussions took place at Kilmarnock early in 1982 with a view to winding up the associate company and transferring all remaining work to Leeds.

Just in time, the pendulum swung again. Eleven more locomotives (nos. 9201–11) were ordered by the Sudan Gezira Board, four 400 h.p. 0–6–0s (nos. 9088–91) by the Nigerian Port Authority, and a new range of cardan-shaft drive sugar estates locomotives brought orders from St Kitts and Fiji. Kilmarnock was reprieved, but the Greenbat works at Armley was later to be mothballed and Leeds manufacture concentrated at Jack Lane in 1984.

Nevertheless, 1982 was a year of mixed fortunes. Events started on 25 January with a change of name from Hunslet (Holdings) Ltd to Hunslet (Holdings) plc (for public limited company) in line with the provisions of the Companies Act 1980. Rae Fryers retired as Chairman and Chief Executive on 13 September but stayed on as non-executive Chairman. Three days earlier, Group President John Alcock's eldest son Peter had been appointed Group Chief Executive, while John's middle son Keith and nephew David Gawthorpe became Joint Managing Directors.

The Nigerian locomotives were despatched between February and May 1982, and the first two cardan-shaft drive sugar estates locomotives departed on 16 April and 3 August. These were, respectively, no. 9086, with a Gardner 6LXB 160 h.p. engine, for the 2 ft. 6 in. gauge lines in St Kitts, and no. 9087, with a 240 h.p. General Motors two-stroke engine, for the Fiji Sugar Corporation's 2 ft. 0 in. gauge system.

Just as, in the 1960s, the Hunslet drawing office had been reluctant to rush headlong into adopting the Twin Disc transmission until it had been thoroughly proven by someone else, its adoption of cardan-shaft drive instead of using rods to mechanically couple the driving wheels was equally cautious. The first step had been to develop a standard wheelset with axle-mounted bevel gearboxes and rubber suspension. A primary gearbox, with high-level input, took the drive from the hydraulic transmission through a double reduction to the rear axle and mid-level output after the first reduction, to lead via another cardan shaft to a single-reduction secondary gearbox on another axle. Different permutations of primary and secondary gearboxes could cater for different wheel arrangements and different transmissions.

These gearboxes were successively used in some eight-ton 54 h.p. locomotives for nickel mines in Canada during 1969, the 90 h.p. mines locomotives and the 216 h.p. and 300 h.p. high-speed B–B mines locomotives; but the St Kitts and Fiji locomotives were the first medium-size surface examples and the first six-wheelers. On the surface examples there was sufficient space to mount a Hunslet designed and manufactured disc brake on the input shaft to give a very cheap and very effective automatic 'spring on, air off' fail-safe braking facility: even if every cardan shaft on the locomotive failed the braking would not be impaired.

The eleven Sudan Gezira Board locomotives, however, retained the traditional coupling-rod drive, mechanically almost identical to the first four supplied to this customer

A standard Hunslet wheel set as used in the later narrow-gauge estate locomotives, complete with final-drive gearbox and disc brake assembly. All was manufactured 'in house' save for the brake cylinder, roller bearings and minor items.

in 1953 but with the latest Gardner 6LXB diesel engine. With over sixty Hunslet locomotives already in service, the Gezira Scheme, lying between the Blue and White Nile rivers some 150 km south of Khartoum, constituted the largest farm under one management in the world, and provided employment for over a million people in the cultivation and processing of cotton.

The project involving the purchase of the eleven locomotives was financed by the World Bank and therefore went out to competitive international tender. Although the Hunslet offer was the dearest it was accepted, albeit for a smaller number than had originally been intended, partly because of the Hunslet product's proven reliability and partly because of the catastrophic failure of a large number of cheaper locomotives purchased from elsewhere in Europe in a similar financing system some years previously.

Sudan Gezira had been one of John Alcock's 'pet' locations: the long non-stop journeys, usually overnight,

on lightly-laid 2 ft. 0 in. gauge track – 100 miles at 20 m.p.h. with a 150-ton train was not unusual – amply bore out his unshakeable faith in the efficiency of a mechanical transmission. There was some poignancy, therefore, in the fact that, as the Jack Lane works closed down for Christmas 1982, John Alcock died at his home in Knaresborough, at the age of 77. Six of the eleven locomotives had been despatched just before, and it was a fitting tribute that the gearboxes and general mechanical design that had been insisted on by the customer showed little noticeable change from his prototype LMS shunter of fifty years before. Old-fashioned? Not a bit of it. John Alcock was stubborn, yes, infuriating sometimes, but he was innovative. His active and enthusiastic involvement in up-to-the minute engineering continued right to the end of his life and was a source of inspiration to many. Horses for courses is one thing: knowing both the horse and the course is another. He got them both right more often than not.

While John Alcock's death was mourned by many, and his well-attended funeral at Knaresborough bore testimony both to his stature and to his popularity, the Board changes of the previous September ensured continuity of the Company's policies, at least for the immediate future, and business continued very much as before.

The newly-developed 150 h.p. rack-and-pinion locomotive was undergoing extensive trials at the National Coal Board's testing centre at Swadlincote, and an important order was received from the Cape Breton Development Corporation (DEVCO) of Sydney, Nova Scotia, for ten low-height 100 h.p. Caterpillar-engined diesel hydraulic flameproof mining locomotives. The Portuguese cement manufacturing company Cimpor ordered a 42-tonne 1665 mm gauge 335 h.p. 0–6–0, and this was to be the first of six locomotives for this customer, including two built in Kilmarnock after surface locomotive work was transferred there in 1989. A 450 h.p. 0–4–0 flashproof refinery locomotive was despatched during May 1983 to the SIBP petrol refinery on Antwerp Docks in Belgium, close to where three other Hunslet standard 0–4–0 shunters had been working since 1973.

In January 1983, Hunslet's Scottish subsidiary Andrew Barclay, Sons & Co. Ltd purchased the 100-year-old Glasgow company Bonar Hugh Smith Ltd. A new company, Hugh Smith (Engineering) Ltd, was formed to operate within the Kilmarnock factory and to continue the Hugh Smith range of heavy machine tools, including vertical

Forty years after the design had first been supplied to the Sudan Gezira Board, the customer was still insisting that the Hunslet 135 h.p. 2 ft. 0 in. gauge 0–8–0 was the only locomotive capable of meeting the long-haul requirements. The final batch of eleven locomotives, nos. 9201–11, is well advanced in this December 1982 view, with no. 9000, the penultimate UK diesel shunter, for British Nuclear Fuels, on the left of the picture, and several 91 h.p. rack locomotives for the National Coal Board in the middle distance, right. The Sudan Gezira Board fleet eventually totalled fifty-two 120–135 h.p. 0–8–0 and twenty-five 88 h.p. 0–6–0 locomotives plus two inspection cars from Hunslet.

The class of '83. Concurrent with the last batch of Sudan Gezira Board locomotives, twelve sets of 'drop-in' 1500 r.p.m. Gardner diesel engines and step-down reduction gearboxes were supplied for 0–6–0 units in Blue Nile Province. They were for installation on site to replace the earlier McLaren engines. Hunslet service engineers Roly Haigh (second from left) and David Walker (far right) pose with the Meringan workshops staff in front of one of the rebuilt locomotives at Barakat in 1984.

plate-bending machines of up to 5000 tonnes capacity, plate edge-planing and milling machines up to 22 metres long and a host of other products. Some idea of the magnitude of these can be judged from the weight of the largest bending machine at 310 tonnes. In the first five months of the acquisition, £1.8 million worth of export orders were received from Egypt, India, Saudi Arabia, the USSR and Thailand. Although centred on, and assembled in, Kilmarnock, the machines incorporated several very large fabrications which provided a good balancing load for the Hunslet welding and heavy machining facilities. Hugh Smith were not strangers to railway work. In *British Railways Illustrated* for June 1995, A. N. Marshall mentions a Hugh Smith plate-edge planer at the St Rollox works of the Caledonian Railway.

For many years the British Army of the Rhine in Germany had been considering a new fleet of locomotives for the rail operations of 79 Railway Squadron, Royal Corps of Transport, and this manifested itself in eleven

locomotives built and supplied during 1984. All were to standard gauge and built in accordance with the requirements of the German State Railways (DB). Five 35 tonne 300 h.p. 0–4–0s were designed and built at Hunslet, while six 600 h.p. 0–6–0s came from Kilmarnock with frames and other components sent up from Leeds. All had Rolls-Royce engines and two-speed transmissions, and acceptance testing took place on the Bicester military railway before despatch to Germany.

The setting-up in the late 1940s of Hunslet Africa (Pty) Ltd, and its subsequent development into Hunslet Taylor Consolidated at the large Germiston Works just outside Johannesburg, had meant that no new Leeds-built Hunslet diesel locomotives had been shipped to South Africa since 1953. Well over two thousand locomotives, from 15 h.p. to 750 h.p., wholly built in South Africa from then on, had carried the Hunslet name, and indeed took numbers out of the Leeds building list in addition to a South African number. By way of examples, taken at random, Hunslet

Hunslet's connection with the British Army started with the Crimean campaign. It ended with the five 35 tonne 300 h.p. locomotives (HE 9221–5) supplied in 1984 to 79 Railway Squadron, Royal Corps of Transport, for work in the depots supporting the remnants of the British Army of the Rhine in West Germany (as it then was).

(Leeds) nos. 7770–7 were also HAD (Hunslet Africa Diesel) nos. 1770–7; this was a batch of 3 ft. 6 in. gauge 250 h.p. shunters for South African Railways built in 1971.

In the early 1980s, changes in the structure of Hunslet Taylor Consolidated, and relaxations in trading policy between Britain and South Africa brought a renewed Leeds interest in what, years before, had been one of its principal markets. John Alcock had always kept in touch both with HTC and with the sugar plantations in particular, where large fleets of both the Germiston-built and the older Leeds-built diesels operated, and this contact was to be useful when the author took over the South African trips in 1980.

The first new locomotives to be supplied after the hiatus were three 20-ton 0–4–0 locomotives, two with 180 h.p. Cummins engines for Sasol Fertilizers and one with a 190 h.p. General Motors engine for South African Railways. They were of Hunslet Taylor design and supplied in kit form for final assembly in the small factory of Prof Engineering Ltd in Alrode, south-east of Johannesburg.

The three kits of parts went away between 21 December 1983 and 1 March 1984. They were followed by a very neat

Leeds-designed 25-tonne 2 ft. 6 in. gauge cardan-shaft drive locomotive for Umfolozi Consolidated Sugar Planters at Mtubatuba. It was powered by a 240 h.p. Cummins engine, and was sent out completely assembled and painted. Umfolozi had a very large fleet of Hunslet locomotives in their plantations close to the estuary of the White Umfolozi River, and the new unit was for the mainline haul to the factory, as a direct replacement for no. 4539, which had been supplied from Jack Lane in 1953. John Alcock had always been well respected by the plantation and factory workers, Zulu to a man, and this new locomotive was to have been named *Mister John* in his memory. Fate intervened: during construction of the locomotive the Umfolozi plantation, and indeed much of northern Natal, was devastated by hurricane Demoina, which severed the road and rail links between north and south. The local superstitions prevailed, and locomotive no. 9237 was despatched on 17 August 1984 carrying name-plates *Demoina* to appease the gods.

The miners' strike against pit closures, which started on 12 March 1984 and was to be rock-solid for a week short

Named after Hurricane Demoina, which devastated northern Natal early in 1984, Hunslet no. 9237/84 stands at the end of the test track shortly before despatch in August of that year. 240 h.p., 25 tons, 2 ft. 6 in. gauge.

The Hunslet erecting shop in the early summer of 1984. This was three months into the miners' strike, and the shop is nearly empty. The first excursion into passenger train work is on the left – the accident-damaged Tyne & Wear Metro unit being repainted after repair. Centre right is one of the British Army of the Rhine 300 h.p. locomotives. Behind stands the cab of Demoina on the frame of a stock 400 h.p. shunter. This frame ultimately became the last Hunslet standard-gauge industrial shunter, no. 9092, despatched to Caledonian Paper Ltd as late as 6 December 1988.

of 12 months, could have killed off Hunslet as it surely did many other companies with mining interests. That the strike did not prove so disastrous was due in no small measure to the strong cash position of the company, aided by the other orders previously mentioned. Just as the term 'industrial action' came to produce exactly the opposite, so did this strike against pit closures succeed in accelerating the demise of many collieries; some never re-opened. The effect on the main-stream mining locomotive business was predictable. Having supplied 162 mining locomotives to British pits in the seven years before the strike Hunslet

despatched only a tenth of that number in the seven years that followed.

Despite a very extravagant international mining exhibition at the National Exhibition Centre in Birmingham in June 1985, at which the large 300 h.p. high-speed man-riding locomotive, the 150 h.p. rack locomotive and a brand-new Greenbat rubber-tyred pony (as in 'pit pony') electric locomotive were exhibited on the company's stand, the market never recovered. Even some of the showpiece pits for which the 300 h.p. locomotives were destined had closed, and four of these units, including the one shown at the NEC, were put into store during July to await regauging and transfer elsewhere as a job for them arose. The remaining rack locomotives, 150 h.p. and 64 h.p., which had been put in hand to stock, sold slowly; three of the older 91 h.p. type were never sold at all in spite of an attempted 'technology transfer' deal with the Chinese coal authority. Other stock batches were curtailed or cancelled and parts used as spares.

However, the corporate Hunslet lap was still capable of being strategically placed. The British Railways Board awarded a contract for twenty-five two-car diesel railbuses to the Falkirk coachbuilding firm of Walter Alexander, with supply of mechanical parts and final assembly by the Hunslet Group. The major portion of the underframes and running gear was in fact supplied from the Hunslet Engine Works, and comprised twenty-five underframe kits and twenty-five complete running chassis, and final assembly took place at Kilmarnock. This underframe work kept the greater part of the Hunslet erecting shop busy through 1985. It was augmented by a major overhaul and refurbishment contract for electric multiple-unit trains operated by the Tyne & Wear Passenger Transport Executive. Three two-car trains were involved: the first was an almost complete remanufacture following an accident at a level crossing; then came the task of bringing two pre-production prototypes into line with the later production fleet.

At the Annual General Meeting of the company held on 21 January 1985, this being the 121st year of business, the Directors' Report and Accounts showed a total turnover up by 10 per cent to £12.5 million, export turnover up 41 per cent to £6.1 million, and a recovery from the previous year's loss of £147,000 to a small profit of £126,000. These figures were the result of trading up to the financial year end of 31 August 1984, and bore the full effect of six months' total loss of National Coal Board business in the first half of the strike. In the circumstances the results were very good indeed.

The slow transition towards a more active role in the passenger transport business developed further when an order was received in March 1985 from Metro Cammell Ltd, Birmingham for eight bogies fitted with Brown Boveri

The ultimate flameproof underground mining locomotive was this 300 h.p. (225 kW) Bo–Bo design capable of taking large numbers of miners several miles to the coal face at speeds of up to 50 km/h. Introduced in 1981 and exhibited at the NEC in 1984, sixteen were built, but events in the mining industry soon rendered them surplus to requirements.

Resplendent in its apple green, red and white livery is No. 10 Yeti, the second of the initial pair of Hunslet diesel-hydraulic rack-and-pinion locomotives, no. 9250/86. (The first was No. 9 Ninian, HE 9249/86.) In this view Yeti nears the end of its journey from the Snowdon summit on 11 September 1996. Llanberis works is in the background. The coach is not coupled to the locomotive: should the locomotive become derailed and allow the coach to proceed on its own, overrun brakes would bring it to a safe if undignified halt. A pair of coach bogies was also produced by Hunslet in 1987. Two further locomotives to the same design were supplied, No. 11 (HE 9305/91) from Kilmarnock (who also gave it Barclays' serial no. 775) and No. 12 (HE 9312) from Leeds in 1992.   ADRIAN BOOTH

Not the ultimate in style but functional if nothing else – and cheap. One of two 2 ft. 0 in. gauge 250 h.p. Bo–Bo locomotives for the Office Nationale de Sucre, Zaïre, in 1986. Hunslet no. 9268/9. Caterpillar diesel engine and hydraulic transmission. A lock, stock and barrel progression from the pre-war Avonside geared steam locomotive.

In 1985 the latest Hunslet standard range of cardan shaft drive diesel-hydraulic locomotives was extended to include a 9¹/₂ tonne version intended to replace the Bord na Mona 'Wagonmasters'. Only fifteen were built for Eire, although variations went to other customers. LM 386, HE 9258/86, arrives at Shannonbridge power station on 5 December 1996 and will stop just short of the tippler seen on the extreme right of the picture. The locomotive will uncouple and be driven through the tippler; the wagons will follow using the feeder mechanism between the rails.   COLIN BOOCOCK

HE 9266/86 was a 60 tonne 603 h.p. radio-controlled shunter supplied to National Gypsum Canada for use at a ship-loading dock at Dartmouth, across the river from the port of Halifax in Nova Scotia. The locomotive was loaded on to an 80 tonne capacity road trailer at the works and carried to Halifax on the Atlantic Container Line's roll-on, roll-off vessel Atlantic Concert whence it was unloaded for the short journey to Dartmouth. In mid-Atlantic it passed 1892-built Hunslet no. 555 returning to its old home in Tralee on board the Concert's sister ship Atlantic Compass.

traction motors and intended for one of the three experimental trains for London Transport's investigation into fleet replacement for the Central Line. A further order for twenty-five bogies came from the Tyne & Wear Metro after examination of existing bogie frames showed up fatigue cracks. The new bogies were used as spares and a 'float' as a programme of modifications got under way.

Overhaul of the Tyne & Wear passenger cars had established a facility for this kind of work, and a 'niche' business began to emerge in the structural repair of accident-damaged vehicles. Several London Underground vehicle repairs, some of them quite extensive, were undertaken in 1985 and 1986. British Rail Engineering Limited ordered auxiliary generators of Greenbat design for the class 150 Sprinters being built at York. These generators were manufactured throughout at Jack Lane, as were all 'Greenbat' electrical machines after the 'mothballing' of the Armley works progressively from 1983.

Also with passenger traffic in mind was an order for two diesel rack locomotives for the hitherto wholly steam-operated Snowdon Mountain Railway in North Wales. Snowdon was operating seven Swiss-built steam locomotives – four supplied in 1895/6 and three dating from 1922/3. Hunslet had a long-established connection with Snowdon, having carried out major overhauls on the steam locomotives for many years.

In so far as the wheels and running gear were concerned, the new diesel locomotives had much in common with their steam ancestors. Snowdon Mountain Railway Nos. 9 and 10, Hunslet nos. 9249/50, were despatched during March 1986 for testing and Railway Inspectorate approval, in preparation for the 1986 tourist season. Two bogies with rack-and-pinion equipment and overrun brakes followed later in 1986 to be used under a new coach body being supplied by East Lancashire Coachbuilders.

The cardan-shaft drive range of locomotives continued to be developed, and several variants were supplied, including

Ten 100 h.p. 13.6 tonne flameproof mining locomotives (HE 9214–9, 9233–6) were built in 1984 for the 3 ft. 6 in. gauge lines of Cape Breton Corporation (Devco) in the north of Cape Breton Island near Sydney, Nova Scotia. Only 4 ft. 0 in. high, to cope with the extremely restricted headroom in the underground roadways, they had Caterpillar 3304 engines and American Funk hydraulic transmissions. The diesel power pack complete with exhaust gas conditioner and all-flameproof equipment was mounted forward of the leading wheels and could be replaced in four hours.

115 h.p. 0–4–0s to Bord Na Mona and to British Steel, Shotton; 140 h.p. 0–4–0s and 240 h.p. 0–6–0s to the Fiji Sugar Corporation; and two 250 h.p. Bo–Bos to the Zaïre Sugar Corporation. A large 60-tonne 603 h.p. radio-controlled coupling-rod drive 0–6–0 locomotive, no. 9266, was shipped to National Gypsum at Dartmouth, near Halifax, Nova Scotia on 16 July 1986. The Nigerian Port Authority ordered four more 0–6–0 shunters similar to the ones supplied in 1982, but fitted this time with the 450 h.p. Rolls-Royce/Perkins CV8T engine – the first locomotive application of this unit. They were nos. 9275–8, and were shipped to Lagos on 26 August 1988.

But the contract that had the greatest impact, in a number of unconnected ways, had to be the one that came in connection with the Channel Tunnel. Hunslet had never been in the mainstream tunnelling locomotive market, since this usually involved very small, usually battery-electric units, which Jack Lane had never been able to

produce economically. Large tunnels, however, were a different matter – as witness *Bernard* in chapter 3; and Greenbat had supplied 31 electric locomotives for the first Mersey tunnel in 1928/9, Hunslet had supplied diesels for the second Mersey tunnel in 1966, and the Channel Tunnel certainly came within the 'large' category.

A wary eye had always been kept on potential contractors throughout the long series of false starts and changes of plan over the years. As early as October 1982, discussions had taken place between Hunslet and Taylor Woodrow on the merits of the rack locomotive principle, with the 150 h.p. National Coal Board design being put forward as suitable for the 1 in 6 drift or adit that was being considered in the preliminary proposal as a means of facilitating tunnelling from a coastal point near Folkestone both seaward in the direction of France and landward towards London. The cliffs on the English side of the Channel determined that the English tunnel portal should

Another design that suffered much from rapid change in the mining industry. The prototype 150 h.p. rack locomotive undergoes test at the NCB test centre at Swadlincote in 1983. Ten were built before Leeds closed; two did sterling work on emergency and ambulance duty during construction of the Channel Tunnel. The design was revived by Kilmarnock in 1997 when examples were supplied to the Chinese Ministry of Coal Enterprises.

be some way inland, at Cheriton, five miles from where the line of route crossed the coast. By 1985 the details were becoming clear. The power requirements for the construction locomotives had increased, and the integration of Greenbat into the Hunslet group was as complete as it was ever likely to be. Taylor Woodrow became a member of the TML (TransManche Link) consortium working on behalf of the operators Eurotunnel. By the time that official invitations to tender for the construction locomotives were issued in the spring of that year, Hunslet already had two designs in mind. Both were battery/overhead-line electrics, to obviate problems with ventilation. The basic frame, running gear and rack-and-pinion features of the Hunslet 150 h.p. flameproof diesel locomotives were enlarged and developed to take two large Greenbat traction motors similar to the ones that had been used in the Mount Isa locomotives mentioned earlier, to produce a 20 tonne 350 h.p. rack-and-adhesion locomotive with the capability of taking more than 100 tons of tools, materials or spoil up or down the eventual 750-yard 1 in 6 adit, and also able to operate if required along the tunnel itself. Complementary to the rack-and-adhesion locomotive was a design based on the cardan-shaft drive diesel locomotive range, whereby one traction motor could be used between two standard gearbox and suspension units to provide a 20-ton 175 h.p. adhesion-only locomotive for the main tunnelling duties.

The first order placed with Hunslet in 1986 for four rack-and-adhesion locomotives was only the second capital order placed by the newly-formed TML (the first was for tunnel-boring machines), thereby emphasizing the importance of these units in the overall scheme of things. Within weeks, four more were ordered, together with twelve adhesion locomotives; and subsequent repeat orders brought the final quantity to seventy-eight, of which eighteen were rack-and-adhesion locomotives, the remainder adhesion-only. The last 40 adhesion locomotives were increased in weight to 25 tonnes, given larger batteries and uprated to 200 h.p.

The first locomotive was ceremoniously handed over by Peter Alcock to the Co-chairman of Eurotunnel, Alastair Morton (later Sir Alastair), on 22 July 1987, and building continued unabated through to May 1990. Two of the

standard 150 h.p. diesel rack locomotives building to stock were supplied as stand-by units in 1988, and one of these was converted for ambulance duties in 1989. A summary of the Hunslet locomotives for Channel Tunnel construction is as follows:

9158–61	Built 1987	350 h.p. rack-and-adhesion electric
9162–73	Built 1987/8	175 h.p. adhesion-only electric
9177–80	Built 1988	350 h.p. rack-and-adhesion electric
9282–3	Built 1988	150 h.p. rack-and-adhesion diesel
9295–304	Rebuilt 1990	Diesel rebuilds of nos. 9166–9 and 9402–7
9400–8	Built 1988	175 h.p. adhesion-only electric
9409–27	Built 1988–90	200 h.p. adhesion-only electric
9428–9	Built 1989	350 h.p. rack-and-adhesion electric
9430–49	Built 1990	200 h.p. adhesion-only electric
9450–5	Built 1990	350 h.p. rack-and-adhesion electric
9459–60	Built 1990	350 h.p. rack-and-adhesion electric

Nos. 9419/20, 9436/7 and 9447 were assembled at Kilmarnock. The 1990 rebuilds used the frames and running gear of ten 175 h.p. electric locomotives together with a Deutz type 413 FW clean-burn ten-cylinder vee engine and hydraulic transmission to give 34-tonne diesel units for use in the landward tunnels where ventilation was not such a problem.

In the midst of this mass of tunnelling locomotives was a contract for three 4 ft. 8½ in. gauge battery/overhead line electric locomotives for works train duties on the Tyne and Wear Metro. These had the same traction motors, disc brakes and axle-mounted gearboxes as the TML adhesion locomotives, and were of roughly the same power and weight. Pantographs were fitted on the roof of the large cabs, and Hunslet rubber-cushioned resilient wheels of the type first used on *Demoina* provided a very silent-running unit.

The full effects of the miners' strike had been reflected in a loss of £896,000 in the 1985 balance sheet; but the 1986 figures, calculated up to 3 August 1986 and announced on 19 January 1987, showed turnover up yet again to £15,696,000 and a return to a modest profit of £360,000. The Channel Tunnel business had been taken at realistic prices and would favourably influence results for the next three years. The worst appeared to be over.

CHAPTER 14

Eclipse of the Family Firm, 1987–1989

At the time of the handing-over ceremony for the first Channel Tunnel construction locomotive, the Hunslet Engine Company, and its direct successor Hunslet (Holdings) plc, had been in the control of just two families for 122 years, with only five men holding the post of what would now be called the chief executive officer.

With a turnover of £15,696,000 in the 1986/7 financial year, the company was grossly under-capitalized with 1,200,000 issued and fully paid-up shares of 25 pence each; or at least this was how the ever more powerful financial institutions of the late 1980s would have viewed matters. But this was no hostage to corporate banking fortune. Sixty-five per cent of the nominal £300,000 share capital was in family hands, and the family would not break ranks, it was said (frequently); no one else had more than a five per cent interest except for one acquisitive shareholder, no larger than a man's hand, who was felt to be under control.

The firm did not borrow, nor did it indulge in rights issues, though such were allowable under a resolution passed at an earlier Annual General Meeting. The many machine tool purchases and the acquisition of other companies had generally been funded out of reserves; dividends were kept comfortably high and published profit figures were consequently artificially low. The aim was always to maintain a steady if unspectacular growth at least in line with inflation, to maintain the Leeds work force at a reasonably constant level, and to spread the risk by having many small contracts at any one time rather than rely on only one large contract. Fifty per cent of work-load on one contract plus, say, ten small contracts was just acceptable; but twenty small contracts would be more palatable.

Advertising was kept to a minimum – just enough to reassure customers that the company was still very much alive. The occasional press release was issued, but only when the contract was of particular interest and never before the customer's firm order had been acknowledged and accepted and payment guaranteed. Only after the Greenbat takeover was there any whisper within the factory walls that might suggest any possibility of chickens being counted before they were hatched. The premature reporting of possible contracts so prevalent throughout the industry in the 1990s would have been anathema within 'Castle Hunslet'.

All this was canniness *in excelsis*, the aim being to maintain a low profile and discourage any adventurous marauders from disturbing family unity. The shares had remained at 25 pence for some time after issue in 1958 and then very slowly risen to bump along at around 150–180 pence throughout the first half of the eighties – not exactly sparkling at a time of very high inflation. Nevertheless, the result was self-determination, maintaining the directors' modest standard of living and providing better than average security for the work force.

Eurotunnel and their construction contractors TML (Transmanche Link) were naturally anxious to promote the Channel Tunnel project, and any equipment contracts placed by them received wide publicity. By the time of Co-chairman Alastair Morton's visit to Leeds on 22 July 1987, the share price had risen to 470 pence even though only small quantities had been traded.

On 21 July 1987 most if not all the national dailies had carried a full-page Eurotunnel advertisement announcing 'EUROTUNNEL TRAINS ARE ALREADY RUNNING ON TIME' and charting the progress of the contract from the receipt of tenders. No one could fail to note the name 'Hunslet', for it appeared five times, once in letters as large as 10 millimetres high.

On 30 July 1987, Peter Alcock issued a letter to all employees advising that 'Hunslet have agreed the terms of an offer to be made on behalf of Telfos Holdings plc to acquire the whole of the issued share capital of Hunslet', adding: 'I am staying on as full time Executive Chairman of Hunslet and I believe that this change of shareholders should be beneficial to us all.' Everything was OK then, and the papers carried the story the following day (Friday), just as the works closed for the annual Bank holiday fortnight.

Peter Alcock's letter had come out of the blue, and the name Telfos was unknown to most of its recipients. The

The first four Channel Tunnel 350 h.p. 900 mm gauge rack-and-adhesion locomotives being assembled at the far end of the erecting shops, early in 1987. The computerized numerically-controlled boring machine in the foreground had been installed only a few years earlier; it was capable of machining the largest fabricated locomotive frames. The larger fabrications for Kilmarnock-built Hugh Smith machine tools were also finished on this machine. Two of these fabrications stand alongside at left.

The first Channel Tunnel rack locomotive, no. LR1 (later RR1), Hunslet no. 9158/87, poses on the test track alongside the old Shepherd & Todd Pearson Street boundary wall immediately prior to Mr (later Sir) Alastair Morton's acceptance visit on 22 July 1987. The 'family firm' had ten days to go. One hundred and fifty years in one picture.

One Mulhauser side-dump muck wagon had to suffice for test purposes. Adhesion-only 175 h.p. Mark I battery trolley locomotives nos. 9162/3 demonstrate tandem working on battery power only in August 1987.

offer was undeniably attractive, however, offering 610 pence cash for each Hunslet 25p share. Alternatively, for each Hunslet share you could have one Telfos 'unit' comprising three new Telfos ordinary shares plus one new Telfos preferred share. The formal document was dated 12 August and remained open until 12 September, by which time it became absolute.

The report in the *Yorkshire Post* for 1 August 1987 said: 'The Telfos directors have every intention of keeping the Hunslet name and expanding the existing business which employs 600 people, mostly in Leeds.' Telfos had at that time a £27 million stock market capitalization. This was said to be a result of its success in turning the Birmingham non-ferrous metals group Charles Clifford round from a situation in which it was losing up to £100,000 per month.

The *Yorkshire Post* account went on to make much of the 'strong financial position' of Telfos, 'with £2m. in cash and

£2.5m. in investment (helped by £3.5m. raised from two rights issues)', but avoids reference to Hunslet also having over £1.5 million in cash and the fact that Hunslet's net assets, at £7.96 million, exceeded those of Telfos, at £7.0 million, even after Hunslet had made a £1.05 million provision against costs to be incurred in rationalizing capacity and in covering 'estimated losses on contracts in excess of the value of related work in progress' and 'estimated liabilities which may arise in meeting contractual commitments'. (From the Hunslet Report and Accounts dated 31 August 1986, published after delivery of the last class 143 'Pacer' railbus and before receipt of the TML orders.)

Background on Telfos is provided by their 1986 Report and Accounts, which shows the parent company as Telfos Holdings plc, an Investment Holding Company, with subsidiaries Charles Clifford Industries (non-trading),

Eurotunnel trains are already running on time.

HUNSLET HOLDINGS PLC
—OF LEEDS—
Eurotunnel trains work schedule

July 86	Receipt by Translink of tenders for locomotives following issue of tender notice in EEC Journal.
August 86	Receipt by Hunslet of Letter of Intent from Translink to enter into a contract for procurement of Rack and Pinion Locomotives.
November 86	Signing of firm contract between Translink and Hunslet for supply of Rack and Pinion Locomotives.
January 87	Commencement of assembly of locomotive main frames.
February 87	Gearbox and transmission assemblies for first locomotive on test.
March 87	First locomotive submitted for initial static electric integrity tests.
April 87	First locomotive on final tests for acceptance by Translink on contractual completion date early May 87.

The Channel Tunnel is planned to open in 1993. But some trains will be running long before then.

In November 1986 Hunslet (Holdings) plc was awarded a £1.2 million contract to supply Eurotunnel with an initial batch of specially designed rack and pinion locomotives.

These won't be carrying passengers. They'll be used to haul spoil from the tunnel to the surface.

There's nearly 4.5 million cubic metres of it. (Enough to fill the National Exhibition Centre three and a half times.)

Of course a contract of this size isn't just great news for Hunslet. It's also a great opportunity for their many suppliers in Yorkshire.

Indeed, for British industry as a whole, there's a lot more coming out of the Channel Tunnel than a few million cubic metres of spoil.

EURO TUNNEL

A breakthrough for Britain.

The Eurotunnel advertisement that appeared in all the quality national daily newspapers on 21 July 1987. Unprecedented publicity with far-reaching consequences.

The first locomotive in the Channel tunnel. Hunslet rack locomotive 9158/87 stands at the foot of the 16% (1 in 6) adit at Shakespeare Cliff, Folkestone, when still almost new. Behind the camera is the first short length of tunnel proper, at the top of the adit the TML workshop and material delivery site.

EUROTUNNEL VIA TML

Charles Clifford Ltd (manufacturers of non-ferrous rod, wire and strip), Metallisation Ltd (manufacturers and suppliers of metal-spraying equipment and materials), Metallisation Service Ltd (suppliers of metal-spraying services) and Stirchley Investments Ltd (investment dealing). The shares in Metallisation Ltd and Metallisation Service Ltd were held by Charles Clifford Industries Ltd. Telfos's ultimate holding company at that time was the Chillington Corporation plc, which was itself a fairly recent merger of Plantation and General Investments and the Anglo-Indonesian Corporation.

The same accounts show a profit on ordinary activities before taxation of £1.5 million on a turnover of £7.9 million from a labour force of 175 people.

Let us just pause awhile at this point to reflect on the British railway rolling stock industry in mid-1987 as it entered the most bizarre and dramatic ten years of its history. Simplistic as it may seem to some, this was still a time when locomotives were built in locomotive works, carriages in carriage works, wagons in wagon works and so on. Diesel and electric multiple-unit trains had increasingly been replacing the traditional locomotive-hauled trains for some years, and were centred on the carriage works, thereby creating workshops that were ultimately to become the 'train builders' of the nineties.

For historical reasons explained in chapter 3, British Railways and its predecessors had been largely self-sufficient in the design and production of their rolling stock. The railway-owned shops reigned supreme and possessed the intellectual property rights for virtually everything that operated on the national system. Private builders had flourished, withered and died in the heyday of the 1950s and 1960s when the Modernisation Plan saw the wholesale replacement of sixteen thousand steam locomotives by diesel and electric units; and only firms with something special to offer, or a strong export base, or those who were also manufacturing electric traction equipment, survived into the 1980s.

So who constituted the industry in 1987? The major player was the collection of state-owned railway workshops of which the major new-build factories, Crewe, York, Derby Carriage and Derby Loco, were destined to be privatized the following year, becoming successively BREL (1988) Ltd, Brel Limited, ABB Transportation Ltd and eventually Adtranz. These works virtually monopolized the BR market but were also beginning to look at London Underground and the export market.

Metro Cammell in Birmingham had an equally firm hold on the lucrative London Underground, Hong Kong Mass Transit and Kowloon–Canton heavy metro markets, but had just scored a significant success with the class 156 Super Sprinter diesel multiple unit for BR suburban services. Metro-Cammell was destined shortly to become part of GEC Transportation Projects Ltd. GEC was a substantial supplier of traction equipment, mainly using BREL for the provision of mechanical parts and final assembly of locomotives. The wagon works portion of the BREL plant at Doncaster was set to be the pilot privatization scheme as RFS Engineering Ltd the following year. There were other firms on the periphery, all contributing their varied skills. Hunslet, Brush at Loughborough, Procor at Wakefield and Thomas Hill at Rotherham were there, to name but four; but the clear indication was that there would, for the foreseeable future at any rate, be two base suppliers of trains (as we could now call them, the distinction between locomotives and carriages having become blurred) for the British railway system.

The thought that there could be a third force in the industry had not escaped us at Hunslet, and the tentative steps towards bogies, railbuses, tube train repair work, etc. had all been directed in the usual low-risk manner towards this end. It was being done by evolution, rather than revolution, and without undue financial risk. As the industrial locomotive market diminished, something had to take its place, and the skills and resources were there.

While the Telfos and Hunslet methods of accounting differed in their areas of creativity, what was clear from the finance-led Telfos figures, as compared with Hunslet's engineering-based traditionalism, was that Telfos expected a higher 'margin' on the sale of products. The quest for higher margins was to exercise the author's mind and the minds of many colleagues over the next few otherwise uneventful months. There was a hiatus, a honeymoon period, while the new owners took stock of the operational works in Jack Lane and Kilmarnock and the mothballed Greenwood & Batley site in Armley, this last to be sold on 10 December 1987 to F. R. Evans of Leeds for £800,000. There was no rush; the order-book was full with work for TML, Nigerian Ports, Cimpor Portugal, Tyne & Wear Metro, British Coal and others. Railways had a future – the British Railways Board forward capital programme and the Channel Tunnel would ensure that, so it seemed – and Telfos had other acquisitions in mind while the Hunslet prices adjusted themselves upwards.

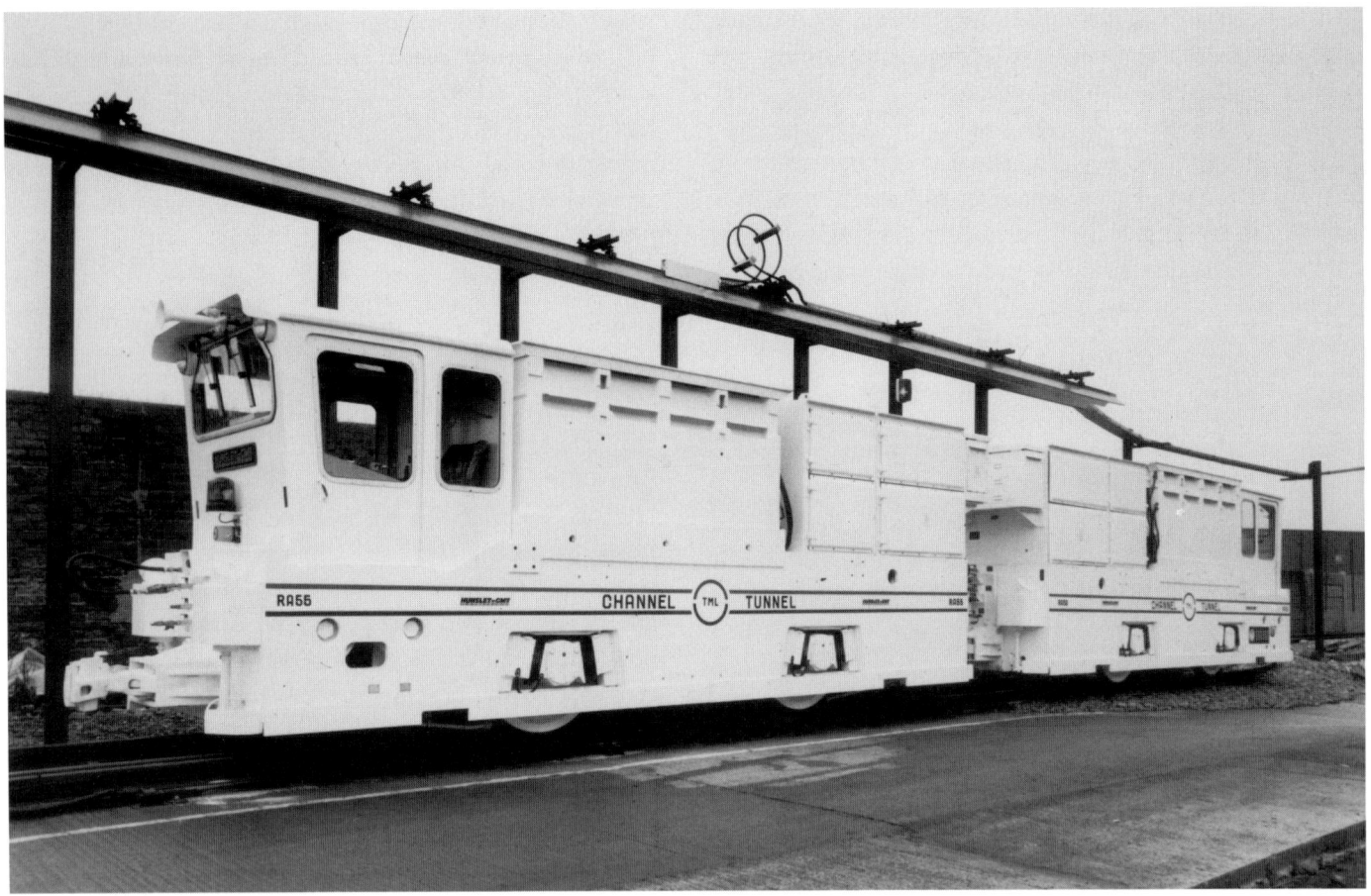

The later Channel Tunnel Mk II battery/trolley adhesion locomotives were uprated to provide tandem pairs weighing 50 tons and producing 400 h.p. RA55 and RA56 (HE 9446/7) stand on the test track just inside the old Railway Foundry boundary wall in November 1989 prior to despatch.

P. E. KEWNEY (HUNSLET BARCLAY LTD)

This was a *Catch-22* situation. Over the years Hunslet's prosperity had been firmly based on a wide-ranging design and manufacturing capability to produce virtually anything the customer wanted in small to medium quantities. This in turn meant higher overheads than in a production-line-based organization, but allowed operation in relatively bespoke niche markets (the rack-and-pinion locomotive was a good example of this) at the expense of high margins. To reduce the semi-bespoke nature of the business would reduce the overhead also, but would weaken the ability to adapt. Herein lies the polarization of attitudes that differentiates between a company viewed as an investment to outperform all other investments and a company viewed as a means of production to give a moderate return in exchange for continuation of the *status quo* and hence security.

Sixty years ago, Col. E. C. Kitson-Clarke, in his book *Kitsons of Leeds*, observed during a dissertation on loyalty between customer and client that 'there are many talented economists who, being outsiders, see most of the game so well that they are vocal beyond their knowledge,' and went on to doubt the wisdom of general standardization and, by inference, the growth of impersonal corporate empires. He feared the loss of markets consequent upon the inability to adapt in times of crisis.

Telfos had stated openly that their objective was to expand 'both organically and by acquisition'. At the end of May and the beginning of June 1988, Hunslet had displayed at a mining exhibition in Turkey and at the IVA 88 international transport exhibition in Hamburg. Even then, almost a year after the Telfos take-over, there was no outward evidence of change in the company approach.

But changes there were. Coincidental with the two exhibitions, Telfos had mounted a £29 million hostile bid for Walter Runciman. Runciman, led by the charismatic

Gary Runciman, a 'latter-day John Alcock', was well known in the City. As a shipowner Runciman took in both the Anchor Line and the Currie Line. It ran a Lloyds syndicate, and had diversified into security services through its subsidiary Tann International, making anti-thermic steel-plated doors and vaults.

The *Sunday Telegraph* of 12 June 1988 described the situation thus: 'this bid for Runciman (very ambitious, it will double its size) is all about assets.'

'Blatantly opportunistic', said Runciman. Absolutely.

In its Telfos *v.* Runciman big fight pot-boiler *The Sunday Telegraph* recapped on the take-over of Hunslet the previous year. In this, the paper said, Telfos had seen not profits but assets, when it paid its £7 million, explaining:

There was £2 million cash in the bank. There was also surplus property, including three small flats in London, now sold, which has raised well over £2 million. In addition there was stock written down in the books at virtually nothing but now reckoned to be worth £4 million. [On top of this] the business, boosted by orders for 70 locos to work in the Channel Tunnel, is set to make profits of £2 million.

Reverting to the Runciman bid, it went on:

Telfos has already snaffled a 24.6 per cent stake... has £10 million in the bank, is backed by deep pockets and has a link to Polly Peck's Asil Nadir.

Comforting words indeed for shareholders in mid-1988. The shareholders needed comfort, for up till then they had been blissfully unaware that any Runciman shares had been acquired by the Telfos board on their behalf, not having been afforded the usual courtesy of a notification document.

On 21 July 1988, Hunslet Holdings plc, acting as a subsidiary of Telfos, entered into an agreement with Matthew Hall plc to purchase the Barnsley-based

The power bogies for one of the prototype tube trains built by Metro-Cammell for London Underground's Central Line replacement evaluation programme were supplied by Hunslet during 1987/8. Metro-Cammell built two trains, BREL Ltd a third. All were four-car trains but with different electric traction packages. This one had Brown Boveri traction motors.

The last standard-gauge locomotives of any kind built in Leeds were three dual-voltage (1500 V d.c. overhead line and 750 V d.c. battery) works machines for the Tyne & Wear Metro, Hunslet nos. 9174–6. They were ordered before the Telfos take-over but not delivered until early in 1989. One of the trio is pictured shortly after its arrival at South Gosforth car sheds, Newcastle.   P. E. KEWNEY (HUNSLET BARCLAY LTD)

manufacturer of mining equipment Qualter Hall and Company for £3.15 million in cash. Qualter Hall had made a loss of £1.13 million on a turnover of £7.4 million in the year ending 31 December 1987, but had a subsisting design-and-build contract for the largest winding shaft in Europe at Maltby colliery, plus other contracts at Lea Hall and Maltby.

Despite an enhanced bid in August 1988 the Runciman deal did not go through, and the holding of 2,629,000 shares already acquired, were to be sold at a profit of £3.3 million some sixteen months later.

One deal that did go ahead at that time was the purchase of GMT for £827,000 partly in cash and partly in the issue of ordinary shares. The name GMT was to feature much in the months to come.

Originally based at Barnsley, Gyro Mining Transport Ltd (GMT) started building non-powered railway equipment in the form of rope-hauled systems utilizing its 'Railhugger' format, this equipment being used underground by the National Coal Board. In 1976 they developed a three-car articulated railcar powered by a linear motor. This never went into general service. After being displayed at various exhibitions, including the 1977 NEC show mentioned in chapter 13, it was sent to the Mining Research and Development Establishment (MRDE) at Swadlincote where it underwent a series of trials. It was decided not to continue with the linear motor concept, and in 1981 the railcar was fitted with a diesel engine and a hydraulic transmission. It remained at Swadlincote, where it was used for testing the new 'rubber' wheels that were being introduced on various underground locomotives.

GMT later moved to new larger premises at Hellaby, near Rotherham, and in 1980 was asked by the NCB to develop a bogie rack locomotive. The specification for two prototype locomotives was that they should be pure rack locomotives and should not be required to work on conventional railway track in the adhesion mode. They were built for the 2 ft. 6 in. gauge, and following completion in 1981 underwent trials on the surface test track at Ledston Luck Colliery.

GMT, unlike most other locomotive builders, did not have a works list for its locomotives; instead, most of the locomotives were named. The two type 85R locomotives were named *Kirstin* and *Stephanie*. After the various trials had been completed, it was decided not to proceed any further with this type and they were returned to Hellaby. *Kirstin* was externally restored and was lent to the National Railway Museum at York for public display.

Seven 160 h.p. rack-and-adhesion locomotives were supplied from 1983 to Gascoigne Wood Mine.

Apart from locomotives, GMT built two railcars for underground use. The first of them was produced in 1980 and was a four-car, battery-powered man-riding set fitted for rack operation. It operated at Kellingley colliery for three years, and was then returned to the works at Hellaby, where it was put into storage.

The second railcar was a five-car unit, an additional intermediate passenger car being added, and was built to operate in the adhesion mode only. After trials had been completed, the railcar was put into service during 1983 at Pleasley Colliery (North Derbyshire Area).

For the International Mining Exhibition held at the National Exhibition Centre, Birmingham, from 10 June to 14 June 1985, GMT introduced a brand-new design of battery-electric locomotive, the type LMC (Locomotive Man Car). After the exhibition had closed, the unit was returned to Hellaby where further trials were carried out. During 1987 British Coal placed an order for this prototype locomotive, for service at a mine in the Selby Coalfield.

The last locomotive to be produced by GMT was a new design of battery locomotive for use on high-speed man-riding trains, and in fact the locomotives built to this design were the largest underground battery locomotives then in use in Europe. Two of them were completed early in 1987 for use at Markham Colliery (Central Area). They were very large locomotives, having a maximum length of 9250 mm, a height of 1830 mm and a width of 1370 mm. These 150 h.p. locomotives were capable of speeds of up to 40 km/h and they weighed 28 tonnes in full working order.

During August 1984, the firm of Underground Mining Machinery Ltd (UMM), which was based at Newton Aycliffe in County Durham, merged with GMT, both firms having similar interests. UMM had been marketing a 'trapped rail' system for use in underground roadways, under the name Mineranger. For this system a number of 40 h.p. locomotives had been built in the late 1960s, most of them using Hunslet flameproof diesel power packs. During 1986, GMT constructed a brand-new locomotive to the original UMM design. This was delivered in March 1986 for underground use at NCB's Royston Colliery (North Yorkshire Area).

The total motive power production of GMT comprised three railcars and thirteen locomotives, all embodying technology of an advanced form that was new to the mining industry.

But let us return to Jack Lane. David Gawthorpe had resigned immediately the Telfos take-over was agreed by the Hunslet Board, leaving Peter Alcock and his younger brother Keith as executive directors for the time being. The Telfos finance director came in as acting Chief Executive.

Early in 1988 Peter Alcock relinquished his position as executive chairman but stayed on in a non-executive part-time capacity for the remainder of the year. Sadly, Keith died in harness in September at the early age of forty-seven, having coped magnificently for two years with considerable distress and painful treatment after collapsing during a flight from Portugal. For some time prior to the Telfos bid, Keith and the author had been involved in a design and marketing investigation into a lightweight shunter/trip locomotive for which BR headquarters at Derby saw a role in the expansion of Speedlink freight services. Another use was seen to be in having one of these locomotives at each end of the weed-killing (or vegetation control) trains. The vegetation control train contracts were due to be farmed out to chemical companies, over a set period of years. In the event, the Speedlink expansion came to naught and the 'shuttle' locomotive concept was stillborn. The vegetation control contracts, after some delay, were let to the Chipman and Schering chemical companies, each to provide one train and between them cover the whole network. In October 1988 both companies placed contracts with Hunslet to provide the power for five years, using surplus BR class 20 diesel-electric locomotives.

The family firm was family no more. The decision was taken to install a new Chief Executive from October 1988 for the Hunslet Holdings component of Telfos, to rationalize the activities of the Leeds and Kilmarnock works, ostensibly to avoid a conflict of interest. The

Hunslet Engine Works would henceforth, it was said, concentrate on underground (mining and tunnelling) locomotives, and would be known as Hunslet-GMT Ltd, while the Andrew Barclay, Sons & Co. works in Kilmarnock would take on all surface shunters and other railway work and would become Hunslet-Barclay Limited. The insignificant contribution brought by the acquisition of GMT and the comparable size of the Hunslet and Barclay works did not provide logic for this decision; but perhaps expansion at Kilmarnock was cheaper than expansion at Leeds. The Channel Tunnel was a one-off flash in the pan, the National Coal Board was winding down, and it was difficult to see how the mining and tunnelling locomotives alone could sustain Jack Lane.

On 6 December 1988, 400 h.p. diesel-hydraulic locomotive HE9092/88 left Jack Lane by road, the last of a long line of standard-gauge shunters directly descended from John Alcock's pioneer work in the early thirties. Its destination was Caledonian Paper plc's new production plant at Irvine in Ayrshire – five miles from Kilmarnock! The class 20 diesel-electric locomotives purchased as surplus to British Rail's requirements for use on the vegetation control trains never came to Leeds but went straight to Kilmarnock.

CHAPTER 15

Down like the stick

The date set for the restructuring of the group and the official launch of the 'Hunslet GMT' and 'Hunslet Barclay' brand names was 9 January 1989.

In the event, Kilmarnock, to its credit and no doubt with considerable feelings of relief, was able to keep its management and work-force intact. In addition to the Hunslet surface locomotive drawings, Barclays inherited the Chipman and Schering five-year Vegetation Control locomotive hiring contracts for British Rail and orders for new locomotives for the Snowdon Mountain Railway, Cimpor Portugal and Cimbrian Unigrain (Uganda), all of which were executed with competence.

The Hunslet works gained nothing in return, although the factory was already busy with the majority of the Channel Tunnel locomotives described in the previous chapter, and the Tyne & Wear Metro works locomotives were on test. Brush Traction Ltd placed subcontracts early in the new year covering mechanical parts for nineteen 1100 h.p. diesel-electric shunters destined for Moroccan Railways, and these were completed by the end of 1991.

A second Brush subcontract, covering body shells for Eurotunnel's Channel Tunnel Shuttle locomotives, originally placed with Hunslet for fabrication at Jack Lane, was transferred to Qualter Hall's Barnsley works.

London Underground Ltd's tunnel cleaning train came in for overhaul later in the year and was subsequently returned to Ruislip by August 1990. The tunnel cleaning train, in effect a large vacuum cleaner, was a five-car unit that had been built at LUL's Acton works between 1972 and 1977. The driving motor cars, one at each end, had been converted from two withdrawn 1938 tube stock driving motor cars, but the three intermediate vehicles

A major contract for Hunslet GMT in 1990/1 was the construction of the mechanical components, including finish-painting, of nineteen Brush-designed 1100 h.p. diesel-electric shunters for Moroccan State Railways.

BRUSH TRACTION LTD

were substantially new. To a casual observer the unit had the general appearance of a tube train, certainly when viewed from the front; and some confusion about London Underground's refurbishment programme arose as a result.

Coincidental with the restructuring came the decision by Telfos to acquire the respected but impoverished Hungarian rolling stock manufacturer Ganz in one of the first West–East commercial link-ups, a move that was loudly praised at the time and which pre-dated the general collapse of communism in eastern Europe by some months. Thus was Ganz-Hunslet formed, an event with no immediate impact on the Hunslet Engine Works in so far as its day-to-day operations were concerned but yet another piece in the complex jigsaw the assembly of which was to prove financially unsustainable.

Hard on the heels of the Ganz acquisition, and reminiscent of Pirandello's *Six Characters in Search of an Author*, came a knock on Hunslet's door from a ready-made rolling stock project team from Metro-Cammell who were fearful of their future in the face of the owning Laird Group's intention to off-load train building from its repertoire. (This was before the purchase of Metro-Cammell by GEC.)

The team was a good one, having handled the successful British Rail class 156 diesel multiple unit programme, and they found enthusiastic foster-parents in Telfos-Hunslet, where they founded Hunslet Transportation Projects Ltd (HTPL) in May 1989, based in the former Charles Clifford offices at Dogpool Mills, Stirchley, Birmingham. Early successes for HTPL were a small number of centre cars for the Glasgow Subway and some of the class 155 to class 153 conversion work for British Rail. Both these projects owed more to the influence of Kilmarnock than to either Leeds or Birmingham, and again had little impact on Jack Lane, although the Glasgow body shells were fabricated there.

These were heady days for Telfos. To the casual observer there was no firm evidence at the time that, financially, the good ship *Telfos* was already fatally holed below the water-line.

On the shop floor at Jack Lane there was little evidence of change as the Channel Tunnel construction locomotives, both rack and adhesion, continued to emerge until the last one was completed on 27 June 1990 – seventy-eight locomotives in just three years. Ten of the earlier locomotives returned to the works to be rebuilt as diesel units for the shorter, landward tunnel at Folkestone between August and November 1990.

In April 1990 was issued the first (and indeed the last) Hunslet Holdings quarterly newsletter. This identified Hunslet (Holdings) plc as a wholly-owned subsidiary of Telfos Holdings plc and listed the members of the Hunslet Group as:

Hunslet-GMT, Leeds:
mining, tunnelling and railway equipment.

Hunslet-Barclay, Kilmarnock:
railway and press equipment.

Qualter Hall, Barnsley:
mine shaft and haulage products.

Ganz-Hunslet, Budapest:
new-build railway engineering products.

Hunslet Taylor Consolidated, Johannesburg:
mining and railway products.

Excil Electronics, Wakefield:
electronic design and manufacture.

Hunslet Transportation Projects Ltd, Birmingham, was shown separately as handling railway project management as a direct subsidiary of Telfos.

The front page of the newsletter was devoted to expansion plans at both Kilmarnock – quite understandable in view of the January 1989 division of responsibility – and Leeds – less understandable in view of the declining mining and tunnelling markets.

Alongside an artist's impression of the proposed new Kilmarnock workshops was the announcement that, as part of the overall expansion, the British Rail sidings to the north of Kilmarnock station and adjacent to the existing Railbus shop had been purchased late in 1989. The new 92,000 sq. ft. factory being constructed would, it said, enable Hunslet-Barclay to tackle any of the large new orders being pursued with Hunslet-TPL or the refurbishment contracts coming out of British Rail over the next few years.

The artist's impression of Jack Lane showed the original grade II listed 1880 office block flanked by a new frontage to the erecting-cum-boiler shop extending over a new 55,000 sq. ft. assembly shop, running parallel to the old one on land previously occupied by Cancel Street and the company car park. The former rail connections into both the original yards were to be reinstated, and a third track would enter the new shop just to the right of the former boundary wall. Effectively the former erecting shop area would be doubled. The newsletter captioned the

The northern boundary of the Hunslet works site on 24 September 1990. Beyond the derelict GMT man-rider is the Channel Tunnel rack locomotive test track incline (at left) with a ruined wall and the main gateway of the original Shepherd & Todd Railway Foundry.   PAUL COTTERELL

The remains of Shepherd & Todds' 1839 enterprise with the rather incongruous TML rack track gradient against the Pearson Street wall, now very much derelict.   AUTHOR

illustration with the statement that 'Hunslet-GMT [sic] is expanding in Leeds for the first time since the 1920s', which was a sanitized, misleading statement – and that the extension was 'in preparation for pending contracts with London Underground and British Rail'.

The remaining three pages of the newsletter were equally upbeat. Ganz-Hunslet's prospects were reviewed, with news of a forthcoming visit by the Prince and Princess of Wales to the Budapest works, where the Prince was to open the newly-refurbished office entrance in May 1990. The Deputy Prime Minister, Sir Geoffrey Howe, visited Jack Lane, where on 8 September 1989 he too unveiled a plaque, commemorating the opening of the refurbished offices there. Qualter Hall of Barnsley were reported as having purchased four companies in six months: mining equipment manufacturers M. B. Wild, Robey Winders and J. & B. Crushers were absorbed into the Barnsley work site, while Excil Electronics, mentioned earlier, remained in their small new factory at Wakefield.

Finally Hunslet Transportation Projects were said to be hard at work after receiving their first major order from the Strathclyde Passenger Transport Executive for vehicles for the Glasgow Subway. 'In addition,' the newsletter continued, 'work is continuing following their quotation for class 323 Electric Multiple Units for BRB's Birmingham Cross City Line.' The report concluded: 'If the quotation [for the class 323] is successful, HTPL will use Hunslet-Barclay's new factory for the assembly of these new rail vehicles.' This was hardly surprising in view of BRB's having issued the invitation to tender to Hunslet-Barclay in the first place.

The public relations machine was rolling, the city analysts were loving every minute of it, and the press was giving more coverage in a week than 'old Hunslet' would have expected in any previous ten-year period. As early as April 1989, only half-way through the Channel Tunnel contract, the newspapers were showing Telfos profits up by 200 per cent to £5.2 million and a share price increase from 170p at the time of the Hunslet acquisition to 219p, adding the rider that the price/earnings ratio of under 10 took little account of the company's very large growth potential. Against a photograph of a Channel Tunnel construction locomotive *The Sunday Times* headlined 'TRAIN BUILDERS ON A £1BN RIDE' and prophesied 'a constant stream of vehicles needed until the turn of the century'.

But those danged chickens wouldn't stay still for the count. On 13 March 1990 the *Yorkshire Post* announced that Telfos, the parent company of Hunslet-GMT, had bought Excil Electronics for £550,000, of which £160,000 was cash and the balance new Telfos shares. Nothing wrong with this *per se*, but the same report went on to say that

the acquisition closely follows a £70m. contract for London Underground won by Hunslet. The contract, which is a joint operation between Hunslet-GMT, in Leeds, and HTPL, in Birmingham, has resulted already in a £2.9m. extension being planned for the Leeds business. Another consequence of the contract is that Hunslet has had to reconstruct its rail link to the British Rail System.

The report of the London Underground contract success was premature and probably the result of a combination of loose PR and misunderstanding, since the tunnel cleaning train was in the Jack Lane works at that time. London Underground was not impressed and apologies were called for. The refurbishment contract went to competitors and the completion of the Channel Tunnel contracts was to result in a severe shortage of work in Leeds. Kilmarnock however was doing well, fitting new transmissions to the BR railbus fleet and converting class 155 Sprinters into single-car class 153 units. However, orders or no orders, the planned extensions at both the main works, Kilmarnock *and* Leeds, were to go ahead as described in the newsletter. On 6 March 1990, with financial grant aid from the Leeds City Council, a contract was let for the proposed factory extension and offices at Jack Lane, at a cost of approximately £1.7 million. The contractor went into liquidation before the works were completed, and on 5 June 1990 a second contract worth £1.6 million was let to another builder who completed the job. Construction of the new factory at Kilmarnock was accomplished without incident at a cost of just over £3 million, but in retrospect the haemorrhage had begun.

Press intimation that something was wrong came for the first time on 6 August 1990, when it was announced that the investment group John Govett was to reduce its 10 per cent stake in Telfos in the aftermath of the latter's £13.85 million rights issue and a proposed £4.5 million flotation on the Austrian stock exchange. The Telfos share price dropped to 153p as a result.

Worse was to come. The Telfos chairman's half-year statement of 2 October 1990 declared a normal interim dividend, but emphasized that the company was 'concentrating its activities on the core engineering businesses and endeavouring to disengage itself from involvement in investment and property' and hinted darkly that 'larger provisions may need to be made at the year end than seemed probable two months ago'.

By the end of October 1990 Telfos shares had fallen to 73p, a 60 per cent tumble in little over a year, and both the chief executive and the finance director had left the company. Much press coverage was given to Telfos's past connection with a Bermuda-based company in which Telfos had a 20 per cent interest against a 60 per cent Polly Peck holding – a deal that had cost Telfos over £5 million. The press also pointed out that much of the £13.8 million rights money raised had been committed to 'a Northern property venture'. This was the derelict 204-acre former Butlins holiday camp site at Filey, which had already been the cause of one property development bankruptcy before Telfos bought it in 1988 against a guarantee from a Swiss bank.

During November the London office of Telfos was closed, the group head office by now being at Jack Lane. By early December the chief executive's seat was again empty. To fill the premier executive positions so dramatically and rapidly made vacant, the managing directors of Hunslet Barclay and Qualter Hall were brought on to the Telfos main board on 10 December and a third, non-executive, director was appointed from outside.

To round off 1990 as Jack Lane's *annus horribilis* it was announced on New Year's Eve that Telfos could not pay a dividend and was, moreover, extending its financial 'year' to fifteen months ending on 31 March 1991. The delay was 'intended to enable the proceeds of planned disposals to strengthen the published balance sheet'.

Some might have argued that 1990 had not been entirely devoid of good news, for, in June, British Rail had awarded the fledgling Birmingham-based Hunslet Transportation Projects Ltd the contract for thirty-seven 'state of the art' class 323 electric multiple unit trains. These were intended to be built at the new Kilmarnock works; they do not at this stage appear to have an impact on the Jack Lane tale, but more of this later.

On 8 January 1991 the Sheffield-based steel castings group William Cook launched a hostile £40 million bid for Telfos Holdings which valued each Telfos share at slightly over 123p. The Telfos board told shareholders that its bankers, Baring Bros., were to review the group's financial position. The review, it said, would 'help focus shareholders' attention on the excellent outlook for those core railway and engineering activities', and it advised the shareholders to ignore the William Cook offer.

On 10 January it became known that Jenbacher Werke, a well-respected Austrian rolling stock manufacturer, was also interested in Telfos's core engineering business and had already acquired a 1.6 per cent stake. They felt that combining Telfos's British and Hungarian activities with their own facilities would create a 'major new force' in Europe's railway engineering industry.

Towards the end of the month Telfos announced that a property sale had fallen through, requiring further provisions in the balance sheet, and that there was the prospect of litigation concerning the group's interest in the Filey site. The chairman of William Cook confirmed that their bid stood, but added: 'A pattern of bad news is emerging which must be causing Telfos shareholders considerable alarm.' On 24 January it was reported that a promised £12–15 million order from Hungarian State Railways (MÁV), crucial to Ganz-Hunslet's future, was likely to be delayed for more than six months. The month ended with the withdrawal from service of the entire fleet of 20 multiple unit trains already supplied by Ganz-Hunslet to MÁV following drive-shaft failures, and with the customer demanding compensation.

Trading in Telfos shares was suspended at 98p on 4 February, Telfos having failed to produce the necessary defence document against the hostile Cook bid. The auditors announced on 6 February that the promised review could not yet be completed, owing to lack of information. A draft was produced, however, which prompted Telfos to make additional provisions of £13.3 million on top of the £8.2 million provided for since November. Shares dropped to 57p, valuing the company at around £18 million. The *Financial Times* relaxed its usually staid prose to comment that the serious money in Europe's railway industry, the likes of GEC-Alsthom, ABB and Siemens, had not needed to wait for the review in order to identify Telfos as 'a pint-size can of worms' with a 'rag-bag of investments', and made the reasonable assumption, later borne out, that William Cook would drop its £40 million bid.

But there were still optimists. Foreign manufacturers wanting a foothold in Britain might find the ready-made business useful once it had divested itself of its liabilities. The class 323 units, worth £81 million, were still to build, and Hunslet had the order from Brush Traction for Channel Tunnel shuttle locomotive body-shells, making £100 million of firm contracts. Barring accidents, there were options on further class 323s. There was a further promise in the offing for electric multiple units for Malaysia worth £65 million.

AEG Westinghouse showed interest in late February, and Jenbacher announced that, while it did not wish to make a full bid for the Hunslet-based group, it saw a way forward for co-operation with the maximum allowable 29.9 per cent stake. But these developments would have to wait until the publication of the 31 March year-end figures. In the meantime Kilmarnock put down a marker with a half-page news item in the *Daily Telegraph* on 28 February announcing, *inter alia*, that 'this 151 year old company [Andrew Barclay, Sons & Company as was] will buck a dismal trend with the commissioning of a £3m. loco [*sic*] plant built in the shadow of its original Victorian works. It will make commuter trains for British Rail use in the Midlands and create 120 new jobs,' these latter being the class 323.

On 29 April the *Yorkshire Post* announced the injection of £14 million into the Telfos coffers by Jenbacher in exchange for the 29.9 per cent shareholding. The last of the remaining original Telfos directors had resigned, the Austrians had representation on the board, and prospects looked much brighter. The only cloud was that the Swiss bank had stopped the clock on the Filey deal, and this soaked up £7.5 million of the Jenbacher funding. The newspaper suggested that a good way of signifying that bad news was a thing of the past would be to change the name from Telfos back to Hunslet Holdings, adding: 'Perhaps a clue to how the board is thinking may be gleaned from the fact that anyone phoning the engine works will be greeted with 'Hunslet Engine Company'. And that is a name from the past.'

But the giant roller-coaster that was the Telfos share price still maintained its white-knuckle potential. The financial report published on 1 August showed a pre-tax loss of £5.8 million, against the £1.0 million that had been estimated in March when Jenbacher had put its £14 million on the line. The loss stemmed from poor operating results at Hunslet-GMT and Qualter Hall, and there was an exceptional charge of £2.06 million for the reorganization of the Leeds factory and the closure of the London office. Also taken above the line was a £3.3 million loss on the sale of an investment. There was no final dividend.

The new chairman, appointed in June 1991, was frank about the situation but felt that the 'horrors' were now in the past. The group was reorganizing the Hunslet Engine Works site and cutting the work-force there. It would now use the spare capacity at Leeds, rather than Kilmarnock, to build the BR class 323 units.

This decision effectively required the Jack Lane site to be split into two distinctly separate operating units with different aims and destinies, connected in theory only in so far as the performance of either might affect the corporate purse.

The corporate purse was further depleted in August when Scarborough Borough Council refused planning permission for a £30 million holiday development on the seemingly-bewitched Filey site, insisting instead that the site be cleared and grassed. On 13 September the Telfos board proposed a reduction of £19.5 million in the share premium account, and an extraordinary general meeting was held to ratify the proposal. The meeting was very sparsely-attended, the mood was sombre, and no attempt was made to hide the urgent need for a 'white knight' rescuer.

Rescue came on 5 November 1991 when the boards of Jenbacher and Telfos announced the terms of a recommended cash offer by Jenbacher for the whole of the share capital, issued and to be issued, that Jenbacher did not already own. The offer of 115p cash per ordinary share valued the issued ordinary share capital of Telfos at approximately £51 million. By 20 November Jenbacher had 57 per cent of the holding and the rest was a mere formality. The gamble was on the performance of the class 323 contract and future orders from British Rail. The Telfos chairman praised the deal, saying that 'it would have been unrealistic for the group to attempt a rights issue after the proceeds of last year's £13.9 m. cash call disappeared in a cloud of smoke.'

To accommodate the class 323 build, the remaining Hunslet-GMT underground locomotive business (renamed Hunslet Engine Company) moved into the former repair shop, taking with it a staff of approximately fifty and a suitable proportion of the machine tools. The balance of the machine tools were either sold or were transferred to Kilmarnock. The machine shop and the Gun Shop were emptied; the former received a new concrete floor and became a large parts store, and the latter was fitted out as a final test bay with a 25 kV shore supply. In December 1991 a new spray booth with two lines of rails was built into and extended beyond the old wheel shop.

Output from the 'reborn' Hunslet Engine Company had been commendable, when one makes allowance for the fact that the work-force was only six per cent of that

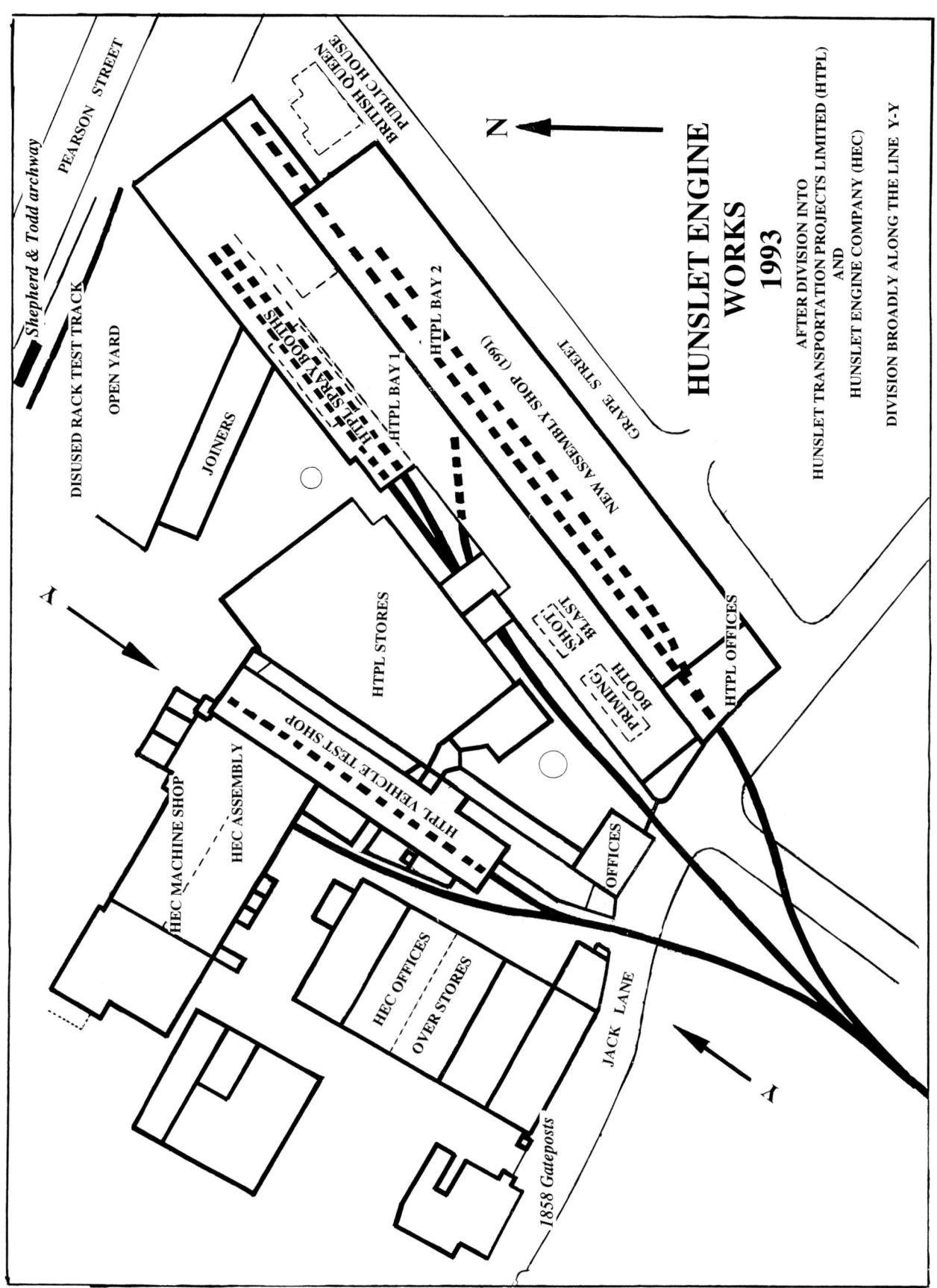

Looking north across Jack Lane during conversion of the works for class 323 production. Realigned original tracks from the main line at left, with a third track to the new shop cutting across the right foreground. The 1882 office block is in the centre.

Re-laying the track into the former Manning Wardle erecting shop in readiness for its conversion into the class 323 test house. The building in the right foreground started life as Manning Wardles' paint shop, then successively Hunslets' accounts and sales office with board room, and finally and fleetingly the Hunslet GMT display room.

The 1882 office block (carrying the date 1864) is flanked by the new curtain wall embracing the original boiler shop and the 1990 assembly shop. The latter commences at the bend in the curtain wall. The date is 29 October 1992.

RABBI WALTER ROTHSCHILD

The view from Jack Lane on 29 October 1992.

RABBI WALTER ROTHSCHILD

employed during the peak of the 1970s. All the Channel Tunnel work had been completed by November 1990 before the workshop reorganization. Four 150 h.p. diesel rack locomotives, nos. 9291–4, a batch that had been put in hand long before the Telfos days, had also gone to the Selby coalfield between April 1990 and 17 January 1991, the last of them being the final locomotive out of the old erecting shop before it was converted for the class 323 electric multiple unit trains.

After the delivery of no. 9294 there was a break of eighteen months before the next locomotive emerged from Leeds. This was inevitable in view of the upheaval, but in the mean time the competitors that remained took full advantage. Clayton Equipment, for example, virtually

cleaned up on British Coal business. The Snowdon Mountain Railway ordered a third 320 h.p. rack locomotive; this was built at Kilmarnock but given the Hunslet number 9305 and despatched on 9 April 1991. A fourth Snowdon locomotive, no. 9312, was later to be one of the first deliveries out of the converted repair shop in July 1992, together with nos. 9310/11, two 150 h.p. battery-electric flameproof bogie units of GMT design for British Coal at Daw Mill and Markham collieries. (The term 'flameproof' was not confined to internal-combustion-engined locomotives.) Nos. 9310 and 9311 were named *Linzi* and *Laura*, thereby continuing the GMT naming practice. The problems of 1990 had severely affected customer confidence at a time when the weak market

Sandra, HE 9320/93, ostensibly brand-new but undoubtedly containing many Channel Tunnel locomotive components, was one of two 900 mm gauge 30 ton diesel-hydraulics (HE 9339/94 was the other) used by the contractor Gama Guris on the construction of the Ankara Metro. Both were out of use by 1997 on completion of the railway. P. E. KEWNEY (HUNSLET BARCLAY LTD)

could least tolerate it, but the three locomotives of 1992 were followed by seven in 1993. These included two 64 h.p. flameproof rack locomotives (nos. 9308/9) for British Coal at Stillingfleet; two 64 h.p. bogie battery-electric flameproof units (nos. 9313/4), also for British Coal at Riccall, both pits being in the Selby coalfield; and a pair of 3½ ton 43 h.p. diesel-mechanicals, nos. 9315/6, for the Uranium Corporation of India Ltd, a steady Hunslet customer in days past. The seventh locomotive was no. 9320, a 900 mm gauge Channel Tunnel look-alike for the Turkish contractor Gama-Guris Construction engaged in tunnelling work for the Ankara Metro. Nos. 9317–9 were pencilled in for a customer in Korea, but the order was not placed.

The order book for 1994 was boosted early in the year by eighteen tiny 2.7 tonne 25 h.p. 2 ft. 0 in. gauge diesel-hydraulics of Jenbacher design with Kubota engines, ordered by the Jan-Pan company in Singapore for plantation use in Borneo. These carried makers' numbers 9321–38. Also in 1994 came no. 9339, another locomotive for Gama-Guris, rebuilt from a surplus Channel Tunnel unit and identical to no. 9320. Also thus rebuilt, but this time to 2 ft. 0 in. gauge, were nos. 9342–5, bought by Sir Robert McAlpine and Sons for use on tunnelling work for the Jubilee Line Extension of the London Underground. Nos. 9340/1 were two further 43 h.p. locomotives for the Uranium Corporation of India.

Six 11 ton 2 ft. 0 in. gauge tunnelling locomotives, nos. 9346–51, were supplied to Balfour Beatty Ltd, also for work on the Jubilee Line Extension, early in 1995, and these were the last locomotives of any kind to be built in Leeds. The highest numbers in the diesel locomotive series, however, were five more 25 h.p. machines of the Jan-Pan type, but increased in weight to 3½ tons (nos. 9359–61) for Balfour Beatty's Jubilee Line contract, and to 4½ tons (nos. 9362/3) for export. These were despatched late in 1994.

Under the headline 'HISTORIC FIRM IS SOLD OFF' the *Yorkshire Evening Post* reported on 9 December 1994 that the Hunslet Engine Company had been sold 'for an undisclosed sum' to the Barnsley-based 'mining and loco manufacturers Qualter Hall'. By this time Hunslet, Qualter Hall, Hunslet-Barclay and Excil Electronics were all directly owned by Auricon Beteilgungs AG, who in turn owned Jenbacher Transport Système.

The staff stayed for a while at Jack Lane, certainly until the Balfour Beatty locomotives were complete; but by June

1995 all had gone to Barnsley, where seven 15 ton locomotives, nos. 9352–8, were built to the order of a Korean company, Ginro, for use in Indonesia.

Also dealt with at Barnsley was the rebuilding of one of the two original GMT-built 150 h.p. battery locomotives; the other had already been rebuilt at Leeds after the closure of Markham colliery, when it was decided to transfer the locomotives to Rossington. The frames for the Indonesian locomotives were machined in the deserted shop at Jack Lane by a Barnsley-based machine operator who travelled to Leeds daily to use the Wadkin plano-milling machine for this sole purpose.

In 1990 Hunslet-GMT had repurchased, at a ridiculously high price, two Hunslet 388 h.p. shunters that had originally been supplied in 1979/80 to the CEGB, and these, nos. 8976/7, had stood around for quite a while before one of them became yard shunter for HTPL handling class 323 units and the other went out on hire. These were the only standard-gauge locomotives to be seen at Jack Lane after the departure of the Tyne and Wear units early in 1989, apart from a similar unit from British Steel, Workington, which arrived for re-engining in early 1995. This was still on site when the other work was transferred to Barnsley, and was finished off in Leeds at the end of June by painters who again travelled up from Barnsley. It was the last locomotive worked on in Leeds.

With the reorganization of 1991 the story of the Hunslet Engine Works went two ways. The first we have described above. The second was the short excursion into passenger train production with the class 323. Contrary to present received wisdom the 323s were not the cause of Hunslet's demise: they were victims of it and of other exceptional circumstances surrounding the railway industry as a whole. Perhaps it was inevitable that the class 323 contract should go to the newly-formed Hunslet Transportation Projects Ltd. In 1989 it was still widely thought that the British Rail investment boom, which had started with the Sprinter diesel multiple units, class 319 and 321 electrics and others, and was continuing with large quantities of the aluminum-bodied class 158, 465 and 466 trains of advanced design, would go on unabated to the end of the century.

Echoing the National Coal Board's concern over monopoly suppliers and its not-too-successful weaning of GMT in 1977, the British Railways Board was also anxious about having too many eggs in too few baskets. There was a desire within BRB to create more competition in the

Full circle. The first Hunslet Engine Company locomotive was supplied for railway construction. So was the last to be built in Leeds: Emma, the first of six 11 ton 2 ft. 6 in. gauge diesel-hydraulics (HE 9346–51) with Deutz air-cooled engines built in 1995, threads her way through subterranean London on Balfour Beatty's Jubilee Line Extension contract.

P. E. KEWNEY (HUNSLET BARCLAY LTD)

Pseudo-Hunslets: the seven 15 ton diesel-hydraulics ordered by the Korean firm of Ginro (HE 9352–8) lined up in the Barnsley works of Qualter Hall prior to shipment to Indonesia.

P. E. KEWNEY (HUNSLET BARCLAY LTD)

388 h.p. diesel locomotive Hunslet no. 8976/80, previously with CEGB at Carrington power station, hauls a class 323 driving trailer vehicle mounted on accommodation bogies across Jack Lane on 29 October 1992. A major disadvantage of the revamped works was the necessity to shunt half-finished vehicles from shop to shop across this busy main road.

RABBI WALTER ROTHSCHILD

388 h.p. diesel locomotive no. 8977/80, the other unit purchased back from CEGB Carrington during 1991, stands on the site of the old weighbridge, while a class 323 peeps out of the test house. 29 October 1992.

RABBI WALTER ROTHSCHILD

The class 323 construction programme was in full swing in the 1990 final assembly shop on 23 September 1993 when units nos. 323217/8 were photographed almost complete. Less than two years later the shop would be empty again.
BRIAN MORRISON

market, and a newcomer with Sprinter management experience was attractive.

The British Railways Board's Regional Railways specification for the class 323 was onerous and the quantity ordered small by previous standards. The initial requirement was for thirty-seven three-car trains for the Birmingham Cross City electrification and the North-West, plus options for two trains for Manchester Airport and four further Birmingham units, totalling forty-three. There was a rather nebulous further option for fourteen trains for West Yorkshire.

There was at that time, or so it would seem, room for a 'niche supplier' for small orders, and HTPL fought hard and won. In Regional Railways they had a demanding customer that was in turn beholden to two equally hard taskmasters, the Passenger Transport Executives of the West Midlands (Centro) and Greater Manchester (GMPTE). Everything had to go right first time.

At the time of tendering there were two perceived major cornerstones in the HTPL bid. One was the track record of its project team and the other was the 'financial stability' of its parent company Telfos. The workers at Kilmarnock, who it was thought would be assembling the trains, had put together the class 143 railbuses, retransmissioned most of the remaining railbuses and successfully rebuilt the single-car class 153 railcars from the earlier class 155 twin units. The performance on these contracts established a *rapport* between Kilmarnock and the Regional Railways engineers in Derby.

The trains for Centro and GMPTE were state-of-the-art three-car 25 kV overhead line suburban units with a top speed of 90 m.p.h. The advanced-technology aluminium bodies were to be made in France under subcontract from Alusuisse of Zürich, and the Dutch firm Holec was main subcontractor for the electrical equipment, including the three-phase drive a.c. traction motors. Holec in turn

A Hunslet 'Lizard' four-way travel reach truck stands between two vehicles of electric multiple unit train no. 323220 at the bottom end of the new shop on 23 September 1993. The fork-lift truck was already five times as old as the workshop itself, yet it would outlive the Leeds rolling stock construction industry.

BRIAN MORRISON

specified a British gearbox, manufactured by David Brown and mounted on bogies from another newcomer, RFS Industries at Doncaster through their subsidiary Specialist Rail Products Ltd. The first electrically-operated power doors on British Rail, and a Dutch-style destination indicator adjacent to the doors, added the finishing touches.

The main build contract for thirty-seven units was signed in June 1990, with the Manchester Airport option (two units) following in May 1991. The four additional units for Centro were ordered in May 1992, by which time production was in full swing at Leeds.

When the initial contract was announced, the delivery date for the first unit was set at April 1992, with full introduction of the Birmingham Cross City service in October 1992. How much delay was caused by transferring the build from Kilmarnock to Leeds is not clear, but when Jenbacher obtained its 100 per cent holding in Telfos during August 1992 a Concession Agreement between Jenbacher and Regional Railways was signed, incorporating a revised delivery schedule, and it was not until September 1992 that type-testing of the first production unit began.

The first impression of the class 323 was, and still is, that of an attractive, stylish, well-engineered train. By November 1992, however, gearbox problems on test resulted in units being sent direct to the Ministry of Defence depot at Kineton for storage. The press, which had been quiet for some time, re-emerged with headlines such as '£90M TRAINS MOTHBALLED' during February 1993.

By June the gearbox problems had been resolved, commissioning had restarted and the British Railways Board submitted a claim for liquidated damages for late delivery. The claim was subsequently reduced as an inducement to meet a revised delivery schedule.

Things went quiet again until just before Christmas 1993, when problems with the line voltage transformer were encountered, resulting in a further modification before testing could begin again. At the same time it was announced that administrative receivers had been called in at RFS Industries Ltd. The problems of divided responsibility, claim and counter-claim were beginning to manifest themselves in profusion. Jenbacher's dream of immediate trouble-free class 323 fleet running, further work from the West Yorkshire PTE in the shape of fourteen more class 323 units, and a batch of class 157 two-car DMUs for Strathclyde, for which a letter of intent

had been signed in May 1993, was shattered. There were to be no more new train orders in the immediate future.

The 1990s phenomenon of political posturing, which was particularly virulent in the van of railway privatization, gathered pace, with party point-scoring and lurid newspaper headlines completely obscuring technical and commercial facts. A headline in the *Yorkshire Evening Post* for 10 February 1994 proclaimed a 'LIFELINE FOR 280 RAIL JOBS' in the wake of Jenbacher's disposing of the design rights for the class 323 (and embryonic class 157), plus the class 155 intellectual property rights which Hunslet-Barclay had obtained from Leyland/Volvo, to Holec Machines en Apparaten BV. Two months later the paper reported on alleged 'dirty dealing' by a German competitor against the Hunslet Engine Company in its bid for tunnelling locomotives for the Jubilee Line Extension of the London Underground. The Opposition spokesman for transport publicly proposed, in the local newspapers, a totally impractical three-point rescue plan for Hunslet which had a hollow ring to it when set against the much more significant problems that were being faced elsewhere in the railway industry – at Doncaster and York to name but two other locations.

It is only fair to say that such squalid involvement of politicians in quoting job losses and alleging foul play, which was not confined purely to the plight of Hunslet, would have been unheard-of only ten years previously. Such was the undignified state to which a major British industry had been reduced.

Further misfortune befell the class 323. In December the fleet was taken out of service again as a result of the sophisticated on-board monitoring unit shutting down when ice was experienced on the overhead line, confusing it with a fault situation. Liquidated damages imposed by the customer were punitive.

By the end of 1995 the trains were working, and in the spring of 1996 Holec was claiming 4.5 million fleet service miles and an ever-improving availability record; but this came too late to help Jack Lane. The combination of an onerous specification, a traction equipment supplier new to the British railway scene, an enthusiastic but inexperienced work-force and a grave financial position only realized too late would have been enough to wound any company. Add to this the reorganization, not only of the company itself but of the whole British Railway market, and the wounds were fatal.

November 1995 and all is quiet. The multi-gauge test track has gone, as have the locomotives awaiting repair. What trackwork is in evidence is the 1991 realignment for class 323's entry into the paint shop. The 1921 extension spans the yard, and the former machine shop wall displays many changes of mind in its patchwork appearance. The original transverse erecting shop was in line with the 'porte-cochère'. All is now but a memory, as Schneider modernizes to house its 650-strong work-force.   AUTHOR

Manning Wardles' 'BOYNE ENGINE WORKS 1858' gate-posts still stand in 1997, but the birthplace of several thousand locomotives has gone.   CHRIS NICHOLSON

The Manning Wardle office block at the south-west corner of the site was used as a canteen by Hunslet Holdings from the mid-fifties. As a listed building it will survive the 1997/8 rebuilding of the works by the new owners. It is pictured here in mid-1997, eighteen months after the closure of Hunslet.   CHRIS NICHOLSON

By the first week in August 1995 the contrast with 1949 was at its most remarkable. The silence was eerie. The offices were empty but for a skeleton staff, and a dozen or so technicians were completing the class 323 build. The last two of these three-car a.c. drive electric multiple unit trains, nos. 323242/3, were in the 1990-built final assembly shop. The first two units, nos. 323201/2, were almost complete after modification and refit following the acceptance test programme. The erecting shop of 1949 (and much earlier), where the author had experienced so much clangour over the years, was empty. The birthplace of locomotives to operate from the Andes to the Antipodes, from the Urals to the Cape, from 15,000 feet above sea level to two miles below ground, was as silent as the grave, from which latter place the spirits of David Joy,

Matthew Murray, John Alcock and countless others would no doubt have had their own very individualistic views.

No. 323202 departed early in August, while nos. 323201 and 323243 were later taken ignominiously by road to Kilmarnock for finishing off; and in these simple acts there came to an end a whole industry stretching over 183 years and affecting the livelihood of every continent. On 8 November 1995 the last vestiges of Hunslet's machinery and equipment were sold at Jack Lane by auction, by coincidence as the last three vehicles were being loaded for transport to Kilmarnock. The move to Kilmarnock had been protracted. The first vehicle of unit 323243 left Jack Lane on 31 August, the other two on 4 September. Two months then elapsed before the three portions of 323201 finally headed north on 9, 13 and 15 November.

The whole of the Jack Lane site was now empty save for a caretaker staff, and was advertised for sale by Wetherall, Green & Smith. Ironically, this firm is a direct successor of the firm of surveyors, auctioneers and estate agents founded by Thomas Hardwicke, the auctioneer who conducted the sale of the Railway Foundry in 1859. The wheel had gone full circle. Defiant to the last stood Shepherd and Todd's archway on the north boundary and Mannings' cast-iron gateposts proclaiming 'BOYNE ENGINE WORKS 1858' facing south.

The run-of-the-mill railway enthusiast was slow to recognize the importance of Leeds as a Railway Town, and the significance in particular of thirty action-packed acres of prolific locomotive procreation. The name Hunslet lives on as a district of Leeds, and by adoption and reputation in Barnsley, Kilmarnock, Johannesburg, etc. *ad infinitum.*

The smoke over Jack Lane, which had changed gradually over the decades from grey to blue as steam gave way to diesel, had finally cleared away altogether; but on Friday 23 August 1996 the *Yorkshire Evening Post* announced that

The former Hunslet railway works is to be transformed into a manufacturing centre for electrical equipment after being bought by the French company, Schneider. Schneider will move its 600 strong workforce from its Merlin Gerin works in Meanwood, Leeds, to the 130 year old locomotive works, which will undergo a £10M modernisation.

The Merlin Gerin works was better known, perhaps, to older generations as Yorkshire Switchgear Ltd. Originally Tramway Supplies Ltd, this firm was established in Leeds in 1907. The link is an encouraging one, the traditional 'iron rail' remaining the common factor; and, while demolition of most of the buildings has been completed,

The dummy weight demonstrates the capacity of the overhead travelling cranes in this estate agent's sales photograph, but the trains had gone. The train maker's dream of a '£1 billion ride' had turned into a nightmare.

WEATHERALL, GREEN & SMITH

Queen Victoria came to the throne in 1837 and was crowned the next year. The British Queen public house, built shortly afterwards, bore silent witness to the railway output of Hunslet from Jenny Lind to the last class 323 – just short of 150 years – and was still standing proud in the shadow of the new but empty shop as the last-named awaited a buyer early in 1996.

WEATHERALL, GREEN & SMITH

the new owners employed an industrial archaeologist 'to ensure that historical information is not lost'. A new landscaped factory estate will emerge, with only the 1990 assembly shop forming an integral working unit. The Railway Foundry arch, the 1880 office block and the Manning Wardle gate posts and office block, all listed items, will be incorporated into the periphery of the site as they stand. Small but nevertheless collectively appropriate memorial, it may be thought, to more than a century-and-a-half of honest endeavour that helped to shape, perhaps sometimes mis-shape, the modern world.

All gone – just about. The long-disused works chimney stands in isolation with a clear view across south Leeds. Compare this with the aerial views in chapters 7 and 10.   MERLIN GERIN (GROUPE SCHNEIDER)

Bibliography

Ahrons, E. L. *The British Steam Railway Locomotive from 1825 to 1925*. Locomotive Publishing Co. Ltd, 1927

Booth, A. J. *Greenwood & Batley Locomotives 1927–1980*. Industrial Railway Society, 1985

Booth, A. J. *A Pictorial Survey of Standard Gauge Industrial Diesels around Britain*. D. Bradford Barton Ltd, 1977

Etherington. A. R. and A. C. Smith, *National Coal Board Flameproof Locomotives*. Industrial Railway Society, 1983

A History of the Middleton Railway – Leeds. Middleton Railway Trust Ltd, 1994

Industrial Railway Society, various publications

Jux, Frank. *Kerr, Stuart & Co. Ltd Works List*. Industrial Railway Society, 1992

Kitson-Clark, Edwin. *Kitsons of Leeds*. Locomotive Publishing Co. Ltd, 1938

Lane, Michael R. *The Story of the Steam Plough Works*. Northgate Publishing Co. Ltd, 1980

The Locomotives of the Great Western Railway. Part three *Absorbed Engines, 1854–1921*. Railway Correspondence and Travel Society, 1956

McLean, A. S. *Locomotives of the North Eastern Railway*. Locomotive Publishing Co. Ltd, 1922 (?)

The Railway Magazine and other periodicals

Redman, Ronald Nelson. *The Railway Foundry, Leeds*. Norwich: Goose & Son Publishers Ltd, 1972

The Reid Report

Rolt, L. T. C. *A Hunslet Hundred*. David and Charles, 1964

Rowlands, David, Walter McGrath and Tom Francis, *The Dingle Train*. Plateway Press, 1996

Taylorson, Keith. *Narrow Gauge at War 2*. Plateway Press, 1996

Townsley, Don. *Trains Illustrated* No. 61. Ian Allan Ltd, 1988

Wear, Russell. *Barclay 150*. Kilmarnock: Hunslet Barclay Ltd, 1990

Index